T0140111

# Knowledge Computing and its Applications

S. Margret Anouncia · Uffe Kock Wiil
Editors

# Knowledge Computing and its Applications

## Knowledge Computing in Specific Domains: Volume II

*Editors*
S. Margret Anouncia
School of Computing Science and Engineering
VIT University
Vellore, Tamil Nadu
India

Uffe Kock Wiil
The Maersk Mc-Kinney Moller Institute
University of Southern Denmark
Odense M
Denmark

ISBN 978-981-13-4102-1 ISBN 978-981-10-8258-0 (eBook)
https://doi.org/10.1007/978-981-10-8258-0


Softcover re-print of the Hardcover 1st edition 2018


Printed on acid-free paper

This Springer imprint is published by the registered company Springer Nature Singapore Pte Ltd.
part of Springer Nature
The registered company address is: 152 Beach Road, #21-01/04 Gateway East, Singapore 189721, Singapore

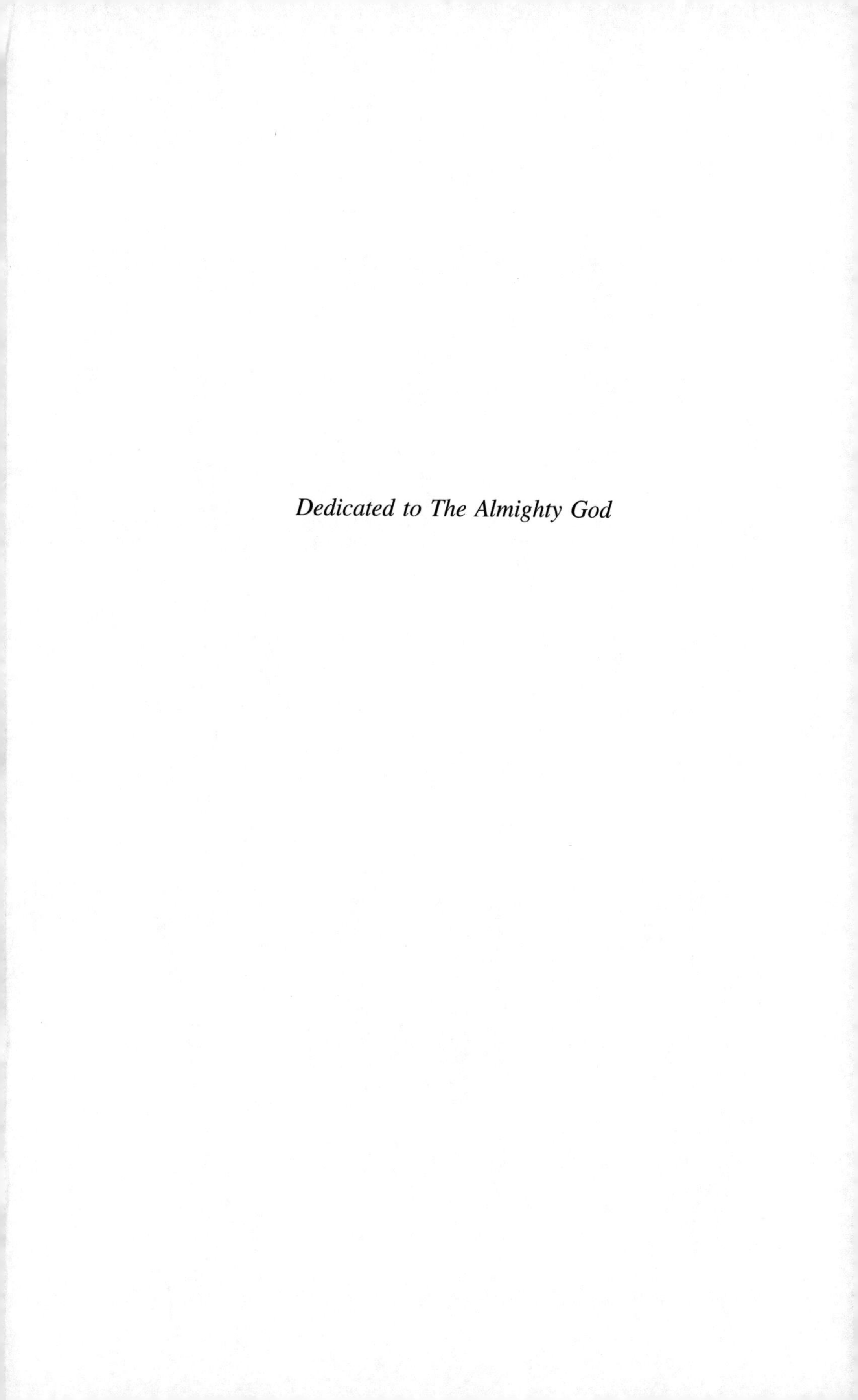

*Dedicated to The Almighty God*

# Foreword

Knowledge computing plays a key role in the current information age. Processing of huge amounts of data and extracting relevant knowledge is a central part of data science today.

A glance at this book and its organization shows that it provides a comprehensive collection of contributions covering current directions, challenges, and applications in the field of knowledge computing. The contributions deal with different dimensions of knowledge processing exemplified by a broad spectrum of applications. The tools and methodologies underlying the contributions provide a state-of-the-art glimpse into modern knowledge computing.

This book consists of five parts divided into two volumes: Volume I: (1) knowledge manipulation techniques for Internet technologies, (2) knowledge computing for large organizations, (3) knowledge computing and competency development; Volume II: (4) knowledge processing in specific domains, and (5) knowledge engineering process in information management. Taken together, these five parts (including a total of 31 chapters) provide an excellent overview of current tools, methodologies, and practice in the field of knowledge computing.

This book can serve as important inspiration for academics, industry experts, business professionals as well as students in their future research endeavors.

Enjoy!

Vijayawada, Andhra Pradesh, India

Dr. Suresh Chandra Satapathy
LMCSI, Senior Member IEEE,
Professor and Head
Department of CSE
PVP Siddhartha Institute of Technology

# Preface

The information society is rapidly migrating to the knowledge economy where knowledge management is renovating to the knowledge ecosystem. Several entities may not provide convenient capabilities directly but can enable capabilities of other products, services, or users. Knowledge economy concerns in substituting knowledge-based products and services within knowledge markets that are built on mechanisms to enable, support, and help in mobilization, exchange of information/knowledge among providers and users.

This Handbook of Knowledge Computing and its Applications is an edited book contributed by seventy of researchers who work in the field of knowledge engineering. The scope of this book includes knowledge manipulation techniques for Internet technologies, knowledge processing tools and techniques in specific domains, knowledge computing in large and international organizations, application of knowledge engineering process in information management, skill, and competency development.

This book is edited as two volumes comprising fourteen and fifteen chapters, respectively.

The first volume focuses on *Knowledge Manipulation and Processing Techniques* comprising three parts such as *Knowledge Manipulation Techniques for Internet Technologies* that include articles dealing with knowledge manipulation, representation techniques that are useful for recent Internet technologies such as Big Data and IoT. The second part on *Knowledge Computing in Large Organizations* focuses the various organizational issues sorted out through knowledge computing methods. The third part deals with the *Knowledge Computing and Competency Development*.

The second volume titled *Knowledge Computing in Specific Domains* has two parts. The first part of volume two delineates on *Knowledge Processing Tools and Techniques in Specific Domains* with articles associating knowledge processing in domains such as health care, manufacturing, multimedia and remote sensing.

The second part deals with the articles concentrating on the *Application of knowledge engineering process in Information Management.*

With the drastic growth of knowledge collections in varieties of forms and in diversified fields, the demand for the knowledge processing and representation increases. Consequently, the claim for the use of knowledge tools and techniques arises. With this motivation, the handbook is planned and is intended for a spectrum of people spanning from academicians, researchers, industry experts, and students. This book also can be beneficial to business managers and entrepreneurs.

The salient features of this handbook include:

- It covers knowledge processing in the developing domains such as Big Data, IoT, health care, and multimedia.
- It also presents methods associated with the knowledge computing and its applications.
- Contributors to this book are researchers and practitioners from various reputed academia and industries within and outside the country.

We thank all the contributors for their relentless effort. Without their support, this book would never become a realization. We express our gratitude to Springer for their support throughout the project.

Vellore, India — S. Margret Anouncia
Odense M, Denmark — Uffe Kock Wiil

# Acknowledgements

First and foremost, the editors would like to wholeheartedly thank the Lord, from the bottom of their hearts, who has been the enduring source of strength and comfort for the completion of this book project. You have given us the power to believe in our passion and pursue our dreams.

The editors would like to acknowledge the help of all the people involved in this project and, more specifically, the authors and reviewers who took part in the review process. Without their support, this book would not have become a reality.

The editors would like to thank each one of the authors for their contributions. Our sincere gratitude goes to the chapter's authors who contributed their time and expertise to this book.

The editors wish to acknowledge the valuable contributions of the reviewers regarding the improvement of quality, coherence, and content presentation of chapters. They also thank the reviewers for their thoughtful, detailed, and constructive comments.

The editors' heartfelt thanks go to all the friends and colleagues who assisted them in all endeavors of assembling this book and their outstanding and generous help in providing recommendations and observations which has been an inspiration for this book.

Finally, the editors would like to acknowledge with gratitude the support and love of their family members for compiling this book. This book would not have been possible without their support.

Vellore, India
Odense M, Denmark

S. Margret Anouncia
Uffe Kock Wiil

# Contents

# About the Editors

**S. Margret Anouncia** is a Professor at Vellore Institute of Technology (VIT) University in India. She received her bachelor's degree in Computer Science and Engineering from Bharathidasan University (1993), Tiruchirappalli, India, and her master's degree in Software Engineering from Anna University (2000), Chennai, India. She was awarded a doctorate in Computer Science and Engineering at VIT University (2008). Her main areas of interest include digital image processing, software engineering, and knowledge engineering.

**Uffe Kock Wiil** received his M.Sc. (1990) and Ph.D. (1993) degrees from Aalborg University, Denmark. He has been a Professor of Software Engineering at the University of Southern Denmark since 2004. His primary research interests are health informatics (clinical decision support systems; patient empowerment) and security informatics (analysis and visualization of crime-related data).

# Abbreviations

| | |
|---|---|
| AALIA | Automatic Anatomic Location Identification Approach |
| ABC | Ant Bee Colony |
| AHELDP | Adaptive Homomorphic Eight Local Directional Pattern |
| ANN | Artificial Neural Network |
| B2C | Business-to-Consumer |
| BCE | Bayesian Consensus Ensemble |
| BDE | Binary Differential Evolution |
| BF | Brevity Factory |
| BLEU | Bilingual Evaluation Understudy |
| BLogReg | Bayesian Logistic Regression |
| CAM | Classification Ability Measurement |
| CART | Classification and Regression Trees |
| CCA | Canonical Correlation Analysis |
| CCS | Classification Computing System |
| CDSS | Clinical Decision Support System |
| CFS | Correlation-based Feature Selection |
| CLBP | Complete LBP |
| CPU | Central Processing Unit |
| CSPA | Cluster-based Similarity Partitioning Algorithm |
| CT | Computed Tomography |
| DEMOIR | Dynamic Expertise Modeling from Organizational Information Resource |
| DIGLIB | Digital Library |
| DL | Description Logic |
| DNA | Deoxyribonucleic Acid |
| DSCFS | Discrimination Structure Complementarity |
| DSS | Decision Support Systems |
| DTI | Diffusion Tensor Imaging |
| EbLDP | Enhanced Local Directional Pattern |
| EC | Evolutionary Computation |

EDM Educational Data Mining
EISM Expertise Information Space Manager
ELDN Eight Local Directional Pattern
EM Expectation Maximization
ER Expertise Recommender
FB Foreign Body
FCBF Fast Correlation-based Feature Selection
FL Fuzzy Logic
FMA Foundational Model of Anatomy Ontology
FMIFS Fuzzy Mutual Information-based Feature Selection
FPRS Fuzzy Preference-based Rough Set
GDP Gradient Directional Pattern
GLSA Generalized Latent Semantic Analysis
GTV Gross Tumor Quantity
HGPA Hypergraph Partitioning Algorithm
HMM Hidden Markov Model
HTML Hypertext Markup Language
ICA Independent Component Analysis
ICT Information and Communication Technology
IDC Incremental Dependency Calculation
IDS Intrusion Detection System
IWSS Incremental Wrapper-based Subset Selection
JAFFE Japanese Female Facial Expression
JCFO Joint Classifier and Feature Optimization
KBCGS Kernel-Based Clustering Method for Gene Selection
KM Knowledge Management
kNN k-Nearest Neighbor
KNN K-Nearest Neighbor
KWS Knowledge Work Systems
LBP Local Binary Pattern
LDA Latent Dirichlet Allocation
LDA Linear Discriminant Analysis
LDeP Local Derivative Pattern
LDNP Local Directional Number Pattern
LDP Local Directional Pattern
LDPv Local Directional Pattern Variance
LGDDP Local Gabor Dominant Direction Pattern
LGDP Local Gaussian Directional Pattern
LGNP Local Gaussian Number Pattern
LLE Locally Linear Embedding
LO Learning Objects
LSA Latent Semantic Analysis
LSDP Local Sign Directional Pattern
LSLS Locality Sensitive Laplacian Score
LTM Landsat Thematic Map

| | |
|---|---|
| LTP | Local Ternary Pattern |
| MAD | Multivariate Alteration Detection TransformationTWOPAC TWinned Object Pixel-based Automated Classification |
| MAE | Mean Absolute Error |
| MASE | Mean Absolute Square Error |
| MB-LBP | Multiblock LBP |
| MIFS | Mutual Information-based Feature Selection Algorithm |
| MIL | Multi-instance Learning |
| MIRI | Multi-instance Rule Inducer |
| MKL | Math Kernel Library |
| ML | Machine Learning |
| MoM | Majority of Majorities |
| MRCPS | Molecular Regularized Consensus Patient Stratification |
| MRI | Magnetic Resonance Imaging |
| MRmMC | Maximum Relevance Minimum MultiCollinearity |
| MSE | Mean Squared Error |
| MSE | Meta-Search Engines |
| NDCG | Normalized Discounted Cumulative Gain |
| NDVI | Normalized Difference Vegetation Index |
| NEETF | National Environmental Education Training Foundation |
| NLDA | Null space based Latent Dirichlet Allocation |
| OCM | Orthogonal Centroid Method |
| OLAP | Online Analytical Processing |
| OPAC | Online Public Access Catalogue |
| PCA | Principal Component Analysis |
| PEAS | Performance Measure, Environment, Actuators, Sensors |
| PGM | Probabilistic Graphical Model |
| PR | Probabilistic Reasoning |
| PSNR | Peak Signal-to-Noise Ratio |
| QCR | Qualified Cardinality Restrictions |
| RDF | Resource Description Framework |
| RDFS | Resource Description Framework Schema |
| RFE | Recursive Feature Elimination |
| RFS | Random Forward Search |
| RLR | Regularized Logistic Regression |
| RMSE | Root Mean Square Error |
| ROC | Receiver Operator Characteristic |
| ROI | Region of Interest |
| RS | Recommender System |
| RTDPCA | Robust Two-Dimensional Principal Component Analysis |
| RWS | Random Walk system |
| S3SVM | Semi-Supervised Support Vector Machine |
| SBE | Sequential Backward Elimination |
| SC | Soft Computing |
| SCAD | Simultaneous Clustering Attribute Discrimination |

| | |
|---|---|
| SERP | Search Engine Result Page |
| SFFS | Sequential Floating Forward Selection |
| SFS | Successive Feature Selection |
| SKOS | Simple Knowledge Organization System |
| SLDA | Supervised Latent Dirichlet Allocation |
| SLogReg | Sparse Logistic Regression |
| SMLR | Sparse Multinomial Logistic Regression |
| SNR | Signal-to-Noise Ratio |
| SOM | Self-Organizing Map |
| SRR | Search Result Record |
| SSM | Simple Selection Method |
| SSS | Small Size Samples |
| SVD | Singular Value Decomposition |
| SVDD | Support Vector Data Description |
| SVM | Support Vector Machine |
| SWRL | Semantic Web Rule Language |
| TEL | Technology-Enhanced Applications |
| TF-IDF | Term Frequency-Inverse Document Frequency |
| TI | Tree Induction |
| TLC | Two-Level Classification |
| TLE | Temporal Lobe Epilepsy |
| TMA | Tissue Microarray Evaluation |
| TSVM | Transductive Support Vector Machine |
| UPFS | Unsupervised Probabilistic Feature Selection |
| URI | Uniform Resource Identifiers |
| URL | Uniform Resource Locators |
| W3C | World Wide Web Consortium |
| WWW | World Wide Web |

# Part I
# Knowledge Processing in Specific Domains

# Probabilistic Graphical Models for Medical Image Mining Challenges of New Generation

**Sridevi Tumula and Sameen S. Fathima**

**Abstract** Probabilistic graphical models (PGM) are one of the rich frameworks. These models are used over complex domains for coding probability distributions. The joint distributions interact with each other over large numbers of random variables and are the combination of statistics and computer science. These concepts are dependent on theories such as probability theory, graph algorithms, machine learning, which make a basic tool in devising many machine learning problems. These are the origin for the contemporary methods in an extensive range of applications. These applications range as medical diagnosis, image understanding, speech recognition, natural language processing, etc. Graphical models are one of dominant tools for handling image processing applications. On the other hand, the volume of image data gives rise to a problem. The representation of all possible graphical model node variables with that of discrete states heads to the number of states for the model. This leads to interpretation computationally obstinate. Many projects involve a human intervention or an automated system to obtain the consensus established on existing information. The PGM, discussed in this chapter, offers a variety of approaches. The approach is based on models and allows interpretable models to be built which then is employed by reasoning algorithms. These models are also studied significantly from data and allow the approaches for cases where the model is manually built. Most real-world applications are of uncertain data which makes a model building more challenging. This chapter emphasizes on PGM where the uncertainty of data is obvious. PGM provides models that are more realistic. These are extended from Bayesian networks, undirected Markov networks, discrete and continuous models, and extensions to deal with dynamical systems and relational data also. For each class of models, the

S. Tumula (✉)
Department of Computer Science and Engineering, Chaitanya Bharathi Institute of Technology (A), CBIT post, Gandipet, Hyderabad, Telangana, India
e-mail: sritumula@gmail.com

S. S. Fathima
Department of Computer Science and Engineering, Osmania University, Hyderabad, Telangana, India



S. Margret Anouncia and U. K. Wiil (eds.), *Knowledge Computing and its Applications*, https://doi.org/10.1007/978-981-10-8258-0_1

chapter describes the fundamental bases: representation, inference, and learning. Finally, the chapter considers the decision making under the uncertainty of the data.

**Keywords** PGM · Probability theory · Graph algorithms · Machine learning Bayesian networks · Undirected Markov networks · Discrete and continuous models

## 1 Introduction

Relating graph theory and probability theory for multivariate statistical modeling, we can build probabilistic graphical models. These models can have a lot of applications such as bio-informatics, speech and image processing, control theory. For computing basic statistical quantities such as likelihoods and score functions, formulation of statistical models has to be in terms of graphs and algorithms. Hidden Markov Models (HMMs), Markov random fields, and Kalman filters are also used as models.

Machine learning models are normally divided into two types such as the discriminative model and the generative model. Examples of categorization or classification of the data, including SVMs, linear/logistic regression, CRFs, are handled by Discriminative models, whereas the generative approach model addresses about generation of the data and various other tasks which are the image generative information. Examples such as HMMs, GMM, Boltzmann machines belong to this model. Deep learning models such as DBMs and variational auto-encoders are again generative model examples. For generative models and delving deeper into them, concepts and thorough knowledge about probabilistic graphical models are essential [1].

The purpose of device learning is to offer no longer handiest computational methods for collecting, enhancing, and renovating information in intelligent structures. Similarly, the learning mechanisms will assist in inducing the expertise from instance or statistics. In cases wherein algorithmic solutions are not to be had, mechanic learning methods are used. They may be notably utilized wherein there is a loss of formal models or wherein the know-how approximately the software program region is poorly defined. In ML studies, numerous medical corporations are concerned and lead the clinical subject to consist of thoughts from awesome regions, inclusive of computational gaining knowledge of idea, synthetic neural networks, statistics, stochastic modeling, genetic algorithms, and sample reputation. ML consists of a wide beauty of techniques that can be sort of categorized in symbolic and subsymbolic in keeping with the character of manipulation taken location at the same time as analyzing [2].

As the state of the art has progressed, graphical models, such as belief propagation and graph cuts, empowered their modeling power for the application of many vision problems for performing approximate inference. For problems such as estimating images, the graphical model is the most convenient model. The relationship of pixels between observed and estimated image is measured conveniently by graph models. Dimensionality of the image is a shortcoming for the graph models. Node in the graph is defined by each pixel of the image or groups of pixels.

For a grayscale image with discrete representation, each gray level of the image is assigned with one discrete state. In the graphical model, 256 states are assigned per node for a given grayscale image and $256^Q$ states are required in case a variable relates to a group of $Q$ pixels. A graphical version computation differs on the order of $q2 \times p$ in which $p$ is represented because the range of nodes in the graph and $q$ represents the wide variety of discrete states in keeping with node. The complexity of inference and the variety of states are constrained by way of the quadratic courting among them for the usage of issues wherein discrete-valued graphical fashions are done. Expressing capability of this undertaking in a graphical model with respect to discrete states is critical for the reason that records of images are non-Gaussian in nature [3]. By way of decreasing the amount of states for each node in the graph inference from the photograph may be traced.

With the advent of artificial intelligence and machine gaining knowledge of, the range of responsibilities that must be computerized will increase. More and more tasks can be developed, employed, examined, and programmed through automation. AI and machine learning can advance in this development provided AI and machine learning recognize and change accordingly. If the automation can identify its inabilities and change accordingly, human interaction can be superseded.

For taking the available information and reaching for conclusions, both about what might be true and about how to act, most tasks require a person or an automated system to reason. In day-to-day life, a doctor needs to take all information about the patient, i.e., symptoms, the test results, the weight particulars, and also the gender of the patient for assessing and reaching to a conclusion about the patients' diseases and the treatment and medicine the doctor has prescribed to the patient. Likewise, a mobile robot also needs to synthesize data from its sensors, cameras, and also other sensors to conclude in which environment it is and also to reach the goal without hitting anything. In the same way, a speech recognition system also needs to take acoustic signal and infer the words spoken which gave rise to it. In this, we can say one could write a special computer program for every domain and every type of question one may answer in principle. Although the resulting system may be quite successful in its particular task, still we can say it is often very brittle. The program requires significant changes when the application changes. Further, in this the general approach is quite and hence difficult to extract lessons from a successful solution and applied to other, which is very different [4].

Based on the concept of declarative representation where one has to construct within the computer, a different approach needed there for a model of system about which one has to reason. In this model, our knowledge is coded for knowing how a system works in readable form. Various algorithms that answer questions based on the model can be manipulated by this method. For instance, our knowledge about various diseases and their relation to a variety of systems and test results can be represented by a model for medical diagnosis. Reasoning the unique algorithm taken, the model and its observations relating to patient and questions and answers pertaining to patients are diagnosed. The characteristics of a declarative representation are separation of knowledge from reasoning. The clear semantics representation is unique, which separates from the algorithm that one can apply to it. So, a popular suite of algorithms may be evolved via making use of any version with a broad elegance, in the areas of both medical analysis and speech reputation. Without enhancing our reasoning algorithms continuously, we will improve the version for a specific utility area. In lots of fields, model-based techniques or declarative representatives are essential additives and models come in many flavors.

## *1.1 Image Mining Challenge*

This is used to mine expertise directly forwardly from an image. The primary section in mining is photograph segmentation, that is, a variety of statistics mining inside the discipline of photograph processing. Image mining handles with hidden information of extraction, image statistics affiliation, and further pattern for which they are no longer truly collected inside the images. This is an interdisciplinary discipline that integrates strategies such as the computer imaginative and prescient, image processing, data mining, statistics bases, gadget gaining knowledge, and artificial intelligence. A totally critical function of the image mining is to generate all large patterns without earlier statistics of the styles [5]. Huge image databases have been adopted through rule mining [6]. A number of researches were carried on. The mining changed into executed according to the integrated collections of pix and its associated photograph statistics [7].

To produce all the information without any loss of the image content is prime insertion of image mining. The pattern types are many and different. Patterns can be classified as classification pattern, discretion patterns, correlation patterns, and spatial patterns. All features of huge databases which include indexing methods, photograph storages, and image retrieval are huddled photograph transferring and are all in image mining system. Organizing an image mining system is a probabilistic method which means numerous techniques ranging from image retrieval and indexing schemes up to facts mining and pattern recognition [8]. Its miles anticipated that a great high-quality image mining device offers customers with a beneficial access into the image storage region. And additionally at the same it also acknowledges information styles and also generates photograph representation. A good pleasant image mining system provides users with a wish to get admission into the photograph garage area, and on the equal time, it recognizes information patterns which generate knowledge in image illustration. Essentially that machine is meant to bring the following capabilities: (1) image garage, (2) image processing, (3) function extraction, (4) image indexing and revival, and (5) pattern and know-how discovery. In addition, machine contemplates a distinct pattern of images as enter whose image-extracted capabilities are much like the image content material. Similarly, the specialty of the mining task, in a great deal is needed to suppose sound and other distractions and their pertinent trouble to positive geometric transformations and robustness. After depicting the photograph content material, the model's rendition of acquired image, we are able to get an appropriate semantic interpretation. For that reason, we get mining outcomes after matching the outline of models with the symbolic description of its complimentary. The consultant description may be just a feature or a set of functions, a detailed verbal description to be able to identify a selected semantic [9]. This presentation of paper depicts a survey on numerous photograph mining techniques stated earlier and also offers a marginal overview for further research and enhancements.

A remarkable variety of researches have been made on photograph mining. Tremendous growths in huge and depicted photograph statistics mines can be stated because of the final results of trends in the vicinity of image acquit ion and storage strategies. Examination to be done with snapshots in databases will provide valuable statistics to human type. In addition, it presents the extraction of hidden statistics, photograph data affiliation, and other unclear styles amassed inside the images.

Mainly, biological information management and mining are set to be the regions of biology studies. The characteristics of Tissue Microarray Evaluation (TMA) device analysis are high considerate and huge facts content material. For making tissue image mining imaginative and quicker, the tissue snapshots want to be indexed, saved, and mined on content material. A pathologist with image examines pattern identification and AI is proposed in this method are vital to exploit. This information which includes disparity or color, are used at image processing and statistics stage to find the output. Pathological gadgets consisting of mobile constituents are diagnosed at semantic degree association, configuration of individual cells into sheets into a tissue image is examined, and supposition of the professional is described on the knowledge level, which is likewise an upper level.

In an image mining method the usage of wavelet transform is said by using [10]. He proposed image mining technique by utilizing wavelet transformation. Wavelet transformation uses not unusual system equal, machine identification, and facts mining fashions with the concept that an actual lifestyles image may be associated with a specific kind, assisting in distinctive predictions and forecasting mechanisms. This method has three steps: first, image collecting; second, learning; and third, classification. This will be utilized for image mining as an opportunity wavelet transformation. The fact that wavelet transformation uses time frequency affiliation is utilized.

The usage of clustering and information compression strategies [11] predicted the image mining technique. In forecasting climate situations, satellite snapshots of clouds are used. Based on the weather surroundings, the tiers in frequency of image acquire now not minute to hour is decided. This could assist in huge series and advent of information ware residence. In this stressful mission, there may be everlasting storage and transmission of snapshots. In their manner of dealing, facts mining clustering method collectively with vector quantization are carried out to cement and compact stable shade photograph [5].

## *1.2 Probabilistic Graphical Models*

Complex systems are distinguished by the existence of multiple interrelated aspects. These are related to the reasoning task. For instance, in a medical diagnosis application, multiple possible diseases exist with dozens or hundreds of symptoms and diagnostic tests. Apart from personal characteristics that frequently form predisposing factors for disease, many other matters also have to be considered.

The characters of these domains are a set of random variables, and the value of the variable defines its property. If for instance we take flu disease, this may be one variable in the domain. It can take a discrete value like yes or no. Another variable like fever may be not discrete rather continuous values. These values (discrete or continuous) for the variables make an important decision making. The task here is to reason out the probabilistic values from the observations. To have a principled probabilistic reasoning, a joint distribution over the space is constructed for some set of random variables $P$. In this model, a large number of questions are allowed to answer such as $P_i$ takes a specific value $pi$ and the posterior distribution of the resultant can be questioned [4].

In pattern recognition application, probabilities play a key role. Theory of probability can be displayed as two rules, i.e., sum rule and product rule. Any probabilistic inference and learning manipulations however complex may be expressed in these two rules/equations only. So, we have to solve the probabilistic models on these algebraic equations. It is found that analysis of these probability distributions can be represented in a graphical/diagrammatical model which is called as probabilistic graphical models [1]. The benefits of those fashions are given as:

(a) It has easy visualization of the shape of the probabilistic model and may be used to devise new models.
(b) Dependencies of the variables and residences of the version may be examined with the aid of going through the graph.
(c) Mathematical expressions are done implicitly to carry out inference and studying in complex fashions and may be articulated in phrases of graphical manipulations.

Every node in PGM represents a random variable or institution of random variables, and the hyperlinks between them are probabilistic relationships between these variables. Joint distributions over all the random variables in the graph may be decomposed right into a product of things each relying handiest on a subset of the variables. One of the directed graphical models is Bayesian networks, wherein hyperlinks of the graph have a particular path. The other class of models is Markov random fields, which may be known as undirected graphical fashions. In those models, hyperlinks do not have directional significance. Causal relationships between random variables may be without problems expressed in directed graphs. Gentle constraints among random variables can be for undirected graph. Directed and undirected graphs represented in an exclusive shape called as issue graph is useful for solving inference issues [1].

## 2 Machine Learning Methods in a Medical Context

Answer for diagnostic and prognostic troubles in an expansion of medical domain names is provided by ML with specific methods, strategies, and tools. For the mining of scientific understanding together with research outcomes, normal patient control, therapy planning and help, ML can be used for the research of the importance of clinical parameters and in their mixtures for evaluation [2].

ML is applied for data analysis data irregularities, continuous data interpretation which comes from intensive care unit, smart alarming for efficient and effective monitoring. With the help of best execution of ML methods/techniques, there can be a combination of computer-based systems in health care. It facilitates opportunities to improve the medical experts' work and also efficiency with quality.

Let us talk few primary ML programs inside the location of healthcare. In [12–14], it is stressed primarily on computer-based structures in reasoning of clinical prognosis. For the era of hypotheses from patient facts, a framework was designed with expert structures and primarily model-based schemes. Rules are formulated from the know-how base of the specialists. But, due to lack of understanding of specialists in hassle solving the understanding base supplied via the specialists changed into no longer sufficient [15]. Used symbolic learning strategies like inductive studying with examples may be used to increase getting-to-know and know-how control to professional systems.

A system is developed known as KARDIO, to interpret ECGs by [16]. In this, a set of clinical cases are used for learning the intelligent system using ML techniques to form a systematic features. These rules are then formulated into a decision tree. It can also be further extended to cases where no previous experience exists for interpreting and analyzing the medical data. Akay et al. [17] has developed a smart system where real-time patient data are taken when the patient is undergoing bypass surgery. With these data, a model is created for both normal and abnormal cardiac structures. It is used for finding the changes in patient conditions. In recent times for research, these models can also be used for initial hypotheses.

Data taken from patient and trying to build a learning algorithm will have a lot of problems, such as the data sets have insufficient data or missing data, noise in the data, sparse data and also problem in selection of parameters. Neural networks which are subsymbolic learning methods can handle these types of datasets. These are frequently used in applications like pattern matching and also applications to improve medical decision making where human-like characteristics such as generalizations and robustness to noise are involved [3, 18–21].

Biomedical signal processing [22–29] is other area of ML application. Because of lack of understanding in the biological systems, essential features and hidden information are not evident. This lack of knowledge affects the inability to not distinguish the relationships between subsystems. The unpredictability of the organic indicators may be because of no delay with inner mechanisms or external impulse, and also with family members. Few of the parameters can be too hard to resolve the use of the traditional techniques. ML techniques rely on the records and can be useful in building model based on the nonlinearity of the data and also extracting the features to develop medical care.

Assistance in medical diagnosis which is the other area [19, 20, 25, 30–36] can be developed by the interpretation of computer-based medical image systems. These systems can be developed by the following doctor's expertise to find the infected/dangerous regions by minimally invasive imaging including computed tomography, ultrasonography, endoscopy, confocal microscopy, computed radiography, or magnetic resonance imaging. The principle aim is to perceive most cancer areas via raising expert's potential and decreasing the interference through preserving accurate prognosis.

There have to be a demand for extra efficient strategies of early detection with people who computer assisted clinical prognosis structures aim to. For the patient, delay in diagnosis, and studying tissue residing in vivo confined type far likely away [37], and for that reason minimizing the shortcomings of biopsies are few topics which can be to be considered. By and large, healthcare environment is vastly depending on computer technology. In most cases, the help of ML technologies is able to provide assistance to the physician by eradicating issues linked with human fatigue and habituation. It also provides a fast method of finding the abnormalities by facilitating real-time diagnosis [2].

## *2.1 Image Mining Challenges*

The motive behind image mining algorithms improves medical image processing applications that are discussed here [38].

The growth of image mining has faced the main difficulty in the lack of understanding the topics and also research results about image mining. Most of the researchers assume that image mining is an application of data mining, and few infer that pattern recognition and image mining are synonym words. Image mining is just exploiting data mining algorithms for images [39]. Information in images, data dependence, and unambiguous patterns stored in the images are all searched using image mining techniques. This field works in two different ways: The first technique goes with the study in a broad variety of independent images and the second technique explores a sequence of linked images [40].

The main aim of medical image mining algorithms is to either extract reckonable features from large volumes of image data or to improve the quality of medical images such as CT scan, X-ray, MRI, having potentially malignant nodules, tumors, or lesions. Various algorithms deal with medical image mining work in a series of three stages.

The first step is the generation of candidates. In this stage, unhealthy Regions of Interest (ROI) from a given medical image is recognized which are apprehensive. These ROIs are usually called as candidates. This step involves an image processing algorithm which searches for ROI in the image with particular variation.

The second step is about feature extraction or calculation of a set of descriptive morphological or texture features for each of the candidates. These are done using innovative image processing techniques.

The third step is classification. Based on candidate feature vectors, the candidate that differs from the rest of the candidates is classified. The classifier goal is to decrease the false positives. Classifier does not include image quantification and enhanced visualization algorithms. These are used by image processing and pattern recognition algorithms for candidate generation and feature extraction.

## 3 Medical Image Mining Applications

PGM technology combined with scientific/clinical application will have those three demanding situations.

First and main is to discover the amount of preprocessing required to study approximately motion and deformation styles. The first rate of capabilities selected quantifies the PGM algorithms:

- The resulting undertaking is to perceive an approach to "dissect" about the case. We recall that exact evaluations of a similar infirmity, for an influenced individual or inward people, can be demonstrated as creative weaknesses of an ordinary condition. It is far speculated that development styles have a place or lie near a nonlinear complex that can be learnt from records by method for nonlinear dimensionality cut charge. Exact issues comprise of picking appropriate perusing strategies, relying on which include improvements to feature, and contrasting new subjects with this learnt portrayal.
- Sooner or later, surveying whether this outline is gainful for the logical programming. In [41], design changes to influenced individual response is valuable for treatment arranging is discussed.

In current years, morphometric sample evaluation using MRI has been broadly investigated for automatic analysis of AD and MCI. Contemporary MRI-based totally completely studies can be categorized into unmarried-template-based totally without a doubt and multitemplate-based totally completely techniques. It is far broadly honored that one-of-a-kind templates can supply complementary statistics that are beneficial for advert and MCI prognosis. Particularly, by way of using each template as a particular view, a few current trends in multiview reading and the use of multitemplate MRI facts had been noted in this economic catastrophe.

### *3.1 Multiview Feature Representation with MR Imaging*

In [41], a multi-view function representation method by means of the use of the usage of MRI facts is advised. Specially, they endorsed a degree mind morphometry through multiple templates and also to generate a wealthy representation of anatomical structures that lets in you to be extra discriminative to separate awesome companies of subjects. In assessment to previous multi-template-based works which test their templates to a common region via deformable registration, they hold the chosen templates in their particular (linearly aligned) areas without nonlinearly registering them to the common area, which will do not forget precise information provided via notable templates. Of their method, affinity propagation is first implemented to pick the maximum particular and consultant templates. Then, subjects from awesome organizations are registered to at least one-of-a-type templates through using HAMMER. The most brilliant local functions may be extracted in extremely good template regions by adopting a function extraction technique.

An ordinary preprocessing approach is hooked up to the MR thought images that are T1-weighted. As a count of first significance, nonparametric unusual bias adjustment (N3) is hooked up to depth inhomogeneity. At that element, a cranium-stripping method is carried out, trailed with the aid of guide audit or adjustment to guarantee easy cranium and dura removal. Cerebellum removal is thus directed via twisting a named layout to every cranium-stripping image. Thereafter, every brain image is segmented into three tissues (gray rely (GM), WM, and cerebrospinal liquid (CSF)) with the resource of the use of FAST; ultimately, all mind snapshots are affine joined with the aid of FLIRT.

This device of acquiring primarily a multitemplate-based illustration of human mind consists of developing a method to select multiple templates. Over a giant variety of first-rate categories (10 for AD, 10 for MCI, and 10 for NC), 30 random templates are decided on. But, the ones (random templates) cannot guarantee to accurately mirror the distribution of the entire population. Moreover, they deliver redundant facts; furthermore, the selection of such unrepresentative images as templates should in addition cause big registration errors. A record-driven template desire technique is used to acquire the maximum extraordinary and representative templates to overcome those boundaries. For this to happen, there ought to be maximum divergence among the decided on templates. Instead, less mistakes in registration is required so that templates need to be representative sufficient to cover the complete populace. At this factor to partition the entire populace (of AD and NC images) into adequate (e.g., $K = 10$) non-overlapping clusters, the affinity propagation set of policies is used. Study shows that, affinity propagation guarantees that an exemplar image that can be a representative image or a template for that cluster, be determined on automatically for each cluster. Further, a template pool may be shaped through combining all the templates from each cluster. Inside the clustering method, a bisection approach is done to discover the best choice charge, and the image similarity is computed as normalized mutual statistics. It have to be stated that, although it is possible to feature greater templates to the present-day set of decided on templates, the ones more templates have to result in the redundant statistics and therefore have an effect at the maximum beneficial illustration of each problem. Due to the fact, MCI is an intermediate phase among advert and NC and is correlated with the tendencies of each advert and NC; only templates from AD and NC subjects are decided on.

The middle steps in morphometric pattern assessment (e.g., VBM, DBM, or TBM) consist of (1) a registration step for spatial normalization of several images proper into a not unusual space and (2) a quantification step for morphometric length. Just like a mass-maintaining shape transformation framework is followed to capture the morphometric patterns of any given state of affairs at the regions of several templates.

First, the capabilities are derived from individual templates and protected together to accumulate a wholesome example. A fixed of areas-of-interest (ROIs) in each template area is first adaptively determined by means of the usage of enforcing watershed segmentation at the correlation map obtained a few of the voxel-smart tissue density values and the class labels from all training topics. Then, to improve each discrimination and robustness of the volumetric feature computed from every ROI and every ROIs further subtle by way of using means of selecting simplest voxels with less high priced illustration energy. In the long run, some assessment is furnished to illustrate the strength of the function extraction approach in obtaining the complementary capabilities from more than one template for representing every situation mind, to reveal the consistency and difference of ROIs obtained in all templates.

The principle downside of contemporary multiview analyzing fashions is that handiest the relationship among samples and their corresponding beauty labels is taken into consideration. In fact, there exist a few amazing essential facts in multiview characteristic illustration the use of multitemplate MR imaging records, as an instance, (1) the connection among a couple of templates and (2) the relationship among super topics. For this reason, expand a dating-triggered multitemplate analyzing (RIML) approach to explicitly version the relationships among templates and among topics.

There are three most vital steps inside the RIML approach: (1) feature extraction, (2) characteristic preference, and (3) ensemble elegance.

In addition, a relationship precipitated sparse (RIS) FS method is proposed underneath the multitask analyzing framework, to version the relationships among templates and among subjects, through the manner of treating the magnificence in each template region as a specific mission.

## 3.2 *Learning and Estimating Respiratory Motion from 4D CT Lung Images*

In unique disciplines of radiology and in numerous clinical programs of radiology, imaging strategies were typically carried out for catching auxiliary and sensible photographs of human frame/organs for evaluation, remedy designing, and drug steerage. Clinical imaging offers snapshots of internal organs that are not straightforwardly open; from those images, docs can collect anatomical and useful information and translate the obsessive states of patients. Using CAD system, quantitative measures of neurotic conditions, as an example, harm, tumor, blood waft, calcification, organ/tissue length and thickness, cardiovascular primary and sensible dysfunctions, and additionally cerebrum community may be amassed to help carry out an correct analysis. Inside the space, in-between images of those varieties of conditions can assist arrange techniques, for instance, intercession, radiotherapy, and surgical treatment? Improvements in imaging hardware have endorsed the development of recent effective and green treatment choices, changing over complicated open surgical tactics to viable negligibly intrusive techniques, shortening sufferers' restoration time, improving chronic solace, and getting rid of the hazard of confusions. Minimally intrusive intervention is presently supplanting all the greater luxurious open surgical treatment in oncology, neurology, pulmonology, and cardiology. Amid insignificantly obtrusive methodologies clinicians rely on imaging sufferers and following of interventional gadgets and consolidating the machine place with their insight into patients' lifestyles systems and neurotic conditions to make bigger a highbrow image to coordinate their sports. Software program fusion mixed with visualization of image and associated devices delivers a powerful tool to help clinical docs play out the technique. In a humdrum interventional technique, initially a diagnostic photograph, for example, computed tomography (CT) or magnetic resonance imaging (MRI) is stuck for planning. Image segmentation with visualization strategies can be carried out at this segment to spotlight the subtle factors of obsessive situations and to prepare the method (for instance, characterizing the passageway component and course manner for biopsy). In radiotherapy planning, tumor(s) are segmented decisively to symbolize the Gross Tumor Quantity (GTV), the scientific goal amount (CTV) and the Planning Target Amount (PTV), and radiation beams are upgraded to amplifying favored radiation dosages to the tumor, at the same time as proscribing the damage to encompassing everyday tissues. At that aspect amid radiotherapy, the making plans information is surely enrolled onto the intra-procedural pix to direct the transport of radiation beams. Normally, in image-guided strategies its miles attractive that a steady imaging method can deliver patients' existence structures facts and song tool location on vicinity. Ultrasound is desired in mild of the reality that its miles ongoing and without radiation exposure. Anyhow, ultrasound may have some policies, for example, limited tissue assessment. Intra-procedural CT or CT fluoroscopy (CTF) are moreover regularly implemented, but they may now not be true sufficient for dynamic organs, for instance, lung and coronary heart imaging.

Comparative issues are furthermore to be trained in radiation treatment, and we are insufficient almost approximately ongoing or sans radiation imaging devices to expose the patients' motion. One-of-a-type regular imaging modalities, as an instance, endoscopy, X-rays, and actual-time MR can likewise be applied for direction. As an instance, bronchoscopy or colonoscopy endoscope cameras are carried out for practical actual-time visualization; in vascular intercession real-time X-ray or fluoroscopy combined with complexity operators is achieved to spotlight the vessels; in real functioning, open MR is applied to get ongoing anatomy images (of the thoughts). On the identical time as endoscopy gives actual-time recordings and fluoroscopy acquires regular second projection images, they depend upon clinicians' aptitudes for exploring the tool to the intention. Troubles for fluoroscopy are that they offer virtually community 2nd projection pix and discover each clinicians and patients to radiation. To fuse greater global records, pre-procedural 3-d CT is probably enlisted with 2nd fluoroscopy images so close-by or second real-time feedback, in addition to a international 3-d instance about in which the gadgets are determined, can be provided. Then again, even though it is possible to perform photograph-guided intervention inner open MR scanners, it is not generally carried out in clinics due to the price of the tool time and the complex scientific setup that goals each other tool to be MRI. Besides the above-said actual-time imaging techniques, tracking of interventional devices is every different project so that you can visualize them in the context of anatomy. The most commonly used tracking techniques of interventional devices embody electromagnetic (EM) monitoring and optical monitoring. For EM monitoring, coil sensors are set up within the interventional gadgets to music their real-time positions within the human frame protected through a magnetic area. Markers can be joined to the tool for optical tracking; to track its feature with a stereo imaginative and prescient framework by way of watching for the machine is rigid. Different tracking strategies are being advanced genuinely so the interventional gadgets may be tracked in the imaging hardware. As an example, primarily coil-based sensors can be brought with recognizing the needle tip and radio-frequency signs and symptoms can be distinguished by utilizing a MR device at the equal time because the intercession technique is finished. It's far on this locale motion damages becomes a noteworthy assignment for authentic guidance: its miles relevant that a real lung movement version can be applied as the guide for coping with the mediation amid every breath-maintaining cycle. Current advances in movement estimation show that via using regression models or device learning techniques, it is far doable to evaluate the lung motion from partially measurable indicators, for example, chest movement indicators. Inside the writing, there are various works that address lung movement, and they may be grouped into three trainings. The number one class is lung movement deformation modeling with registration; however, they just increase the respiration styles and do not accomplish movement estimation. The second one class employs an person affected individual's dataset, for example, first extracting lung movement in pre-procedural 4D outputs and for a while using this model amid remedy (radiotherapy or intercession), in which 4D filtering is needed. At ultimate, a factual model can be related to consolidate every gathering and individual statistics for movement demonstrating. Breathing styles organized from a large quantity of subjects can be executed to control the estimation of dynamic images of people, however while the 4D CT images are not available for them. As an instance, you can make use of the movement of an affected person's outer factors in combination with a movement model to make up for indoor breathing movement. The test is that the limited measurement of respiration sensor symptoms will not reflect the excessive-dimensional lung motion exactly, and in spite of the fact that they display respiration tiers nicely, they will no longer gauge precise lung tumor place and form dynamics. With new system analyzing and estimation improvements, the noncontact vision-based totally absolutely movement monitoring devices may be performed for recognizing the excessive-dimensional chest ground motion, and it is promising to degree dynamic lung movement through measurable model-based totally prediction among immoderate-dimensional chest ground and lung movement vectors. In this phase, we acquaint a manner with gauge dynamic lung snapshots from the 3-d CT and immoderate-dimensional chest ground symptoms. This estimation machine includes of levels: the education and the assessing degrees. Inside the training section, subsequent to appearing longitudinal registration of the 4D CT statistics of each education concern, the respiration movement fields are ascertained. At that element, each one of the images, motion fields, and chest floor actions of each mission are adjusted onto a template area for education a statistical model. The relationship of some of the trunk floor movement and lung breathing movement is then settled within the template area via manner of the use of expectation or relapse calculations. Amid the motion estimation prepare, the trunk floor motion alerts and the 3-d CT of a affected character is probably stuck and changed to the template area through photograph registration, and we are able to exercise the movement prediction model professional above to measure affected person-specific lung place motion from the chest ground signs and symptoms. At very last, the assessed dynamic lung photographs are modified decrease once more to the affected individual photograph place. For example, the usage of CT lung image motion restores as we hooked up that its miles viable to estimate the dynamics of the lung by means of using available real-time chest movement symptoms. But for the vital clinical image processing strategies required for motion compensation, we tended to the overall query, i.e., given the changeability of the complete shape, the way to estimate the shape from particularly regarded data. Truth be recommended many engineering troubles fall into this estimation elegance, and one-of-a-type prediction and regression fashions can be done. In this element, a big PCA version or K-PCA version can be utilized to extend the immoderate-dimensional alerts and fields onto low-dimensional regions. At that factor, Bayesian estimation or a manual vector device is applied for estimation. At prolonged final, utilizing a CTF-guided mediation, as an instance, we constitute how motion estimation may be related in a systematic software program. Future works contain assessing how the estimation model may be more energetic and dependable to cope with numerous sizes and breathing sorts of the sufferers, and the way to put in force them regularly in a real-time estimation of location and nation of the tumor for intervention and radiotherapy [41].

### 3.3 *Hierarchical Parsing in Medical Images Using Machine Learning Technologies [41]*

In a few clinical imaging bundles, from simple body-issue ID to lung knob recognition, one significant undertaking is to restrict and translate specific anatomical frameworks in clinical photographs. For an expert master, this test is consistently finished through taking advantage of the "life systems marks" in logical pictures—the fitting structure or appearance elements of the life systems of interest. Truth be told, inside the course of logical preparing and exercise, a restorative master learns heaps of those "life structures marks" from seeing a tremendous type of therapeutic photographs. In the meantime as tablet supported calculations have rise as progressively understood inside the logical imaging region, it isn't astonishing that the investigations organize has become keen on developing calculations to naturally limit and decipher a few life structures in clinical depictions. The viability of those calculations is continually depending on huge kind of life system marks if any might be pleasantly extricated with the guide of method for way of the calculations. At the indistinguishable time as we look often as possible, it has been made specific photograph channels to extricate life structures marks, a more noteworthy ultra-current research form demonstrates the predominance of picking up learning of-based genuinely techniques for two thought processes: (1) machine becoming acquainted with period have created to determine real overall issues. (2) Greater clinical photograph datasets are to be needed to encourage the "investigation" of entangled restorative imaging data.

In [41], a talk on a breaking down-based absolute system can "investigate" life system marks and limit different anatomical structures in logical pictures. In this system, it has a layered shape and parses human life structures in a harsh to-great style. Particularly, our device begins from the littlest and the most number one anatomical element—anatomical points of interest, and this is then drawn out to the recognizable proof of life systems pressing compartments (locale of leisure activities), coarse organ division, and sometime reaches to decide and characterize the organ limits. On one attitude, calculations created on each layer may also furthermore advantage tremendous picture evaluation bundles; for instance, jumping challenge identification of knee meniscus empowers to mechanize attractive reverberation (MR) imaging making arrangements. As an open door, calculations progressed on diminish layers moreover give middle results to the higher layers. As a case, milestone identification gives essential photograph look that prompts to coarse organ division. Its miles appropriately important that, as a result of reality calculations on this structure, motivation to separate life structures marks by means of becoming more acquainted with, they might be presently not confined to exact organ slants. Due to this the entire structure, it is versatile to interesting imaging modalities notwithstanding organ frameworks.

In the meantime as conventional procedures attempt to format channels or layouts for particular point of interest more prominent to a great degree, current research and practices have utilized gadget examining strategies for an answer adaptable to remarkable milestones, particularly imaging modalities. Advisor works include which lease probabilistic boosting tree and arbitrary timberlands to break down the coming inclinations of points of interest.

Programmed discovery of anatomical pressing holders has moreover been radically inquired about. An introduction to basic difficulty is the best approach to manage the entry fluctuation of particular thing. The thing identifiers for various postures are completely performed, and the adjusted stance is the representation that relates to the indicator which triggers the best reaction. To appraise protest postures utilizes close-by introduction evaluation. Nowadays, negligible space considering (MSL) has approved promising impacts in assessing rough adjustments of anatomical structures in logical pictures. The basic idea is to partition the change parameters into locale, introduction, scale, etc. What's more, prepare indicators to assess each of them successively. On account of the way that it treats the aim life structures (e.g., a coronary heart chamber or the knee menisci) all in all, this approach has the heaviness of radical becoming more acquainted with, especially for the primary occurrence of breaking down (region finders).

The displaying of frame priors decides the viability of coarse organ division. Following the developing utilization of the dynamic frame demonstrate, one-of-a-kind strategies have been proposed to improve the productiveness of shape ahead of time displaying in a few conditions. Inside the nonlinear shape adaptations through a blend of Gaussians and complex considering, it made utilization of incomplete enthusiastic frame display (ASM) to make the frame form solid to lacking constraints. A meager most imperative perspective assessment (PCA) is gotten sparser modes and bring near orthogonal added substances. It separates the frame demonstrate into a few autonomously displayed added substances for various leveled shape case. Each of those works inside the course of gaining little shape forms and holds frame data. For exact organ division, deformable designs have been impressively contemplated, especially inside the region of logical photograph division. The impressively analyzed vitality of deformable models originates from their ability to stage anatomic frameworks through misusing requirements got from the photograph actualities (base up) joined with prior data around the ones frameworks (apex down). The disfigurement strategy is by and large detailed as an advancement issue whose reason trademark comprises of an outside (photograph) era and an internal (shape) day and age. While internal power is intended to hold the geometric advancements of the organ beneath take a gander at, the outside power is characterized to transport the model toward organ impediments. In popular, the outside power term regularly originates from angle records, for instance, picture inclination. In bleeding edge day years, additional exertion has been contributed at the total of different photograph abilities, as a case, close-by measurements and surface models. By joining particular photograph skills as the outside power, deformable models have completed to a great degree great accomplishment in various logical practices. Gadget picking-up information of procedures has given an ultra-present-day course to a more prominent regular outside power design. Through the use of learning-based absolute strategies, limited capabilities can be put from tutoring realities. Through this way, the "design" of outside quality transforms into information or information pushed and can be drawn out to consider stand-out imaging modalities.

### *3.4 Anatomy Landmark Detection [41]*

Anatomical historic points are naturally broad components found in exact organ frameworks. They will be characterized to give a clarification to the morphological homes of anatomical frameworks and straightforwardness discussion among researchers inside the fields of science and cure, etc. In the meantime as most anatomical points of interest develop as noticeable in vivo in the logical imaging system, the ones anatomical milestones assume critical parts inside the elucidation of clinical depictions. Since the geographic historic points control voyagers in investigating the earth, in like manner anatomical milestones offer strategies to explore the clinical pictures. For instance, anatomical substance material interior a restorative photograph can be chosen with the helpful asset of the utilization of the areas of anatomical historic points. Likewise, a couple of anatomical historic points moreover offer imperative hints to analyze sicknesses. For instance, the back intersection part of a lumbar vertebra and the spinal wire might be utilized to analyze spondylitis and test the sickness level. At long last calculations that can mechanically discover anatomical points of interest can immediately profit a few restorative utilize times. But the immediate impact on restorative utilize examples, programmed discovery of anatomical milestones in addition, clears the way for specific logical pictures assessment obligations. For instance, to introduce deformable models for organ division, you may utilize routinely recognized historic points. They can likewise offer a preparatory change for photograph enlistment. Because of the perplexing appearance of a few anatomical milestones, we utilize a becoming more acquainted with based absolutely technique for individual historic point location. Because of its data-driven nature, contemplating-based practical strategies besides make adaptable to exact imaging modalities, for example, figured tomography (CT), MR, positron emanation tomography (residential pooch), ultrasound. Its miles detailed to milestone recognition as a voxel-cunning sort issue. For the most part, voxels near the point of interest are mulled over as outstanding specimens and voxels a long way from the historic point are contemplated as poor ones. To check a chose point of interest identifier, first explain the milestone in preparing set photographs. The colossal and lousy examples/voxels are chosen principally in view of the separations among the clarified point of interest and the voxel. For each preparation test (voxel), a firm of clean capacities is removed in its system. All acclaimed capacities are then bolstered appropriate directly into a course wonderfulness system as checked. The course system is intended to address the exceedingly uneven effective and horrendous specimens. Actually, astounding voxels round historic points underneath investigate are positives and whatever is left of the voxels are negatives; the proportion of positives to negatives is regularly a ton considerably less than 1. The instruction set of standards is "one-sided" to positives, with the end goal that each unnecessary astounding should be proficiently arranged; however, the negatives are permitted to be misclassified. The ones misclassified negatives, that is, fake positives, can be what's more used to prepare the ensuing falls. At run-time, on an indistinguishable time from positives are relied upon to experience all falls, most negatives can be dismissed inside the initial a few falls and do not require promote assessment.

In [41], a framework to parse human anatomies in scientific pix is added. The framework is designed in a difficult-to-excellent style to localize and perceive human anatomies from landmarks to unique organ limitations. In each layer, numerous gadget getting-to-know generations, which encompasses AdaBoost, sparse instance, iterative clustering, and so on, are used to investigate "anatomical signatures." In this manner, our framework may be prolonged to numerous scientific programs in one in every of a kind trust modalities and can be without trouble tailored to new programs. No matter the truth that the specific layers independently advantage particular clinical use times, similarly they help every different from a technical component of view. For instance, the land marking set of rules offers appearance cues for coarse organ segmentation and in exceptional manner coarse organ segmentation paves the way for specific segmentation.

## *3.5 Machine Learning in Brain Imaging [41]*

Brain images genomics has pulled in increasing attention these days. It is far a transpiring studies field that has emerged with the advances in excessive-throughput genotyping and multimodal imaging strategies. Its significant venture is to take a look at the connection among hereditary markers, for example, unmarried nucleotide polymorphisms (SNPS) and quantitative features (QTS) extricated from multimodal neuro-imaging information (e.g., anatomical, beneficial, and subatomic imaging filters). Given the first-rate significance of imaging and genomics in the brain ponder, crossing over those factors and investigating their associations can in all likelihood supply an advanced robot comprehension of normal or harassed mind capacities. Likewise, adjustments in imaging phenotypes normally go earlier than those in disease reputation and mental consequences and are prevalent to be nearer to the fundamental hereditary devices. Along these traces partner hereditary statistics with imaging phenotypes, rather than illness reputation, is exceedingly encouraging for revelation of powerful hereditary design and can in all likelihood help uncover the soonest brain adjustments for anticipation.

Early endeavors in imaging genomics were usually pair sensible univariate examinations for quantitative broad affiliation considers (GWASS), which were completed to relate high-throughput SNP statistics to significant scale photo QT facts. A fundamental relapse display, as an instance, that actualized in plink, turned into regularly used to look into the introduced substance impact of each single SNP on every single imaging QT. Those fundamental relapse fashions have been regularly combined with idea checking out, in which the significances of relapse coefficients were found out at the identical time. Pairwise univariate investigation changed into applied as part of conventional affiliation studies to hastily give vital association statistics among SNPs and QTs. However, both SNPs and QTs were dealt with as self-sufficient and disconnected units, and as a result, the hidden connecting shapes among the units become disregarded. One-of-a-kind relapse fashions had been later acquainted with examines the multilocus influences on imaging phenotypes. Rather than locating singular SNPs with crucial affects, multivariate models have empowered the recognizable proof of SNPs that together have an effect on phenotypic alternate. GCTA, a typical heritability examination tool, makes use of a direct combined relapse show and has efficiently exhibited that $\sim$45% of the phenotypic trade for human tallness may be clarified by using the joint impact of simple SNPs. Likewise thinking about the inter-correlated nature and high-dimensional placing of imaging and hereditary facts, insufficient relapse models are specifically supported. These models can deal with the relationship issue, as well as help to distinguish few evidently significant hereditary markers for easy understanding. Additionally, numerous modern-day endeavors have been committed to in advance learning-guided relapse fashions, and lots of evaluations have affirmed the advantageous a part of in advance data structure in catching extra particular imaging genomic connections.

Bimultivariate affiliation examination has as of late gotten increasing attention for research of complicated multi-SNP-multi-QT courting. Existing bimultivariate fashions widely applied as part of imaging genomic research can via and massive be arranged into two types: well-known relationship examination (CCA) kind and dwindled rank relapse (RRR) type. In view of the presumption that a real imaging genomic flag on the whole includes few SNPs and QTs, both CCA and RRR have their scanty paperwork. These scanty models are intended to higher healthy the imaging genomic contemplate as they yield inadequate examples for easy translation. Instance contemplates makes use of these techniques in imaging genomic programs. Earlier mastering has moreover been analyzed in bimultivariate fashions as of past due, and its useful component has likewise been accounted for in many critiques.

Advancement examination has been normally taken into consideration in first-rate expression facts investigation and has as of late been changed to interrupt down GWAS statistics to separate organic bits of understanding in the light of realistic comment and pathway databases. As of past due, it has been reached out to the imaging genomics area, to discover extraordinary state affiliations in view of in advance getting to know, which include crucial great units (GSS) and brain circuits (BCs), which usually incorporate various characteristics and numerous QTs, for my part. By way of at the same time thinking about the difficult connections between the interlinked hereditary markers and corresponded imaging phenotypes, imaging genomic enhancement investigation (IGEA) offers greater strength to extricating organic bits of understanding on neuro-genomic dating at a frameworks science degree. Within the maximum current decade, real and system studying has been assuming a primary element in imaging genomic thinks approximately and has efficiently superior the disclosures of biologically essential biomarkers and further fundamental affiliation designs. Potkin et al. utilized plink to research the hereditary premise of hippocampal decay in Alzheimer patients and efficiently outstanding five hazard characteristics required inside the path of protein debasement, apoptosis, neuronal misfortune, and neurodevelopment. In preference to auxiliary changes, another gathering as of late found out a GWAS result among brain availability and hereditary versions, in which one threat fine SPON1 become diagnosed and moreover recreated in a self-sufficient accomplice.

In multilocus, affects were researched in opposition to the temporary projection extent using the flexible net approach, wherein SNPs in traits RBFOX1 and GRIN2B had been, one after the other, determined to be profoundly contributory genotypes. Via making use of a variation of PCR over all qualities and an extensive database of voxel-clever imaging facts, outstanding 10 vital SNPS in GRB-related proscribing protein 2 qualities (gab2), to essentially connect to all voxels. So to make use of RRR, inside the wake of approving its major execution on a huge scale dataset, to take a look at the multilocus influences over the voxel-wise imaging statistics but using their longitudinal adjustments. Their discoveries affirmed the important thing a part of APOE and TOMM40 in AD and highlighted some other ability relationship too. However, these version degree discoveries, in advance, record guided strategies additionally assist to find competitor pathways, whose trouble can also doubtlessly set off adjustments in imaging phenotypes. As an instance, a protecting bunch tether punishment introduced to RRR changed into utilized to demonstrate the pathway having an area with SNPs and a few pathways were accounted for to be related with longitudinal shape modifications in mind, as an instance, the insulin flagging pathway, chemokine flagging pathway, and Alzheimer's illness pathway. With the help of S2CCA, numerous to-specific connection among APOE and all inclusive amyloid collections, and detailed limited amyloid examples encouraged by using the joint impact of APOE SNPs are inspected. In the interim, a comparable check was carried out utilizing KG-SCCA, in which a transcriptome coexpression device becomes related as an earlier, and comparative amyloid collection examples have been diagnosed [41].

## *3.6 A Connectome-Based and Machine Learning Study [41]*

With the advances in computational neuro-imaging, its miles now able to investigate complete mind maps of structural connectivity it's far possible to chart the employer of white count connectivity throughout the complete mind. With the resource of combining segmented gray remember tissue statistics with white rely fiber tractography from diffusion tensor imaging (DTI) MRI from weighted magnetic resonance imaging (MRI), i.e., structural mind connectome. The mind connectome gives a super diploma of information approximately the business organization of neuronal network shape. Its miles each at a regional degree and moreover concerning the whole brain community [41] because of which the mind connectome has these days grow to be instrumental inside the investigation of neuronal systems organization. And this dating of fitness and ailment, with appreciate to neurological conditions together with epilepsy, has come to be trivial.

Epilepsy is a neurological disorder. Its miles at once associated with pathological adjustments in thoughts community corporation. Maximum varieties of epilepsy are understood to stand up from epileptogenic pastime growing from localized mind areas. With lot of evidence, it suggests that focal seizures are in reality the result of hyperexcitation of localized networks than remote cortical areas. Focal epilepsy is appeared to be an illness of neuronal networks in which there is a deviation from structural and functional business enterprise of neuronal systems which leads to unabated neuronal hyper-excitability and seizures.

Histopathological adjustments were first of all measured to be restricted to isolated thoughts areas. There can be growing frame of evidence from neuro-imaging to suggest that more subtle abnormalities are pervasively allocated across more than one brain areas with the hippocampal sclerosis in people with temporal lobe epilepsy (TLE). In TLE, functionally or structurally associated areas with the hippocampus showcase gray depend on mobile loss. It additionally affects white count lack of the sensory input from part of the body and pathological network structure reorganization. Due to these motives, epilepsy is now nicely concept-out to be a circumstance that emanates from unusual neuronal structures instead of remote brain regions. The idea of epilepsy as a neuronal network sickness features considerably inside the maximum cutting-edge epilepsy. Global league toward epilepsy seizures the class from it. It is not completely understood approximately the scientific relevance of pervasive network abnormalities concerning seizure control. So the more superb the disbursed networks, the greater numerous clinical endophenotypes regarding seizure manipulate. Connectome primarily based studies is in its initial levels. But it has established a constant pattern of intra and further temporal connectivity abnormalities. Moreover, individualized styles of network abnormalities may be straightaway related to diagnosis [41].

## 4 Limitations with Medical Images

The obstacle concerning technical regard is the restrictions of guide capability manage human organs. The restrained access, constrained visualization of tissues and also restricted diagnostic facts of the tissues. To overcome those troubles worldwide technology traits have created new methods and technologies which entails robotics and endoscopic interventions with smart sensor devices. These sensor-driven gadgets offer upward push to improvement and spread of wise structures in scientific area by way of the use of ML strategies. Therefore, the need for brand spanking new imaging standards such as laser scanning microscopy and fluorescent imaging with PGMs is excessive. Even though medical results are looking ahead to, technically tendencies are very hopeful. The endless research in this factor will make clean the right possibility of this technology. Moustakis and

Charissis' work [3] discussed the position of ML in medical applications wherein it becomes insisted that became shifting far from the accuracy measures as sole evaluation criteria of gaining knowledge of algorithms. The opportunity trouble of comprehensibility can be very trivial and additionally cautiously considered in the evaluation.

## 4.1 Future Guidelines

Image mining is a diffusion of records mining approach. Many image processing algorithms include photograph mining. So, photograph mining is attracted via many researchers to discover new answers to the issues inside the recent years. The subsequent investigations may be protected with the implementation of the Bayesian networks for relevant feedbacks. Extra investigations are to be completed to locate the hidden relationships among images [5].

Theoretically, it is essential to enhance the information of ML algorithms and additionally to discover the mathematical justification for his or her residences. This is required considering an expertise and answering of essential questions with the behavior of ML methods as observed. Conversely, few problems that are worried inside the procedure of getting-to-know expertise in positioned to exercise are the visualization of the learned knowledge. There may be a need for the algorithms to mine in addition to discover the noise and outliers within the information [6].

Few other problems which upward push up in PGM packages are over becoming and scaling houses at the same time as achieved with large datasets, immoderate-dimensional information (input (characteristic), and output (training-lessons) areas).

Because technical, organizational, and social problems end up intertwined inside the issues like the requirement for directness of the mastering final results with respect to guidelines while choosing the algorithm within the clinical context which integrates the patient records and position of smart systems in health care turns into very trivial, the location of health care relies upon at the successful combination of the generation with the organizational and social context within which it is far finished. From the preceding studies and know-how, the ones issues are complicated because the achievement universal performance of information structures [42, 43] and choice help structures specially [44, 45]. Since moral issues are worried, clinical data of affected individual inclusive of analysis and remedy will become trivial. Few of those problems are gadget-focused which inherently associated issues of the ML research. Therefore, they have a look at of the emerging demanding conditions and moral troubles from a human-targeted perspective with the aid of thinking about the motivations and ethical dilemmas of researchers, builders, and scientific users of ML strategies in scientific programs which have turn out to be important [2]. With the improvement of brain imaging and genotyping techniques, it has opened amazing alternatives finding the underlying diseases and its mechanisms. More complicated fashions may be building to capture the actual and the hidden information with the genome sequencing, longitudinal imaging, and mind connectome. Scalability is the foremost challenge as the existing techniques are a success simplest when worked on small statistics sets. Models based on regression and association with dimensionality reduction is needed for the near future.

Massive facts are promising destiny into the brain imaging genomics subject to overcome the computational complexity. Due to the fact the conventional GWAS is based totally on supercomputing techniques, attempts are made to consciousness at the overall performance of GWAS. Some of the excessive overall performance tools are rapid-LMM, EPIGPU, and GBOOSt. To reinforce up the SCCA, an effort modified into made by way of combining intel Math Kernel Library (MKL) and offload version for intel many included middle (MIC). its miles located that a reliable twofold velocity was accelerated without any code change. A tenfold development became accomplished by using coupling map/lessen framework with random wooded area for associating SNPs and photograph phenotypes. The capacity of the large statistics techniques had a contribution of the initial steps. In future, greater such techniques are required for mind and genome great research [41].

HAFNI framework can be advanced via the usage of superior algorithms including sign sampling techniques with dictionary learning and sparse representation techniques. To handle FMRI "big facts," the newly evolved HAFNI-enabled massive scale platform for neuro-imaging informatics (HELPNI) or hadoop-/spark-based structures is carried out. HAFNI framework can be used to brain dynamics, to pick out community-based totally neuro-imaging biomarkers to symbolize, subtype, and diagnose different neurological/psychiatric issues (e.g., autism, schizophrenia, post-traumatic strain disorder, and many others) [41].

## 5 Conclusion

Developing these probabilistic graphical models requires some critical points for large-scale adoption of probabilistic graphical models in an area where there is a lot of uncertainty about the capacities of computerized systems. Important point to be observed is that just by being more accurate systems are not successful. The impact of it can be measured only when the radiologists are able to diagnose the patients with the assistance of software developed. So there is a need for validation with respect to radiologist's accuracy. Next point to be observed is that research innovation should be specific to data domain. Changing from SVM to the neural network or probabilistic graphical models will not improve the performance of the system in a major way.

It is important to analyze, assume, and evaluate appropriately according to the problem. The relationship between image processing and probabilistic graphical models in image mining is critical. Image processing algorithms will extract the features that are given to probabilistic graphical models for successful analysis.

In-depth basic/fundamental knowledge behind these probabilistic graphical models will lead to a good understanding of statistical issues that are vulnerable to the classifier.

In preliminary approaches, a large number of image mining algorithms or probabilistic graphical models are designed which have not performed up to the mark in the real world. So there should be coordination between subject experts and model developers. Finally, the end user should benefit out of the model developed by the researcher. In this chapter, a review on a variety of mining techniques that were proposed by researchers was presented.

## References


1. Christopher, M. *Bishop pattern recognition and machine learning*.
2. www.dcs.bbk.ac.uk/~gmamoulas/ACA199_workshop.pdf.
3. Moustakis, V., & Charissis, G. (1999). Machine learning and medical decision making. In *Proceedings of Workshop on Machine Learning in Medical Applications, Advance Course in Artificial Intelligence ACAI99, Chania, Greece* (pp. 1–19).
4. Fung, G., Krishnapuram, B., Bi, J., Dundar, M., Raykar, V., Yu, S., et al. (2009). Mining medical images. In: *Third Workshop on Data Mining Case Studies and Practice Prize, 15th Annual SIGKDD International Conference on Knowledge Discovery and Data Mining, Paris, France*.
5. Sudhir, R. (2011). A Survey on image mining techniques theory and applications. *Computer Engineering and Intelligent Systems, 2*(6).
6. Ordonez, C., & Omiecinski, E. (1999). Discovering association rules based on image content. In *Proceedings of the IEEE Advances in Digital Libraries Conference (ADL'99)*.
7. Megalooikonomou, V., Davataikos, C., & Herskovits, E. (1999). *Mining lesion-deficit associations in a brain image database*. San Diego, CA, USA: KDD.
8. Missaoui, R., & Palenichka, R. M. (2005). Effective image and video mining: An overview of model based approaches. In *MDM'05: Proceedings of the 6th International Workshop on Multimedia Data Mining* (pp. 43–52).
9. Fernandez, J., Miranda, N., Guerrero, R., & Piccoli, F. (2007) *Appling Parallelism in Image Mining*. www.ing.unp.edu.ar/wicc2007/trabajos/PDP/120.pdf.
10. Sanjay, T., et al. (2007). Image mining using wavelet transform. In *Knowledge-based intelligent information and engineering systems* (pp. 797–803). Springer Link Book Chapter.
11. Pattnaik, S., Das Gupta, P. K., & Nayak, M. (2008). Mining images using clustering and data compressing techniques. *International Journal of Information and Communication Technology, 1*(2), 131–147.
12. Kralj, K., & Kuka, M. (1998). Using machine learning to analyze attributes in the diagnosis of coronary artery disease. In *Proceedings of Intelligent Data Analysis in Medicine and Pharmacology-IDAMAP98, Brighton, UK*.
13. Zupan, B., Halter, J. A., & Bohanec, M. (1998). Qualitative model approach to computer assisted reasoning in physiology. In *Proceedings of Intelligent Data Analysis in Medicine and Pharmacology-IDAMAP98, Brighton, UK*.
14. Bourlas, P., Sgouros, N., Papakonstantinou, G., & Tsanakas, P. (1996). Towards a knowledge acquisition and management system for ECG diagnosis. In *Proceedings of 13th International Congress Medical Informatics Europe-MIE96, Copenhagen*.
15. Bratko, I., Mozetič, I., & Lavrač, N. (1989). *KARDIO: A study in deep and qualitative knowledge for expert systems*. Cambridge, MA: MIT Press.

16. Hau, D., & Coiera, E. (1997). Learning qualitative models of dynamic systems. *Machine Learning, 26,* 177–211.
17. Akay, Y. M., Akay, M., Welkowitz, W., & Kostis, J. B. (1994). Noninvasive detection of coronary artery disease using wavelet-based fuzzy neural networks. *IEEE Engineering in Medicine and Biology*, 761–764.
18. Lim, C. P., Harrison, R. F., & Kennedy, R. L. (1997). Application of autonomous neural network systems to medical pattern classification tasks. *Artificial Intelligence in Medicine, 11,* 215–239.
19. Lo, S.-C. B., Lou, S.-L. A., Lin, J.-S., Freedman, M. T., Chien, M. V., & Mun, S. K. (1995). Artificial convolution neural network techniques and applications for lung nodule detection. *IEEE Transactions on Medical Imaging, 14,* 711–718.
20. Micheli-Tzanakou, E., Yi, C., Kostis, W. J., Shindler, D. M., & Kostis, J. B. (1993). Myocardial infarction: Diagnosis and vital status prediction using neural networks. *IEEE Computers in Cardiology*, 229–232.
21. Miller, A. S., Blott, B. H., & Hames, T. K. (1992). Review of neural network applications in medical imaging and signal processing. *Medical & Biological Engineering & Computing, 30,* 449–464.
22. Nekovei, R., & Sun, Y. (1995). Back-propagation network and its configuration for blood vessel detection in angiograms. *IEEE Transactions on Neural Networks, 6*(1), 64–72.
23. Neves, J., Alves, V., Nelas, L., Romeu, A., & Basto, S. (1999). An information system that supports knowledge discovery and data mining in medical imaging. In *Proceedings of Workshop on Machine Learning in Medical Applications, Advance Course in Artificial Intelligence-ACAI99, Chania, Greece* (pp. 37–42).
24. Pattichis, C., Schizas, C., & Middleton, L. (1995). Neural network models in EMG diagnosis. *IEEE Transactions on Biomedical Engineering, 42*(5), 486–496.
25. Phee, S. J., Ng, W. S., Chen, I. M., Seow-Choen, F., & Davies, B. L. (1998). Automation of colonoscopy part II: Visual-control aspects. *IEEE Engineering in Medicine and Biology*, 81–88.
26. Pouloudi, A. (1999). Information technology for collaborative advantage in health care revisited. *Information and Management, 35*(6), 345–357; Rayburn et al. 1990.
27. Yeap, T. H., Johnson, F., & Rachniowski, M. (1990). ECG beat classification by a neural network. In *Proceedings of the 12th Annual International Conference of the IEEE Engineering in Medicine and Biology Society* Philadelphia, Pennsylvania, USA (Vol. 3, pp. 1457–1458).
28. Coppini, G., Poli, R., & Valli, G. (1995). Recovery of the 3-D shape of the left ventricle from echocardiographic images. *IEEE Transactions on Medical Imaging, 14,* 301–317.
29. Hanka, R., Harte, T. P., Dixon, A. K., Lomas, D. J., & Britton, P. D. (1996). Neural networks in the interpretation of contrast-enhanced magnetic resonance images of the breast. In *Proceedings of Healthcare Computing, Harrogate, UK* (pp. 275–283).
30. Innocent, P. R., Barnes, M., & John, R. (1997). Application of the fuzzy ART/MAP and MinMax/MAP neural network models to radiographic image classification. *Artificial Intelligence in Medicine, 11,* 241–263.
31. Karkanis, S., Magoulas, G. D., Grigoriadou, M., & Schurr, M. (1999). Detecting abnormalities in colonoscopic images by textural description and neural networks. In *Proceedings of Workshop on Machine Learning in Medical Applications, Advance Course in Artificial Intelligence-ACAI99, Chania, Greece*, pp. 59–62.
32. Veropoulos, K., Campbell, C., & Learmonth, G. (1998). Image processing and neural computing used in the diagnosis of tuberculosis. In *Colloquium on intelligent methods in healthcare and medical applications*. UK.
33. Zhu, Y., & Yan, H. (1997). Computerized tumor boundary detection using a Hopfield neural network. *IEEE Transactions on Medical Imaging, 16,* 55–67.
34. Delaney, P. M., Papworth, G. D., & King, R. G. (1998). Fibre optic confocal imaging (FOCI) for in vivo subsurface microscopy of the colon. In V. R. Preedy & R. R. Watson (Eds.),

*Methods in disease: Investigating the gastrointestinal tract*. London: Greenwich Medical Media.
35. Anderson, J. G. (1997). Clearing the way for physician's use of clinical information systems. *Communications of the ACM, 40*(8), 83–90.
36. Lane, V. P., Lane, D., & Littlejohns, P. (1996). Neural networks for decision making related to asthma diagnosis and other respiratory disorders. In *Proceedings of Healthcare Computing, Harrogate, UK* (pp. 85–93).
37. Ridderikhoff, J., & van Herk, B. (1999). Who is afraid of the system? Doctors' attitude towards diagnostic systems. *International Journal of Medical Informatics, 53,* 91–100.
38. Sudhir, R. (2011). A survey on image mining techniques theory and applications. *Computer Engineering and Intelligent Systems, 2*(6).
39. Pearl, J. (1988). *Probabilistic reasoning in intelligent systems: Networks of plausible inference*. Morgan Kaufmann.
40. Boykov, Y., Veksler, O., & Zabih, R. (2001). Fast approximate energy minimization via graph cuts. *IEEE Transactions on Pattern Analysis and Machine Intelligence, 23*(11), 1222–1239.
41. Yousofi, M. H., Esmaeili, M., & Sharifian, M. S. (2016). A study on image mining; its importance and challenges. *American Journal of Software Engineering and Applications, 5* (3–1), 5–9.
42. http://pgm.stanford.edu/intro.pdf.
43. Machine Learning and Medical Imaging 1st Edition Academic Press. (2016). eBook ISBN: 9780128041147; Hardcover ISBN: 9780128040768.
44. Tappen, M. F., Russell, B. C., & Freeman, W. T. (2004). Efficient graphical models for processing images. In *IEEE Conference on Computer Vision and Pattern Recognition (CVPR)*.
45. Wainwright, M. J., & Jordan, M. I. (2008). Graphical models, exponential families, and variational inference. *Foundations and Trends® in Machine Learning, 1*(1–2), 1–305. https://doi.org/10.1561/2200000001.

# Pipeline Crack Detection Using Mathematical Morphological Operator

**A. Prema Kirubakaran and I. V. Murali Krishna**

**Abstract** A pipeline crack is a major hazard to any type of liquid transportation. Oil industry depends mainly on human detection of these cracks, which leads to many problems from health to environment disaster. To detect a crack, pipelines' inner layers are X-rayed, and these X-rays were later manually evaluated for cracks and holes. This technique evolves lot of time and resources. This proposed research work helps to diminish this problem by analyzing the cracks and holes through a computerized solution. A pipeline with crack is analyzed using image analysis and processing which comprises of various pattern recognition techniques. Image analysis and processing is one of the most powerful innovations in today's world. It brings all kinds of pattern recognition together and solves the problem of data identification and misuse of data. This is achieved by applying the method of pattern analysis and recognition. As a result, this technique of image analysis and processing is used to detect the holes and cracks which occur in a pipeline that carries any type of liquids and gases. This paradigm helps in environmental safety. As in the pipeline industry there are many man-made equipments and methods, computer application to carry out these process is lacking. To make a shift over to the computerized image recognition, high-frequency filter (HFF) with Gate Turn off thyristor (GTO) using unsupervised-based learning algorithm is implemented. Mathematical morphological operator and edge detection principles are used for image evaluation. Initially a digital camera with fiber optic cable is passed inside a pipeline to capture the cracked images. These images are converted as raster images and stored as bits. Later, these images are processed to view for hidden points using unsupervised cluster algorithm; after evaluating the hidden points, the crack has to be measured for its length and to identify the location where it occurs and this is achieved by developing mathematical morphological operator. Images are always

A. Prema Kirubakaran (✉)
Department of Computer Science, Annai Violet Arts & Science College, Chennai, India
e-mail: unjanai@gmail.com

I. V. Murali Krishna
IIT Madras, Chennai, India

I. V. Murali Krishna
Remote Sensing, NRSA, ISRO, Hyderabad, India



S. Margret Anouncia and U. K. Wiil (eds.), *Knowledge Computing and its Applications*, https://doi.org/10.1007/978-981-10-8258-0_2

not clear, so the dataset formed is always unclear; in order to smooth the images, the edge detection principle is applied. The captured images are read as pixel groups, after converting into raster images. If the pixels grouped as clusters are clear with no zero bit values, it denotes that the pipeline is without any crack else even a small relapse in any one of the pixel will make the image vague. The blurred picture denotes that there is a defect in the pipeline. This helps to locate the image with defect, which is rectified and thus it results in an effective, defect-free passage that will carry any type of liquid or gas. The extraction of hidden patterns from a large quantity of data recognition activities seems to be a major conflict in image detection. Usually, the size of a pipeline is 22 m; in order to avoid the problem of large data, the length to capture an image can be reduced to 10 m and later the rest. The safety of the environment and the manual operator is very important during the transportation of any substance in a pipeline; this major security concern has made image detection to become a popular component in the area of image analysis and processing. In traditional manual detection systems, the manual operator needs to spend much time in analyzing the data and also it generates high false data rates. There is an urgent need for effective and efficient methods to discover both the unknown and unexpected novel image display over a pipeline network from those that are extremely large in size and high in dimensionality and complexity. So the pattern recognition-based image analysis detection systems have been chosen which are more precise and require less manual processing time and input from human experts. This research focuses on solving the issues in image detection communities that can help the system operator to make processing, classification, labeling of data and to mitigate the outcome of image data. The system administrator finds it difficult to preprocess the data. Even though it has been done successfully, the overwhelming output of the images makes the task a failure and even sometimes images go unidentified. To overcome this situation, frequent updating of data is needed. In order to reduce the workload of the administrator, four major image analysis and processing techniques involving pattern recognition task have been introduced. Image detection datasets have been used in this research, and the proposed algorithms will be implemented in MATLAB. In this research, for classification of network data, several existing algorithms like Kohonen-cluster algorithm, Canny's edge detection algorithm, and mathematical morphological operator algorithm for the simulation of images are proposed. The crack is measured using mathematical morphological operator. Mathematical morphology is evaluated using the concept of geometric measurements. Set theory is applied to evaluate morphological-based geometric measurements. An important technical goal is to provide sufficient information so that the readers can apprehend and possibly implement the technique that has been derived. The result of this study will build a system, for identifying the defects in an oil pipe, by matching to an image database. This system can be implemented or replaced with the existing manual one. **Need of this topic**: Engineers pursuing mechanical stream and who would like to have a career from normal mechanical to oil pipeline can refer the topic through this book for their career enhancement. The chapters on erosion, corrosion, and dilation will help them to make a more effective study on cracks and holes which can be applied through robotics. This book will focus more on detecting the cluster cracks that are neglected during an inspection of an oil pipeline. This concept of detecting a hole or a crack can be applied to any type of pipeline that is going to transport any type of substances. The topic of the book can be the same or it can be altered according to the needed definition of the engineering society. The opportunity to write these chapters will help to learn more about the mathematical operators to detect a clustered crack.

# 1 Introduction

## *1.1 Image Analysis and Processing*

Image analysis is the process of taking out required information from digitized pictures or images by means of the technique known as digital image processing. Increasing digitization has paved way for many new techniques, and this leads to data complexities which are continuously posing new challenges. These challenges are conquered with various techniques and solutions. Big data concepts termed as new era of data analysis and processing helps to handle with volume of data from various structures and sources. Data is coming in variety of forms like structured, semi-structured, unstructured, and sensor data. An excellent image analysis tool is our own eyes, which extracts high-level information. Proper mechanism is required to handle decision-making situations. Digital image processing is the only practical technology for

- Classification
- Feature extraction
- Pattern recognition
- Projection
- Multi-scale signal analysis

## *1.2 Image Coordinates*

Signals are also used for image processing. These image outputs are treated as 2D signals, and standard signal processing techniques are applied for image detection. Image analyzed will be considered to be a function of two real variables as $A(X, Y)$, where $A$ will be considered as amplitude defined as the brightness of the image at the real coordinate position $(X, Y)$. This chapter will deal with three important categories of manipulating an image: image processing, image analysis, and image understanding.

# 2 Pipeline

Crude oil and natural gas over land transporting are safe and efficient through pipelines. Each and every day, oil companies transport enough crude oil and petroleum products to fill 15,000 tanker truck loads and 4,200 railcars. Pipeline network transports three million barrels of oil every day. Everything from water to crude oil even solid capsule is being transported across millions of miles.

## 2.1 *Pipeline Design*

This chapter to begin with needs to learn about the designing of a pipeline system. Factors like distance to be traveled, quantity of liquid or gas that is to be transported, and the quality of the substances to be traveled have to be considered.

NACE defines the following for the pipeline designers' concentration before designing a pipeline:

- Sensitive areas nearby
- Communities located nearby and their concern
- Temperature and climatic conditions
- Public safety concern and any other risk that is posing near
- The terrain facility for the construction of a pipeline (Fig. 1)

In case of pipelines underwater, the condition is still hectic at the time of inspection. To overcome these problems, a computer-based monitoring system was proposed to set up and as a part to make it efficiently work, image analysis with mathematical morphological operators to detect the defects and edge detection principle to remove the shadow images is being implemented (Figs. 2 and 3).

## 2.2 *Causes for Pipeline Damage*

Though there are many reasons, the main cause is due to corrosion. This occurs because the pipeline runs for long meters, and for many years together, they get damaged due to rust and the residue formation in the inner wall of the pipeline, and this results in the cause of cracks and holes thus leading to oil leakage affecting the environment badly. A crack in the pipeline can be identified as given in Fig. 4.

**Fig. 1** Ground-level pipelines

**Fig. 2** Pipeline under water

## 2.3 Pipelines Monitoring—Smart Pigs

Monitoring of pipelines should be done with weekly, bi-weekly or monthly inspection by the operators and they inspect the pipelines with all relevant safety measures. Manually, internal inspection is carried out by gadgets like high-resolution inspection tools also called as intelligent pigs or pipeline intervention gadget, to detect the dents, damages, and corrosions.

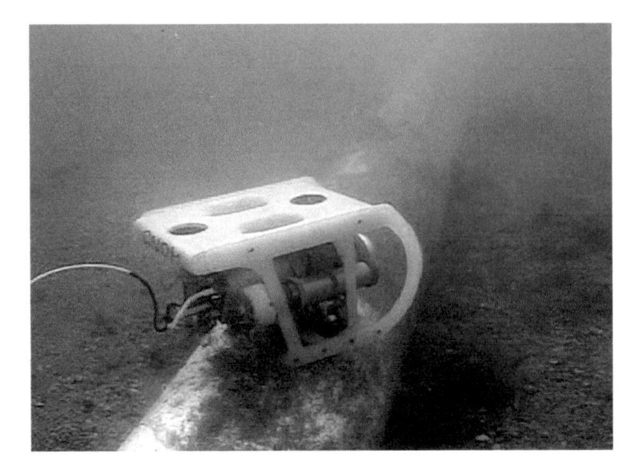

**Fig. 3** Manual pipeline intervention in side water level

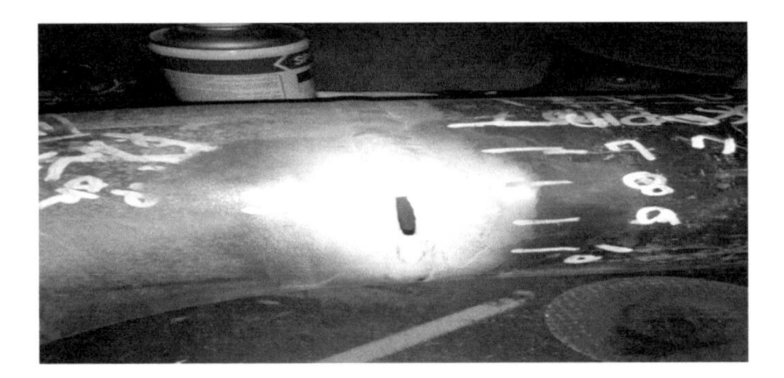

**Fig. 4** Picture showing a crack in the pipeline

## *2.4 Drawbacks of Existing Pipeline Models*

One of the most extensive representations is smart pigs (pigs) pipeline intervention gadget. Earlier smart balls and X-ray machine detection were used.

**Drawbacks**

1. Routine leak surveys, including pinhole-sized leaks are required.
2. Emergency leak location, minimizing product loss evaluation is mandatory, and cleanup costs are high.
3. Validation of alarms generated by CPM systems (with leak location) is not successful.
4. Acceptance testing of new pipelines failed.

## 2.5 *Image Analysis and Processing Model of a Pipeline*

Image defects encountered can be accessed by the sizing methods in the piping industry. Although the focus of this chapter with research has been on image analysis and processing of pipeline applications, the technology can be extended for many other applications also, including the inspection of stainless steel and aluminum-based products. Defects in a pipeline are identified using a computerized program in this chapter. As discussed earlier, a camera is sent inside the pipeline path that captures the images and sends to the system.

## 2.6 *Characteristics of Image Analysis Model*

After reading the theory of pipeline, the analysis process of an image has to be considered, and the size of the digitized camera used to capture and send the images to the data section for image analysis should be sized conveniently according to the inner diameter of the pipeline. The miniature cameras come with different specifications, but the right one has to be chosen depending on the diameter of the pipeline that helps to select the right camera for image capturing. Choosing a camera depends on the diameter of the pipeline that is used for transporting different substances.

## 2.7 *Implementation of Digitized Camera with Fiber Optic Cable*

The rapid deployment of fiber optic technology to enhance the transportation of images through light sources helps to capture the picture from the camera and sent to the system that is prepared to monitor the pipeline image techniques. Impact technologies like LLC help to explore the possibilities in attaching a fiber optic cable to a digitized camera to capture the pictures from inside the pipeline. This helps to have a good knowledge of the type of image processing technique involved in taking pictures. A study with the help of any cameraman will helps us to select the perfect one for picture capturing inside the pipeline.

## 2.8 *Pipeline Evaluation*

Pipelines have to be evaluated before choosing the fiber optic cable (FOC) and the camera for further processing. This is carried out with the help of an inspection engineer who guides to tell the structure, material, and the inside coating of the pipeline that are buried underneath. Pipelines are drained, and this process is termed as a "**shut-down operation**," where the path is closed for evaluation before the next process commences. In this duration, pipelines are checked for cracks, holes, or any other defects, and these are replaced or repaired to carry out the next assigned task. This may take 10–30 days, and numerous people evolve in this process because even a minor negligence will cause oil spills that results in heavy loss to the industry and also to the environment. After the study of the pipeline characteristics, computerized methods to detect the defects are carried out with the insertion of the digitized camera inside the pipeline to find and locate the damaged areas in the pipeline.

## 2.9 *Fiber Optic Cable*

A pipeline runs for meters together; a camera travels or stretches to cover this distance with an attached fiber optic cable. To achieve longer distance travel of data without any loss and to resist electromagnetic interferences, optical fibers are widely used. This novel procedure of camera attached to a fiber optic cable to capture pictures is the first step in this image analysis process (Fig. 5).

The jacket is used to hold the cable without movement while traveling inside the pipeline to capture a jerk-free image. Microbend clamping device is used to hold the jacket that has an optical photo detector, and this is connected to the laptop or desktop or to an IPAD through the optical electrical converter. This design of the system helps to capture and send the pictures for image analysis and processing.

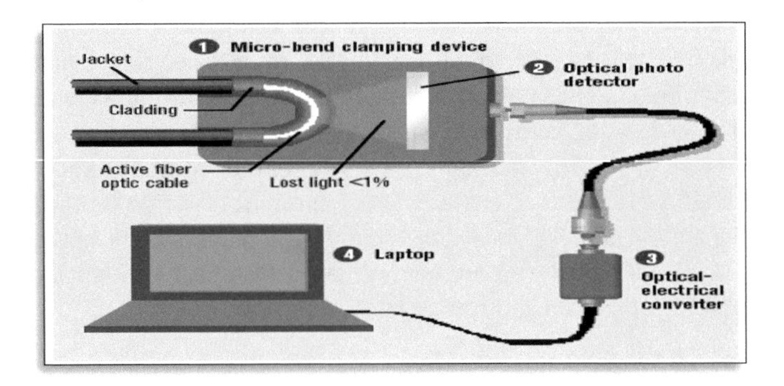


**Fig. 5** Overall structure of the FOC clamping

# 3 Image Detection Techniques

## 3.1 Gate Turn-Off Thyristors

A GTO is implemented here because it acts as rectifiers, switches, and as a voltage regulator. PNPN technique is applied in these GTOs which helps to detect repeated occurrence of 0s.

## 3.2 High-Frequency Filter

A high-frequency filter is applied here to filter the images and give the required section of an image for further image processing. Proposed cluster algorithm enhances the design to perform image modifications and noise reductions using high-pass and low-pass filters (Fig. 6).

A cluster image developed from the defected image is sent from the camera. Visual images on a computer are manipulated by computer graphics software. There are two categories of computer graphics raster and vector graphics. Focus on either vector or raster graphics is performed by many graphics programs, but there are a few that combine them in interesting ways. The concept of raster graphics is instigated for image analysis and processing. Raster graphics is set up as raster images that hold the bit images. This option in raster graphics helps to develop the bit manipulation concept in this chapter. The bits transmitted are then grouped into sets for mathematical morphological implementation (Fig. 7).

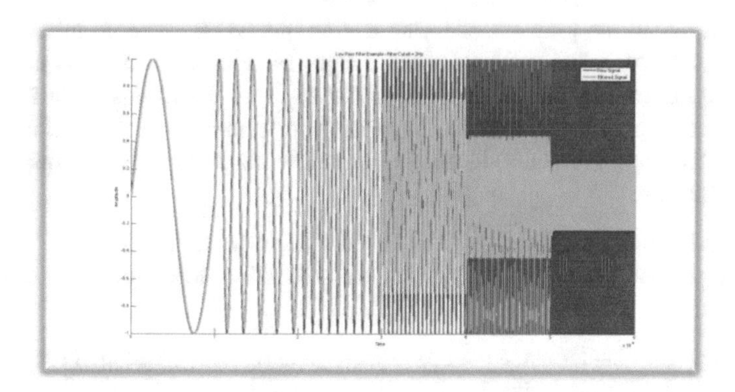

**Fig. 6** HFF domains (Picture Courtesy SADARA)

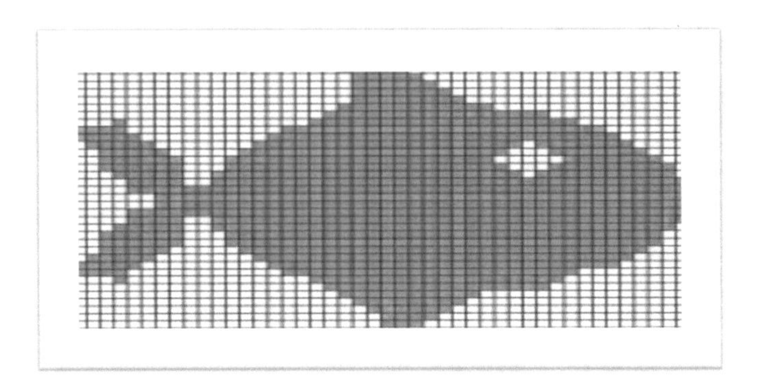

**Fig. 7** Picture showing bitmap image (Picture Courtesy Computer Graphics by Hearn)

## *3.3 Image Analysis Using Bitmaps*

Digital images can also be represented as raster image, and it is also called a bitmap. .gif, .jpg, .bmp are the various formats for storing images in the system. Red–Green–Blue (RGB) is one of the best methods to identify the image with pixel notifications.

## *3.4 Methods to Identify Hidden Structures in an Image*

Implementation through machine learning for hidden structure images, a precise concept of unsupervised learning algorithm plays a major role, unsupervised learning refers to the problem of trying to find hidden structure in unlabeled data. Since the samples taken to identify the images are unlabeled, there is no error or reward signal to evaluate a potential solution. This discriminates unsupervised learning from supervised learning. Unsupervised learning is closely related to the problem of density estimation in statistical application. Approaches to unsupervised learning include:

i. clustering (e.g., *k*-means, mixture models, hierarchical clustering),
ii. blind signal separation using feature extraction technique for dimension reduction

Through this chapter since the bits are analyzed without predefined set of results, *k*-means clustering algorithm is used to detect the hidden structure in the image. Images classified into bitmaps are applied through k-means structure to classify the hidden bit images. Images can be as clusters or can be as a single image, these are converted to bitmap image and then through the *k*-means, the defected parts are identified. Length and the exact position are calculated using the mathematical morphological operators.

## 3.5 Filtration of an Image

A filtering of an image is useful for many applications, including smoothing, sharpening, removing noise, and edge detection. A filter starts its process work from a kernel which is a small array applied to each pixel and its neighbors within an image. Filters are always applied to images for clear view, and this process is known as convolution and this is achieved by applying either in spatial or frequency domain. Image domains are used to overview transformations between images. This helps to make a study of filtered images. The high-frequency filter design is used here to refine the image before bit valuation. The design is very important because the anomaly caused during image perfection is rectified with this design. The decision whether to use a high-pass filter or low-pass filter should be taken with all properties in concern. With these reviews, the images taken by the digitized camera with a fiber optic cable connection are implemented to get the resultant structure as required for the detection faults in the pipeline.

## 3.6 Characteristics of Cluster Algorithm Model

Cluster analysis used for image processing is a general task to be solved. The clusters are represented as groups with small distances within the cluster members and data space dense areas. Clustering can therefore be originated as a multi-objective optimization problem, and an approximate process based on cluster algorithm can be opted depending upon the density threshold, maximum occurrence of projected clusters that depends on individual dataset, and the results to be expected. Cluster analysis is an iterative process of knowledge with trial and failure involvement and not automatic. Thus, it is required to modify data preprocessing and parameters, till the result with desired properties is achieved.

## 3.7 K-Means Algorithm

To implement in a best way *k* centroids are declared, each representing a cluster. All centroids should be planned and placed because they are received from different location causing different results. The better choice is to place them as much as possible far away from each other and take each point that belongs to a given dataset and relate it to the nearest centroid for further analysis. When no point is pending, the first step is completed and an early group age is done. At this point k new centroids are used as centres of the clusters resulting from the previous steps has to be recalculated, a knowledge discovery approach helps to clarify the different methods used in *k*-means algorithm in case of clustering large datasets. To read the values in a much faster method, the technique of learning fast nets has been applied.

# 4 Morphological Image Processing System

In this chapter, the method for background of the theory is discussed and evaluation of the theory used to identify the defects is identified. Based on the mathematical theories of sets and to logical notations, its principle supports in studying the morphological properties (shape, size, orientation, and other forms) of the object (patterns) through nonlinear transformations as associated with reference object (restructuring element). Through these continued processing, image processing to detect the cracks is achieved. The techniques used to perform these tasks are as follows (Fig. 8).

This is a vital and initial step to keep the pipeline infrastructure in proper shape (quality-wise); the pig (pipeline intervention gadget) used needs a lot of manual operation and there are differences between the image photographed and the picture marked manually. Thus, automated system implementation is required. The basic task for automated operation is to find out the cracks, holes, joints, and fissures in the images taken by the camera. These cracks have their own specific patterns (images described as such), matching the Gaussian profile on which pattern recognition has to be performed. Mathematical morphology is used to extract image components with regard to geometric features. An image is not taken under assumption; the features are extracted from the image that is used for representation and description of the image. This is possible by taking the values from the image domain, and this can be used as semantic information.

## 4.1 Implementation of the Operator Tool

### 4.1.1 Design of the Algorithm

Morphological operations are used mainly to detect expanding and shrinking image to a given structuring element. When the images are operated in the initial state, it appears only in black and white called binary images and later after technical development, they are applied for color images called grayscale images.

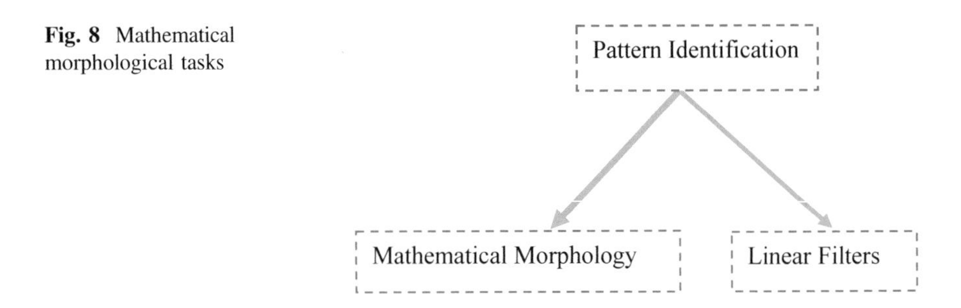


**Fig. 8** Mathematical morphological tasks

The image domain is mapped as 2D coordinates from minimum to maximum range, which defines the possible intensity values of the image. The values for a binary image (black and white) range from [0–1], for grayscale images (taken as 8 bit), it ranges from [0–255]. Mathematical morphology operates on nonlinear structures and thus uses totally different type of algebra, required for 2D attributes than the linear algebra. As defined the 2D image $F$ is defined as follows:

$$F = Z^\wedge 2|\square\ [I(\mathrm{Min}); I(\mathrm{Max})] \tag{1}$$

$I$ image intensity represented as bits
$Z$ bit error recovery

That maps 2D coordinates to the range [$I$(min) to $I$(max)] that defines the possible intensity values. If an image is black and white, the dark colors are considered as background and white are parts of the image. But now to locate a perfect crack position, white parts are placed as background and dark ones are parts of an image. When the image is photographed and sent to the system, with the leak software, the data is read for pixel image and it is analyzed for picture perfection. Leak software is a tool that helps to find the leak values with leak detection methods.

In case of grayscale image, to differentiate the colors, homogeneous technique is followed, where same colors are grouped together as white and the remaining as dark color (Fig. 9).

Picture showing the crack being locked with a redline indicating the location of a crack in the pipe (Fig. 10).

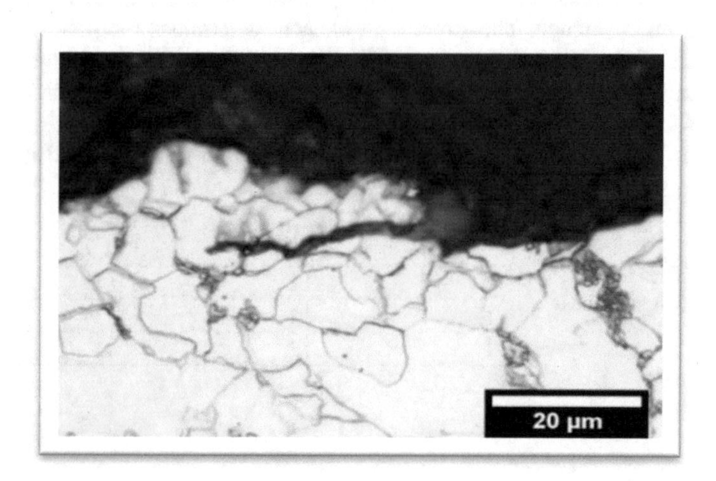


**Fig. 9** Cracks in branches (picture showing the white and dark color image)

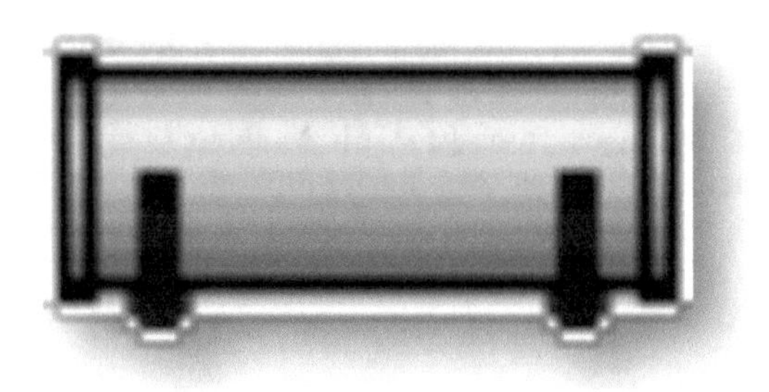

**Fig. 10** Portion marked for repair (Picture from Shutterstock.Com)

## *4.2 Analysis with the Bitmap Set*

Dilation is a morphological expansion operation—this operation is used both in binary and grayscale images. In binary images, structural element is moved over the image where the dilation takes on binary images. Two pieces of data are taken as data inputs.

### 4.2.1 Data Gathering Procedure

First the image to be dilated is collected from the images sent by the camera (a digitized fiber optic cable is connected). Then the fiber optic cables carry the communication signals using pulses of light. The second step is to collect the samples from the image. The structuring element determines the precise effect of the dilation on the input image. Thus, the data is gathered by the images that are sent by the camera (for 1 s minimum 500 images). Foreground pixels are represented by 1s and background pixels by 0s. An example of a structuring element with a $3 \times 3$ squares and origin in the center is shown in Fig. 4, and then these images are scanned for pixels (Table 1).

**Table 1** Values for evaluation of $3 \times 3$ pixels

| | | |
|---|---|---|
| 1 | 1 | 1 |
| 1 | 1 | 1 |
| 1 | 1 | 1 |

## 4.3 *Mode of Analysis—Detection of the Crack with the Morphological Operator*

### 4.3.1 Types: Erosion and Corrosion

After the procedure of grayscale binary value application, now the image should be reduced using erosion. Erosion is a morphological shrinking operation: In binary images, structural element is moved over the images; here pixel is taken to the result image, and it is written as:

$$F - E = \{XEZ^\wedge 2|E * \mathrm{CF}\} \tag{2}$$

CF correction factor (calculated from the available bit set)

where $F$ is an image and $E$ is a structured element for evaluation. In case of gray scale, each pixel touched by $E$ is considered and the minimum intensity for all the pixels is calculated.

Dilation plays a vital role, and this operation mostly uses a structuring element for probing and expanding the shapes that is present in the input image. It is one of the most prominent operations in mathematical morphology.

### 4.3.2 Opening and Closing Techniques

Dilation and erosion can be combined to form two important higher order operations. The opening technique is used to remove small objects from the image and closing removes small holes (for original images). For defected images, closing removes small objects and opening removes small holes.

## 4.4 *Proposed Method to Implement the Image Analysis Model*

Now with all these techniques, cracked image is to be located. First a subtraction is performed to obtain the defected image as follows:

$$\mathrm{OI} - \mathrm{CI} = \mathrm{UDI} \tag{3}$$

where OI stands for original image, CI stands for cracked image, and DI stands for undetected image. So the remaining dilation (defected image as called) is taken for morphological evaluation. If there are two or more cracks (multiple), then mathematical morphology deals with two image processing techniques. In this method, two images are considered, in which the first one is taken as input and uses an isotropic structural element, for a dilation or erosion process. This process is same for any number of cracks. Now to proceed further with multiple cracks geodesic reconstruction technique is applied, where as stated earlier the first one is taken as input and the second one is used to confirm the result. This process is repeated (iteration) until the stability of the image is obtained. By applying this process, the original image is either reduced or expanded in size by one pixel, for every iteration. Such an image called as the marker image is then confirmed by a mask image. The number of iterations gives a measure for the distance of the pixels in a cracked image. Repetition of dilation and erosion takes place until the defected image is successfully evaluated.

## 5 Edge Detection of an Image

Edge detection is a process that is used to make the image of an oil pipeline to appear bright and clear for further analysis, after smearing mathematical morphology for curves, error detection must be used, to read the pixel values and eliminate the shadow images that appear. It reads the coordinate values and builds accurate pixel values using this method. It also mainly helps to remove the shadow images if it occurs during the camera intervention inside the oil pipeline. The line or part where an object or area begins or ends is an edge (Fig. 11).

### 5.1 *Methods to Detect Edge Defects*

Edge detection has many methods, but they are grouped under two categories such as search-based and zero crossing-based methods. Smoothing filters' application helps in differentiating the way of computing edge strength. The different types of

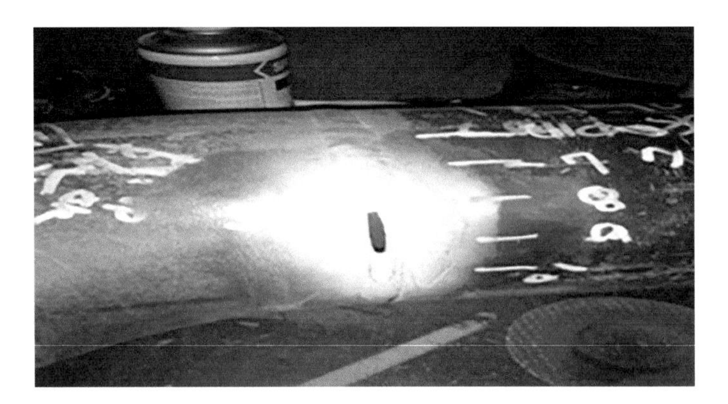

**Fig. 11** Pipeline showing crack with measurement

edge detection algorithms are gradient-based algorithm includes Laplacian-based algorithm and Canny's edge detection algorithm. Canny's edge detection algorithm is the best one when compared to other edge detection algorithms.

### 5.1.1 Smoothing of the Detected Image

In smoothing, the data points of a signal are modified so individual points (presumably because of noise) are reduced, and points that are lower than the adjacent points are increased leading to a smoother signal.

Smoothing is used in two important ways that can aid in data analysis

(1) It must be able to retrieve more data after smoothing process
(2) It must be able to give analysis details that is both easy to access and robust.

## 6 Conclusion

The purpose of this chapter is to develop an algorithm to ease the work of a manual operator in identifying the cracks in an oil pipeline, with better accuracy and reduced false detection of image recognition. Our proposed approach is also suitable for any type of computerized pipeline intervention. The study aims to construct a computerized mobility model which is useful to an environmental protection service. This model handles the problem of image analyzes interruption on the computing during the transportation of any form of liquid in a pipeline, where the substance travels from one location to another. The solution might be the boundary of convex and non-convex environment, which is successful. Furthermore, mathematical morphological operators and edge detection algorithms applied here are evaluated, thus giving an appreciated result.

The performance of existing X-ray image analysis and smart pigs for pipeline analysis over this computer-based monitoring system (CBMS) is compared and analyzed. Based on the performance analysis of the existing models, the developed CBMS shows a good result, and it is faster and the time duration is limited when compared to the other two existing models. These two models, X-ray and smart pigs, take 2 days and a day, respectively, for image analysis. CBMS gives the result within seconds and gives approximately 90% result. Thus, human errors and time delay in detecting the cracks or holes or any other defects caused due to natural calamities or due to human errors is avoided. The overall performance of CBMS may vary depending on the type of algorithm chosen to remove clusters, but the ultimate output remains the same with a good record of result.

## References


1. Aldenderfer, M. S., & Blashfield, R. K. (1988). Sage university paper on quantitative applications in social science. *Cluster Analysis, 36*(1), 1–13.
2. Wardhani, A. W., & Gonzalez, R. (1998). *Automatic image structure analysis*. e-prints (pp. 1–9).
3. Ledda, A., & Luong, H. Q. (2006). *Image interpolation using mathematical morphology* (pp. 12–49).
4. Bhoite, A. (2011) Automated fiber optic cable end face field inspection. *Technologic*, 3–8.
5. Blashfield, R. K., & Aldenderfer, M. S. (1978). The literature on cluster analysis. *Multivariate Behavioral Research, 13,* 271–295.
6. Busch, C., & Eberle, M. (1995). Morphological operations for color-coded images. In F. Post & M. Göbel (Eds.), *Proceedings of the European Association for Computer Graphics* (pp. 193–204).
7. Can, F., & Ozkarahan, E. A. (1984). Two partitioning type clustering algorithms. *Journal of the Association for Information Science and Technology, 35,* 268–276.
8. Can, F., & Ozkarahan, E. A. (1989). Dynamic cluster maintenance. *Information Processing and Management, 25,* 275–291.
9. Canny, J. (1986). *A computational approach to edge detection* (pp. 3–6).
10. Crouch, C. J. (1988). A cluster-based approach to thesaurus construction. In *11th International Conference on Research and Development in Information Retrieval* (pp. 309–320). New York.
11. Domingo, J., Benavent, X., Vegara, F., & Pelechano, J. (2005). *A new approach for image colour morphology* (pp. 234–245). Tech. Rep., Institute of Robotics, University of Valencia.
12. Dougherty, E. R., & Astola, J. (1994). *Introduction to non-linear image processing*. Bellingham, Washington: Spie.
13. El-Hamdouchi, A. (1987). The use of inter-document relationships in information retrieval. Ph.D. thesis, University of Sherfield.
14. Ester, M., Kriegel, H.-P., & Sander, J. (1996). *A density based algorithm for discovering clusters in large spatial database with noise* (pp. 226–231).
15. Everitt, B. (1980). Cluster analysis. In F. Khalvati (Ed.), *Computational redundancy in image processing* (pp. 4–6, 3–57, 2nd ed.).
16. Gonzalez, P., Cabezas, V., Mora, M., Cordova, F., & Vidal, J. (2010). Morphological color images processing using distance-based and lexicographic order operators. In *Proceedings of the 29th International Conference of the Chilean Computer Science Society*.

# Efficiency of Multi-instance Learning in Educational Data Mining

**S. Anupama Kumar and M. N. Vijayalakshmi**

**Abstract** Educational data mining (EDM) is one of the emerging technologies in recent years. The various changes in the process of teaching and learning have brought in a lot of challenges to the stakeholders to understand the learners toward the different methods of teaching and the way they perform in various teaching environments. This chapter is an application of Baker's taxonomy in an educational dataset to predict course outcome of the learners during the middle of the course. The experiment is conducted using different single and multi-instance-based learning algorithms. The efficiency of the single and multi-instance learning algorithms was measured using the accuracy rates and the time taken to build the model. In single instance algorithm, decision stump tree was found very effective and in multi-instance learning, the Simple MI method was found very effective. The precision of the instance-based learning algorithms is calculated using Wilcoxon rank method, and multi-instance learning algorithm is found to be more accurate than the single instance learning techniques.

**Keywords** Prediction · Single instance learning · Multi-instance learning
Accuracy · ROC · PRC · MCC · Rank

## 1 Introduction

In recent years, the technical advancements and global education scenario have changed the means in which people teach and learn. The technical advancements empower the teachers with a variety of teaching aids to teach through traditional or virtual learning methods. Students are able to learn through virtual learning, distance learning, integrated computer technologies, collaborative learning, etc.

S. Anupama Kumar (✉) · M. N. Vijayalakshmi
Department of MCA, R V College of Engineering, Bengaluru, India
e-mail: anupamakumar@rvce.edu.in

M. N. Vijayalakshmi
e-mail: vijayalakshmi@rvce.edu.in



S. Margret Anouncia and U. K. Wiil (eds.), *Knowledge Computing and its Applications*, https://doi.org/10.1007/978-981-10-8258-0_3

It becomes vital in this era to understand the need of new-generation education tools and the methods to teach the students and understand the behavior of the student community toward those methods. Equally, the behavior the various stakeholders have toward these tools should also be understood to bring out betterment in education.

EDM is an emerging discipline in the field of data mining that is used to develop methods for exploring the unique types of data available in the educational departments. This analysis will help the stakeholders to understand the students better and their process of learning.

The three main objectives of EDM are:

1. To assist the learners to improve their performance in the academics, let tutors to design an efficient tutor material, etc., are termed as pedagogic objectives of EDM.
2. Optimizing the educational infrastructure and maintaining resources, request for more courses in the institution, etc., constitute toward the management objectives.
3. The commercial objectives create segmentation in the market facilitating toward the recruitment of students in colleges.

## *1.1 Educational Applications*

The major applications of EDM are listed below:

**Communicating to Stakeholders**:

Communicating with stakeholders includes helping the administrators and educators in analyzing students' behavior toward a new course and understanding their interest toward the course. This can be achieved by applying exploratory data analysis through statistical analysis and visualization, process mining, etc.

**Maintaining and Improving Courses**:

EDM models are designed to help the course administrators and educators in determining how to improve the course contents using information about student usage and learning. Association rule mining, classification, and clustering can be used to develop EDM models to help the administrators in this area.

**Predicting Student Grades and Learning Outcomes**:

The EDM models can be applied to predict the grades the student would get or predict their learning outcome at every stage of their studies. Classification, relationship mining, and clustering can be applied to achieve the goal.

## *1.2 Baker's Taxonomy*

Baker's taxonomy is the basic dictionary followed by all the EDM researchers. The different data mining techniques that can be applied on educational data mining are classification/prediction, clustering, association mining/relationship mining, distillation of human judgment, and model discovery.

1. **Prediction**

Prediction can be defined as development of a model that can be used to infer a predicted variable from a set of predictor variables. The prediction techniques are used to predict the course outcomes of the learners, their behavior in the class during various courses/toward the different learning environments, the behavior of the tutor toward the learner under various environments, the learner's behavior toward different learning environments, etc.

2. **Clustering**

Grouping of data using a set of data points refers to clustering. The technique can be applied on the dataset when the miner is aware of the familiar categories. This technique can be applied to group the students according to certain criteria and infer new knowledge out of it. They can also be used to investigate the similarities and dissimilarities in the behavior of the students toward an environment/course.

3. **Relationship Mining**

This method is applied when the miner wants to find the relationship available between the variables in a dataset. They can also be used to find the strong and weak relationship between the elements in a dataset.

4. **Distillation with Human Judgment**

Human judgment plays a vital role in making inferences about the data that is presented in an appropriate manner. It is very difficult to prepare a model using the human judgment, but they can be used as a visualization tool. It can be combined with any data mining techniques to create and implement models educational data. The main purpose of applying this method is classification and identification. The refined data helps the miner to identify the well-known patterns and develop models using the patterns.

5. **Discovery with Models**

Discovery with models is an analytical model using which we can find the learning material most opted by the students, analyse their behaviour towards the material and the tutor impact on the behaviour of the student.

This chapter aims at describing the applications of the classification techniques in predicting the student course outcome at an early stage in a higher educational domain. The application of single and multi-instance learning algorithms in predicting the student course outcome at an early stage is discussed in the further sections. The various algorithms in both the single instance and multi-instance paradigms are implemented and compared using various metrics. The accuracy of single and multi-instance learning algorithms is then compared and assessed using

Wilcoxon ranking technique. The results proved that MIL is efficient in obtaining accurate models when compared to SIL for predicting the performance of the student during the middle of the course. Section 2 gives an introduction to the problem domain, the experiment setup, and the metrics used in analyzing the efficiency of the solutions.

## 2 Problem Domain: Predicting Student Course Outcome at an Early Stage

Predicting course outcome plays a vital role in any higher education program. The course outcome can be predicted using measurable and non-measurable attributes. Attributes which are measurable can be easily predicted. Measurable attributes include the marks obtained by the students, the level of their understanding the course, etc. The non-measurable attributes can be student's creativity, team building capability, leadership skills, analyzing capacity, etc. Predicting the course outcome helps the institution to build strategic programs to improve students' performance during their period of studies in the institutions [1]. The outcome of the course is predicted for various reasons. In [2], the author examined the relations among students' perceived usefulness of the student learning outcomes, the different measures adopted in learning, rating the self-confidence in psychology, and evaluating the department's helpfulness in students' skill development. In [3], the authors have explored the relationship between the performance of the students in 23 online courses and quality area. The authors concluded that the results obtained infer positivity toward the quality of the interpersonal interactions and student grades. The author in [4] discussed the outcome of the different learning outcome variables in the course. The different challenges involved in predicting the course outcome are

(i) **Phases of Prediction**

Predicting the course outcome at the right time is very important for any educational institution. Prediction of course outcome at different stages of education can be interpreted in various ways. Predicting the course outcome during the middle of the course or at the beginning of the course will help the tutor to analyze the performance of the students and help them to train the students to perform better in semesters and exams. This will help them to identify the students who tend to fail and train them better to pass the course.

(ii) **Range of Attributes**

Selecting a suitable parameter is an important task in predicting the course outcome. The authors in [5] have discussed the importance of selecting different attributes that are suitable to forecast the results of the students and implemented it using different data mining techniques.

(iii) **Selection of Appropriate Algorithm**

Prediction of student's course outcome has been carried over by a variety of researchers using different data mining techniques. The authors in [5–8] have applied classification techniques like decision trees, rule-based, and naïve-based algorithms to predict the student grades and analyze the performance of the students under various environments. In [9], the authors have used a priori rules to classify the student interest and predicted the performance of the students.

From the studies made, it is understood that single instance representation will be more suitable for a traditional learning environment. For a multi-learning environment where the course has different types of components and each student carries out the components in multiple ways, the multi-instance learning (MIL) will be more appropriate representation of the model. The different instance-based learning algorithms are discussed in Sect. 2.3. This experiment is carried out using both traditional supervised learning systems and multiple instance learning techniques. Section 2.1 gives a description of the dataset used in the experiment to predict the course outcome.

## 2.1 *Description and Selection of Dataset*

A dataset is a collection of data. The data can be in a structured or unstructured form. The data can be stored in a flat file, simple database, database table, etc., depending on the form of the data. The dataset may comprise data for one or more members, representing it in the form of rows. The structure and properties of the data define the number and types of the attributes or variables, and various statistical measures applicable to them, such as standard and kurtosis [10]. For predicting the course outcome, National Environmental Education Training Foundation (NEETF 2000) divides the factors that influence learning outcomes into the following categories:

- External (such as gender, parents' educational background, student's prior educational data),
- Internal (psychological factors involving the behavior of the student),
- Social, curricular, and administrative data.

The varied learning styles of the tutors will also play a vital role for the better achievement of the student (Klavas 1994; Thomas et al. 2000). Students in varied environments show interest toward learning, and with parent's involvement in their studies, they tend to get better grades and test scores, better attitudes and behavior (Brown 1999; Peterson 1989, etc.).

**Table 1** Dataset description

| Component | Name | Description |
|---|---|---|
| Assignment | No_as | Number of assignments taken by the student |
| | No_sta | Number of students assigned/assignment |
| Quiz | No_qz | Number of quiz |
| | No_qzt | Time taken to solve the quiz |
| Lab sessions | Lab | Number of laboratory sessions |
| Tests | No_t1 | Number of students participated in test 1 |
| | No_t2 | Number of students participated in test 2 |
| Marks | Mark1 | Marks scored by the students in test 1 |
| | Mark2 | Marks scored by the students in test 2 |
| Total Marks | TM_t | Total marks scored in test 1 and test 2 |
| | Tm_a | Total marks scored in assignments |
| | Tm_q | Total marks scored quiz |
| | Tm_lab | Total marks scored in laboratory |
| | TotalE | Total marks excluding the laboratory |
| | TotalF | Total of all the above |
| Result | Result | Pass/Fail |

The authors in [1, 11–13] have collected and applied different data mining techniques like classification and clustering in predicting the course outcome of the students using different attributes. The course outcome has been predicted using two class variables, namely Pass and Fail.

The data of semester I students of a particular postgraduate program is collected in a flat file. The noise in the dataset has been removed, and the missing values are replaced by the average score of the class during the preprocessing. Table 1 describes the data used in this **work**.

The following section gives an introduction about the classification task, its types, general approach of classification to solve a problem, the different metrics that can be used to analyze the classification algorithms.

## 2.2 *Introduction to Classification Technique*

Classification is the task of learning a target function *f* that maps each attribute set **X** to one of the predefined class labels **y**. Classification prediction comprises two levels: classifier model construction and the usage of the classifier constructed [14]. Figure 1 explains the classification model.

The classification model can also be called as classifier which predicts the class value from other explanatory attributes.

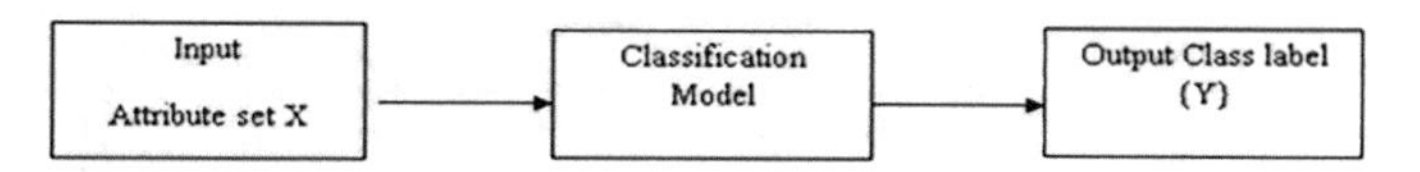


**Fig. 1** Classification model

### 2.2.1 Types of Classification Techniques

The classification models are built to predict class of an object on the basis of its attributes [15]. Classification models can be used for two different purposes via (i) descriptive modeling and (ii) predictive modeling.

Descriptive modeling is an illustrative tool to discriminate objects of different classes. It can be used to visualize/summarize the data and explain the characteristics of the data.

Predictive modeling is used to forecast the class label of unknown records. It consigns a class label to the attribute set of unknown records and predicts the target variable of binary form.

### 2.2.2 Solving a Classification Problem

A classification technique is a systematic approach in building classification models from an input dataset. To build a model, learning algorithms are applied on the dataset and the algorithm with the highest accuracy can be employed as the best fit for the relationship employed between the attributes and the class label. Therefore, the algorithm should be capable of predicting the class labels of unknown records. The learning algorithm should learn the model with the help of the training set (which contains more records so as to train the algorithm) and apply the model to the test dataset which contains the required input data. The dataset can be divided using cross-validation method or training and test set. The training set is fed into the learning algorithm, and the algorithm learns and builds a model using the data. Then the model is applied on the test data to receive appropriate output.

### 2.2.3 Classification Accuracy

The performance of the model can be assessed using the following metrics

1. **Confusion Matrix**: The matrix shows the number of test records correctly predicted against the incorrectly predicted records. The output can be expressed in the form of a matrix explaining the true and false rates of the test records. Table 2 shows the entries in a confusion matrix.

Each entry $f_{ij}$ in this table denotes the number of records from class $i$ predicted to be class $j$. From the table, it is understood that total number of correct predictions made by the model is $(f_{11} + f_{00})$ and the total number of incorrect predictions is $(f_{10} + f_{01})$.

**Table 2** Format of confusion matrix

| | | Predicated class | |
|---|---|---|---|
| Actual class | | Class = 1 | Class = 0 |
| | Class 1 | $f_{11}$ | $f_{10}$ |
| | Class 0 | $f_{01}$ | $f_{00}$ |

(i) Accuracy

It describes the number of correct predictions against the total number of predictions.

$$\text{Accuracy} = (f_{11} + f_{00})/(f_{11} + f_{00} + f_{10} + f_{01}) \quad (1)$$

(ii) Error rate

It explains the number of wrong predictions against the total number of predictions.

$$\text{Error rate} = (f_{10} + f_{01})/(f_{11} + f_{00} + f_{10} + f_{01}) \quad (2)$$

The efficiency of algorithm is evaluated using the accuracy class rate:

True positive (TP), false positive (FP), recall, *F* measure, and receiver operator characteristic (ROC) values.

2. **True Positive Rate (TP)** is the ratio of the true positive value against the total true positive and false negative.

   True positive rate: (TP)/(TP + FN)

3. **False Positive Rate (FP)** is the ratio of false positive value against the total false positive and true negative value.

   False positive rate = (FP)/(FP + TN)

4. **Precision** is the ratio of true positive against the total true positive and false positive values.

   Precision = (TP)/(TP + FP)

5. **Recall** is the ratio of true positive value against the total true positive and false negative values. Recall = (TP)/(TP + FN)
6. ***F* Measure** is defined as: *F* measure = (2(precision * recall))/(precision + recall)
7. **ROC Value**: The ROC curve is a metric that is plotted from the origin which moves for each true positive and false positive values [16]. The vertical and horizontal axes show the percentage of true positives and false positives. Any ROC curve value holding between 0 and 1 is highly appreciative for a better accuracy.

### 2.3 Instance-Based Learning

Instance-based learning refers to storing and using the instances to improve the performance of the classifier. The instance-based classifier can be either single instance-based learners or multi-instance-based learners. All the existing supervised learning algorithms like decision trees, rule-based algorithms, distributed networks fall under single instance-based learning where the examples are complete for understanding. The author in [17] has implemented various single instance algorithms to study the application of classification techniques over student data. The author has applied on one algorithm from each tree, rule, and naïve Bayes technique over a student set. Multiple instance learning (MIL) algorithms work with incomplete knowledge about labels of training examples in contrast with the learning algorithms. A variety of supervised learning techniques like decision trees and rule-based techniques have been modified and implemented for MIL. The author in [18] has implemented both single instance and multi-instance algorithm to predict the performance of student in a virtual learning environment and found that MIL has better accuracy than the single instance learning methods. Section 3 discusses the implementation of the dataset using the various single and multi-instance-based learning algorithms.

## 3 Experimentation and Results

The dataset is experimented using single instance learning and multi-instance learning algorithms. All the experiments are carried out using tenfold stratified cross-validation.

Initially, the experiment is carried out with a variety of single instance learning algorithms, and the accuracy of the algorithms is compared using accuracy, accuracy rates, and time taken to evaluate the model.

Secondly, the dataset is then experimented with the various multi-instance learning algorithms and compared with each other using the same parameters as discussed in single instance learning techniques.

### 3.1 Implementation of Single Instance Learning Classification Algorithm

The various single instance learning algorithms that are included in this experiment are decision tree-based algorithms, rule-based algorithms and naïve-based algorithms. The various single instance learning algorithms based on trees include decision stump tree, J48, random forest, and random tree. Methods used on rules include rule-based algorithms, tree-based algorithms, and naïve Bayes algorithms.

This section explains the implementation of various single instance learning algorithms and its efficiency in predicting the course outcome.

### 3.1.1 Rule-Based Algorithms

The different rule-based algorithms used in this experiment are Zero *R*, One *R*, and Ridor. The rules were generated based on the two different class variables: Pass and Fail. The Zero *R* algorithm generated one rule with 88.3% of correctly predicted instances and 11.69% of incorrectly predicted instances. The number of rules generated by Zero *R* algorithm is not sufficient to predict the performance of the student.

The One *R* algorithm on the other hand generated two rules, one on each target class. The rule generated is stated below:

Rule 1: TotalE< 39.5-> Fail

Rule 2: TotalE >= 39.5-> Pass

From the rules, it is understood that the rules were generated using the total marks obtained by the students excluding the laboratory component and that has played a major role in building the target variable. The correctly classified instances are 95.9% and incorrectly classified instances are 4.0936%.

The Ridor algorithm generated four rules with 96.4% of correctly predicted instances and 3.5% of incorrectly predicted instances. The rules generated are given below:

**Rule 1**: Result = Fail (171.0/151.0)
**Rule 2**: Except (TOTALE/100 > 39.5) and (quiz2 <= 9.5) => Result = Pass (95.0/0.0) [46.0/0.0]
**Rule 3**: Except (quiz1 > 8) => Result = Pass (5.0/0.0) [2.0/1.0]
**Rule 4**: Except (T1/30 <= 1.5) => Result = Pass (2.0/0.0) [1.0/0.0]

The NNGE algorithm has generated rules using multiple combinations in the dataset and is given below. Rules generated:

**Rule 1**: class Pass IF: 3.0<=quiz1<=10.0 ^ 2.0<=quiz2<=10.0 ^ 10.0<=t1<=50.0 ^ 6.0<=T1/30<=30.0 ^ 0.0<=t2<=50.0 ^ 0.0<=T2/30<=30.0 ^ 9.0<=a1<=10.0 ^ 1.0<=a2<=10.0 ^ 40.0<=TOTAL/100<=96.0 ^ 36.0<=lab 50<=50.0 ^ 76.0<=total 150<=146.0 (147)
**Rule 2**: class Fail IF: 4.0<=quiz1<=6.0 ^ 0.0<=quiz2<=6.0 ^ 5.0<=t1<=17.0 ^ 3.0<=T1/30<=11.0 ^ 0.0<=t2<=11.0 ^ 0.0<=T2/30<=7.0 ^ 0.0<=a1<=10.0 ^ 0.0<=a2<=10.0 ^ 12.0<=TOTAL/100<=39.0 ^ 0.0<=lab 50<=50.0 ^ 12.0<=total 150<=89.0 (17)
**Rule 3**: class Pass IF: quiz1=4.0 ^ quiz2=0.0 ^ t1=0.0 ^ T1/30=0.0 ^ t2=0.0 ^ T2/30=0.0 ^ a1=0.0 ^ a2=0.0 ^ TOTAL/100=4.0 ^ lab 50=0.0 ^ total 150=4.0 (3)

Even though the accuracy rates of NNGE and Ridor are same, the Ridor algorithm takes 0.05 s to generate the model while NNGE takes 0.06 s to generate the model. All the rule-based algorithms are compared in terms of accuracy and efficiency. The Ridor and NNGE algorithms are found effective in terms of accuracy where One $R$ is better in terms of the time taken to generate the rules. The NNGE algorithm takes more time in generating the rules when compared to Ridor even though their accuracy is same. Therefore, Ridor performs better in accuracy and One $R$ performs better in time.

### 3.1.2 Tree-Based Algorithms

The different tree-based algorithms used in this experiment are decision trees including J48 and decision stump, random forest, and random tree. In [19], the author has implemented decision tree, neural networks, and $K$-means nearest neighbor algorithms to predict the performance of the students based on the academic performance and the social background of the students.

The various classification techniques were used by [20] in their research work to predict the performance of the students in a distant learning course and C4.5 proved to be very effective for their dataset. The different tree classification algorithms like C4.5 REP tree and random forest tree have been used by [21], and the author concluded that the accuracy rate of C4.5 is 84% when the dataset is considerably high and the random forest could perform better with 92% of accuracy.

In our experiment when the algorithms are implemented, it is found decision stump tree and random forest tree classified 97.06% records accurately. The decision stump tree generated the tree in 0.02 s, whereas the random forest tree has taken 0.08 s to generate the output. The random tree classified 95.9% records accurately, and C4.5 algorithm classified 88.3% of records accurately. The random tree algorithm has taken 0 s to generate the model, and C4.5 has taken 0.05 s to generate the model. From all these outputs, it is understood that decision stump tree performs better in terms of accuracy and efficiency among the tree-based algorithms.

### 3.1.3 Naïve Bayes Algorithms

Two naïve Bayesian algorithms are implemented in this experiment, namely naïve Bayes and naïve Bayes simple. The naïve Bayes classification approaches need only one scan of the training data [22], i.e., the approach does not need many examples to classify. In [8], the authors have implemented naïve Bayes techniques to predict the performance of student results and found that to be an effective technique. For the given dataset, the naïve Bayesian algorithm was found effective with 86.4% of records classified correctly when compared with naïve Bayes simple algorithm with 82.4% of records classified correctly.

### 3.1.4 Comparison of Single Instance Learning Algorithms

The algorithms are compared based on their accuracy rate, ROC value, and the time taken to build the model. The accuracy class rates like true positive (Sensitivity), false positive (Specificity), true negative, and false negative values of the Pass and Fail target values are considered when any two algorithms reported same accuracy values. Table 3 reports the weighted average of the various algorithms implemented during the experiment.

The following observations can be made from Table 3.

1. Decision stump algorithm is very efficient than all the algorithms in terms of Sensitivity and Specificity. The Sensitivity of the algorithm is the percentage of instances correctly classified meeting the target value (positive instances—Pass value), and the Specificity is the number of correctly classified instances meeting the target value (negative instances—Fail value).
2. The ROC values of the algorithms are compared by keeping the true values in $X$-axis and false values in $Y$-axis. The random forest algorithm holds good when compared to all other algorithms in terms of ROC value since it tends to 1 very closely. Since the other measures of the random forest are not very efficient, the tree is not suitable for this dataset.
3. Figure 2 shows the comparison of the algorithms based on the accuracy rate and time taken. From the graphs, it is clear that decision stump algorithm performs better in terms of accuracy rate and time taken.

The following section explains the implementation of the multi-instance learning algorithms in this experiment.

**Table 3** Weighted accuracy class—single instance learning

| Algorithm | Sensitivity | Specificity | Precision | Recall | *F* measure | ROC |
|---|---|---|---|---|---|---|
| Zero *R* | 0.883 | 0.833 | 0.78 | 0.833 | 0.28 | 0.497 |
| One *R* | 0.959 | 0.136 | 0.96 | 0.959 | 0.959 | 0.912 |
| Ridor | 0.965 | 0.091 | 0.967 | 0.965 | 0.966 | 0.937 |
| NNGE | 0.965 | 0.222 | 0.964 | 0.965 | 0.963 | 0.872 |
| Naïve Bayes | 0.865 | 0.061 | 0.93 | 0.865 | 0.884 | 0.953 |
| Naive Bayes simple | 0.825 | 0.067 | 0.922 | 0.825 | 0.852 | 0.95 |
| Decision stump | 0.971 | 0.047 | 0.974 | 0.971 | 0.972 | 0.933 |
| Random forest | 0.971 | 0.091 | 0.972 | 0.971 | 0.971 | 0.99 |
| Random tree | 0.959 | 0.136 | 0.96 | 0.959 | 0.959 | 0.96 |
| J48 | 0.883 | 0.883 | 0.78 | 0.883 | 0.828 | 0.497 |

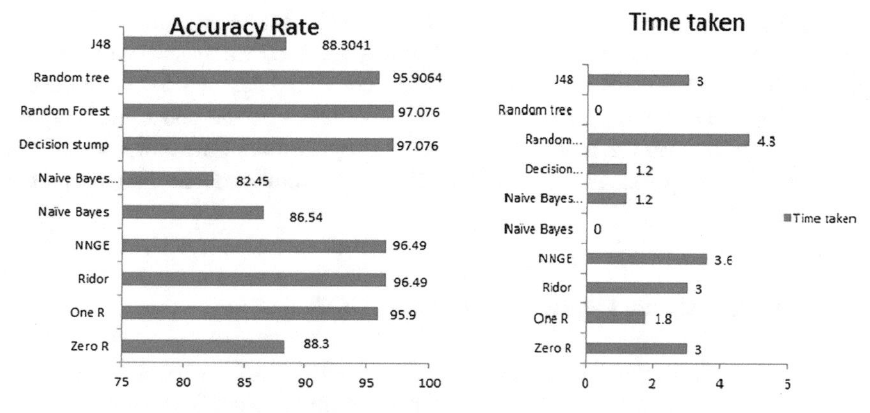


**Fig. 2** Comparison of single instance learning

## 3.2 *Implementation of Multi-instance Learning Classification Algorithms*

Multi-instance learning (MIL), which is defined by Dietterich et al. [23], is a variation on the standard supervised machine learning scenario. Every example in MIL consists of a multi-set (bag) of instances. Each bag has a class label, but the instances themselves are not directly labeled. The learning problem will form a model based on given example bags and predict the future bag. Several MIL methods are based on probabilistic models.

In [24], the author has discussed the different single instance and multiple instance learning methods and applied the algorithms to various experiments. He has also discussed the different methods using which the algorithms can be compared and the efficiency of the methods can be justified.

### 3.2.1 Construction of Multi-instance Dataset

The dataset used in the experiment is initially preprocessed and converted. The propositional dataset is converted into multi-instance dataset with relational attributes. The student_id is considered as the Bag_Id which is always a nominal attribute, and the other attributes are considered as instances and the result attribute is the target value. All the missing values are removed during the preprocessing phase so that the dataset is ready for implementation.

### 3.2.2 Rule-Based Algorithms

Two multi-instance rule-based algorithms are used in this experiment. They are multi-instance boost algorithm with PART rules, multi-instance wrapper algorithm with PART rules, MI rule inducer (MIRI), Simple MI with PART which are implemented over the dataset. The MI boost algorithm, wrapper algorithm, and Simple MI predicted 98.7578% of the instances correctly, and MIRI predicted 97.5155% of the instances correctly and the time taken by the boost and wrapper algorithm to develop the model is also same. But when the time efficiency of the algorithms is compared, Simple MI with PART algorithm performs better than the other algorithms.

### 3.2.3 Multi-instance Logistic Regression (MILR)

The MI Logistic regression algorithm predicted 98.75% of instances correctly in 0.08 s. The algorithm uses standard or collective multi-instance assumption, with linear regression. In this experiment, arithmetic mean has been used.

### 3.2.4 MI Tree-Based Algorithms

The various tree-based algorithms like MI tree induction (TI) and MI two-level classification (TLC) are implemented to predict the course outcome. Both the algorithms are implemented using the C4.5 classification model to build two classifier instances. The MITI algorithm predicts 97.5% of the instances correctly while the TLC algorithm predicts 98.75% of the instances correctly.

### 3.2.5 MI Meta-learning Algorithms

The other algorithms implemented were MI nearest neighbor, MI SMO, and MI optiball algorithms. When the correctly predicted instances of all the three algorithms were compared, the nearest neighbor algorithm was found very ineffective predicting only 71% of the instances correctly, optiball with 93%, and SMO with 91% of the instances predicted correctly. The following section represents the comparison of the MI algorithms in detail.

### 3.2.6 Comparison of the Multi-instance Learning Algorithms

All the MI algorithms were compared using the accuracy rate, ROC, PRC value, and the time taken to build the model. The accuracy class rate like true positive, false positive, true negative, and false negative values of the Pass and Fail target values are considered when any two algorithms reported same accuracy values. Table 4 depicts the weighted average accuracy of the various algorithms implemented during the experiment. PR curves plot the precision versus recall values.

**Table 4** Comparison of MIL algorithms

| Algorithm | Sensitivity | Specificity | Precision | Recall | *F* measure | MCC | ROC | PRC area |
|---|---|---|---|---|---|---|---|---|
| MBI boost with PART | 0.988 | 0.130 | 0.988 | 0.988 | 0.987 | 0.920 | 0.996 | 0.997 |
| MILR | 0.988 | 0.066 | 0.988 | 0.988 | 0.988 | 0.922 | 0.992 | 0.989 |
| MINND | 0.739 | 0.800 | 0.834 | 0.739 | 0.782 | 0.043 | 0.469 | 0.836 |
| MI optiball | 0.938 | 0.329 | 0.938 | 0.938 | 0.938 | 0.609 | 0.804 | 0.919 |
| MIRI | 0.975 | 0.132 | 0.975 | 0.975 | 0.975 | 0.844 | 0.922 | 0.965 |
| MITI | 0.975 | 0.132 | 0.975 | 0.975 | 0.975 | 0.844 | 0.922 | 0.965 |
| MISMO | 0.913 | 0.913 | 0.834 | 0.913 | 0.872 | 0.000 | 0.500 | 0.841 |
| MI wrapper with PART | 0.988 | 0.130 | 0.988 | 0.988 | 0.987 | 0.920 | 0.906 | 0.970 |
| MITLC | 0.988 | 0.130 | 0.988 | 0.988 | 0.987 | 0.920 | 0.906 | 0.970 |
| Simple MI | 0.988 | 0.130 | 0.988 | 0.988 | 0.987 | 0.920 | 0.906 | 0.970 |

Precision–recall curves are better to highlight differences between models for highly imbalanced datasets. The Mathews Correlation Coefficient (MCC) is calculated between the observed and predicted binary classifications. The coefficient lies between −1 and +1. A coefficient of +1 represents a perfect prediction, 0 no better than random prediction, and −1 indicates total disagreement between prediction and observation.

The following observations are made from Table 4.

1. The algorithms MI boost with PART and MI wrapper with PART differs only using PRC area.
2. The MILR algorithm performs better than the other algorithms when compared using the accuracy rate.
3. The nearest neighbour algorithm was inefficient.
4. All the accuracy metrics when compared to other algorithms. From the MCC and PRC, it is clearly understood that the algorithms were very efficient in predicting the outcome and suitable for such kind of datasets. Figure 3 shows the comparison of the algorithms using the weighted accuracy rate and time taken.

From Fig. 3, it is clearly understood that even though the accuracy rate of the Simple MI, MITLC, MI wrapper, MILR, and MI boost are equal, the time efficiency of the algorithms differs in building the model. The time efficiency of

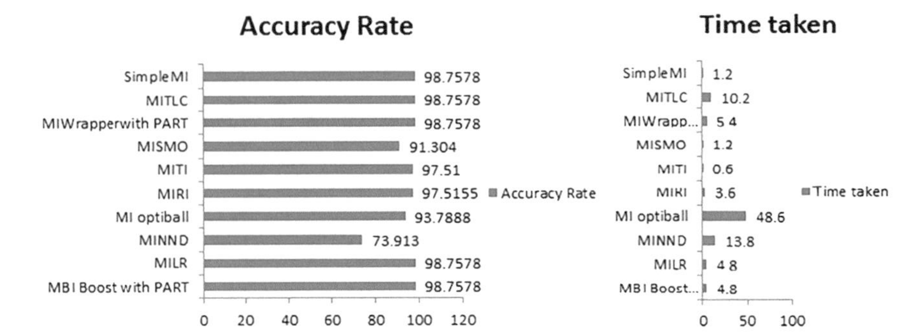


**Fig. 3** Comparison of MI learning

Simple MI and SMO is same, but the accuracy rate of Simple MI is more. Therefore, Simple MI is very efficient when compared to all other algorithms. The following section explains the comparison of single and multi-instance learning algorithms.

## 4 Comparison of Single Versus Multi-instance Learning Algorithms

The implementation of various single instance algorithms and multi-instance algorithms were discussed in the previous sections. The results of the algorithms were compared using different metrics like weighted accuracy rate, Sensitivity, Specificity, ROC values, and the time taken to generate the model were considered and analyzed. From the comparisons, it is clear that decision stump performs better in single instance algorithms and MI Simple performs better in multi-instance learning algorithms. When these two algorithms are compared, Simple MI is more effective than decision stump. The different algorithms implemented in the previous section should be analyzed to find out the best method to find the course outcome of the student. The authors in [18, 24] have discussed the Wilcoxon rank sum test which can be applied on the accuracy levels obtained from two different classification models and analyzed. This test evaluates the performance between the two methods using Mean rank and Sum rank. The scores are ranked from lowest to highest values. Figure 4 depicts the output of the tests.

From ranks calculations, the mean accuracy rate of SI is 92.45 and MI is 94.78. Therefore, it is clearly understood that selecting multi-instance learning algorithms to predict the performance of the students in the course is more effective when compared to selection of the single instance algorithms.

**Fig. 4** Implementation of Wilcoxon rank sum test

| Attribute_Y | | Attribute_X | | Statistical test | |
|---|---|---|---|---|---|
| Accuracy SI | | Accuracy SI | | Measure | Value |
| Avg | 92.453249 | Avg | 92.453249 | Used examples | 0 |
| Std-dev | 5.463323 | Std-dev | 5.463323 | Sum ranks + (T+) | 0.0 |
| | | | | Sum ranks - (T-) | 0.0 |
| | | | | E(T+) | 0.000000 |
| | | | | V(T+) | 0.000000 |
| | | | | Z | 0.000000 |
| | | | | Pr(>\|Z\|) | 1.000000 |
| Accuracy MI | | Accuracy SI | | Measure | Value |
| Avg | 94.782030 | Avg | 92.453249 | Used examples | 10 |
| Std-dev | 7.764212 | Std-dev | 5.463323 | Sum ranks + (T+) | 46.0 |
| | | | | Sum ranks - (T-) | 9.0 |
| | | | | E(T+) | 27.500000 |
| | | | | V(T+) | 96.000000 |

# References


1. Quadri, M. N., & Kalyankar, N. V. (2010). Drop out feature of student data for academic performance using decision tree techniques, GJCST computing classification H.2.8 & K.3.m. *Global Journal of Computer Science and Technology, 10*(2).
2. Marsh, P. A., & Poepsel, D. L. (n.d.). Perceived usefulness of learning outcomes predicts ratings of departmental helpfulness. *Teaching of Psychology*, 1532–8023. 0098-6283. https://doi.org/10.1080/00986280802374245.
3. Jaggars, S. S., & Xu, D. (n.d.). Predicting online student outcomes from a measure of course quality. CCRC Working Paper, 57. Retrieved from http://ccrc.tc.columbia.edu.
4. Mitchell, T. M. (2010). Generative and discriminative classifiers: Naive bayes and logistic regression. In *Machine learning*. Mcgraw Hill.
5. Kumar, A. S., & Vijayalakshmi, M. N. (2011). Efficiency of decision trees in predicting student's academic performance. In *International Conference on Computer Science, Engineering and Applications (CCSEA 2011)*, Chennai (pp. 335–341). 2231-5403.
6. Pagallo, G., & Haussler, D. (1990). Boolean feature discovery in empirical learning. *Machine Learning, 5,* 71–99.
7. Weiss, S. M., & Indurkhya, N. (1991). Reduced complexity rule induction. In *International Joint Conference on Artificial Intelligence*, 678–684.
8. Kumar, A. S., & Vijayalakshmi, M. N. (2012). *Inference of naive bayes techniques on student assessment data* (pp. 186–191). Berlin Heidelberg: Springer.
9. Kumar, A. S. (2016). Edifice an educational framework using educational data mining and visual analytics. *I.J. Education and Management Engineering, 2*, 24–30. 2305-3623.
10. Principles of data mining and knowledge discovery. (1999, September). In *Third European Conference, PKDD'99*, Prague, Czech Republic (pp. 15–18). 978-3-540-66490-1.
11. Qasem, A. A., Emad, M., & Mustafa, A. I. (2006). Mining student data using decision trees. In *International Arab Conference on Information Technology, ACIT.*
12. Crain-Dorough, M. L. (2003). *A study of dropout characteristics and school-level effects on dropout prevention* (Unpublished doctoral dissertation). A Dissertation Submitted to the Graduate Faculty of the Louisiana State University and Agricultural and Mechanical College.
13. Ayesha, S. (2010). Data mining model for higher education system. *European Journal of Scientific Research, 43*(1), 24–29. 1450-2165.

14. Danso, O. S. *An exploration of classification prediction techniques in data mining: The insurance domain* (Unpublished doctoral dissertation). A Dissertation Presented to the School of Design, Engineering, and Computing. Bournemouth University. http://www.comp.leeds.ac.uk.
15. Kaufmann, M. (1993). *C4.5: Programs for machine learning*. San Francisco, CA, USA: Inc.
16. Ramesh, V., Thenmozhi, P., & Ramar, K. (2012). Study of influencing factors of academic performance of students: A data mining Approach. *International Journal of Scientific & Engineering Research, 3*(7).
17. Rajeshinigo, D., & Jebmalar, P. J. (2017). Educational mining: A comparative study of classification algorithms using WEAK. *International Journal of Innovative Research in Computer and Communication Engineering, 5*(3), 5583–5589.
18. Zafra, A., Romero, C., & Ventura, S. (2010, October 25). Multi-instance learning versus single-instance learning for predicting the student's performance. In *Handbook on educational data mining*. CRC.
19. Frederick, U. N., & Christiana, O. C. (2016). Evaluation of data mining classification algorithms for predicting students performance in technical trades. *International Journal of Engineering and Computer Science, 5*(8), 17593–17601. 2319-7242.
20. Kotsiantis, S. B., Pierrakeas, C. J., Zaharakis, I. D., & Pintelas, P. E. (n.d.). Efficiency of machine learning techniques in predicting students' performance in distance learning systems. *Recent Advances in Mechanics and Related Fields*, 297–305.
21. Shah, N. S. (2012, January). Predicting factors that affect students' academic performance by using data mining techniques. *Review of Business Pakistan*, 631–668.
22. Dole, L., & Rajurkar, J. (2013). A decision support system to predict student performance. *International Journal of Innovative Research in Computer and Communication Engineering, 2*(13), 7237–7273.
23. Dietterich, T. G., Lathrop, R. H., & Lozano Perez, T. (n.d.). Solving the multiple instance problem with axis parallel rectangles. *Artificial Intelligence, 89*, 31–71 (Elsevier Science).
24. Shah, N. S. (2012). Predicting factors that affect students' academic performance by using data mining techniques. *Pakistan Business Review*, 631–668.

# Ontology-Driven Decision Support Systems for Health Care

**S. Shridevi, V. Viswanathan and B. Saleena**

**Abstract** Decision support systems (DSSs), as means of diffusing clinical guidelines, are powerful software system that will result in an improvement of medical practices. However, they are not invariably efficient and may suffer from limitations among which are lack of flexibility and weaknesses in the integration of many clinical guidelines for the management of patient's details. Recent research efforts resulted in a vital range of semantic reference systems enriched with vocabularies, thesauri, terminologies, and ontologies. The intensive use of ontologies is included in a new approach to create modern intelligent systems, reusing and sharing pieces of declarative information that plays a significant role in a DSS. A lot of effort has been made to produce standard ontologies for medical knowledge representation. This chapter brings an overview of semantic knowledge representation frameworks such as RDF and OWL for developing ontology and presents a DSS that is enabled by ontology for healthcare domain. A clinical use case is illustrated highlighting the role of ontology in medical DSS.

**Keywords** Ontology · RDF · RDFS · OWL · Medical ontologies
DSS

## 1 Introduction

Knowledge illustration presents a vital downside in today's science, notably if this knowledge has to be effectively used for reasoning as a part of the decision support systems (DSSs). Medical domain is defined by the abundance of existing

S. Shridevi (✉) · V. Viswanathan · B. Saleena
VIT University, Chennai, India
e-mail: shridevi.s@vit.ac.in

V. Viswanathan
e-mail: viswanathan.v@vit.ac.in

B. Saleena
e-mail: saleena.b@vit.ac.in



S. Margret Anouncia and U. K. Wiil (eds.), *Knowledge Computing and its Applications*, https://doi.org/10.1007/978-981-10-8258-0_4

professional knowledge, and practically each of its specializations includes a constantly growing and interacting range of relevant guidelines. A long goal is illustration of this knowledge in a form which can be used by systems, supporting medical decision making. An approach is critical which can change systematic illustration of various kinds of medical knowledge that may be used for various reasoning, ranging from offline and online warning systems to planning tending activities.

In this chapter, our aim is to give an overview of different metadata structures and schemes proposed for knowledge depiction, storage, and management of information. Any knowledge representation system must have styles for (i) representation and (ii) inference. Based on that, a brief discussion on two vital Semantic Web ontology representation formalisms—namely RDF-based metadata framework *and* OWL-based metadata framework—is presented. The frameworks are described and illustrated with rules and query languages for handling metadata and repositories. The chapter is structured as follows. Section 2 will discuss RDF-based metadata framework constructs; Sect. 3 will discuss OWL-based metadata framework constructs. Section 4 demonstrates a case study in the application of ontology-based DSS for clinical system by exploiting popular ontological resources of medical domain. Section 5 contains the conclusions and summary.

# 2 RDF-Based Metadata Framework

The RDF particulars were outlined starting from the earliest stage as a dialect for metadata about Web resources. The metadata model of RDF was proposed by the W3C and comprises of the below specifications:

- RDF/XML in terms of XML namespaces is used as specifications for RDF in XML syntax, XML information sets, and XML-based specifications.
- RDF statements are instances or extensions to the specifications for RDF schema (RDF(S)), which is utilized to characterize RDF vocabularies or models.
- The SPARQL is a query language and protocol for retrieving RDF instances from RDF information stores or data stores.

## *2.1 Resource Description Framework*

The Resource Description Framework (RDF) [2] offers a basic yet valuable semantic model based on graph structure. Generally, RDF is a meta-language for characterizing declarations or statements about resources and relations or links among them. Uniform Resource Identifiers (URIs) are utilized for recognizing such statements and connections. RDF gives constructs for characterizing resources and properties using classes. Statements are framed using these classes that declare facts about resources. RDF utilizes its own constructs (RDF schema or RDFS) for designing a schema for a resource. RDF is less expressive than RDFS, and RDFS incorporates subclass/superclass connections and restrictions or constraints on the resources complying with the schema. The motive of RDF's theoretical model is to break down data, and every little piece has unmistakably characterized semantics in such a way machine can comprehend it and derive meaning with it. Presently, utilizing RDF's constructs follows below rules:

**Rule #1: Information should be considered as statements; each statement must be in the form of Subject–Predicate–Object, and change of this triple order is not allowed**.
**Rule #2: Uniform Resource Identifier (URI) must be the way to name or identify a resource and it must be universal**.

RDF utilizes URIs rather than words for identifying resources and properties. Besides, all the URIs in such a vocabulary typically share a namespace prefix for this vocabulary as mentioned below:

**http://www.w3.org/1999/02/22-rdf-sentence structure ns#**

RDF vocabulary is summarized in Table 1:

The rest of the RDF Syntax names are: datatype, Description, parseType, ID, about, resource, li and nodeID with rdf: prefixed for all. Hereby, rdf:name is utilized to demonstrate RDF vocabulary constructs and the URI of rdf:type is given as below:

**http://www.w3.org/1999/02/22-rdf-sentence structure ns#type**

### 2.1.1 Syntax and Examples for Basic RDF Constructs Follows

Consider a basic sentence: The name of "sickperson1" is "Alex" which has the below parts:

- http://www.hospital.org/Sickpersons/sickperson1 as Subject or Resource,
- Name as Predicate or Property,
- "Alex" as Object value.

Representing a sick person using RDF expects all resources to be considered as URIs. The doctor of this sick person is also a resource. Doctor name is Kishore with kishore@clinic.org as his mail id, and he treats http://www.clinic.org/Sickpersons/

**Table 1** RDF constructs

| | |
|---|---|
| Classes | rdf:Statement, rdf:Property, rdf:XMLLiteral, rdf:Seq, rdf:Bag, rdf:List, rdf:Alt |
| Properties | rdf:type, rdf:predicate, rdf:object, rdf:subject rdf:first, rdf:rest, rdf:_n, rdf:value |
| Resources | Rdf:nil |

sickperson1. This could be taken as “http://www.clinic.org/Sickpersons/sickperson1” is dealt by somebody with name Kishore and email as kishore@-clinic.org. This can be taken as two sentences: The person identified as http://www.clinic.org/Physicians/physician1 has name Kishore with kishore@clinic.org as his mail id. The sick person http://www.clinic.org/Sickpersons/sickperson1 was dealt by this doctor. From the aspect of data integration, RDF promotes three different container objects such as Bag, sequence and alternative where URIs are used to identify same nodes and RDF graphs can be merged using the same.

### 2.1.2 RDF Serialization

Multiple syntaxes are followed in serializing RDF data model. Among them, RDF/XML and triples-based syntax are the two commonly used syntaxes. Consider the example discussed earlier. The XML serialization of that example is as follows.

```
<?xml version=“1.0” encoding=“utf-8” ?>
<rdf:RDF xmlns:rdf=“http://www.w3.org/1999/02/22-rdf-syntax-ns#”>
<rdf:Description rdf:about=“http://www.hospital.org/Sickpersons/sickperson1”>
<treated-by rdf:resource=“http://www.hospital.org/Physicians/physican1”/>
</rdf:Description>
<rdf:Description rdf:about=“http://www.hospital.org/Physicians/physician1”>
<Name>Kishore</Name>
<Email>kishore@hospital.org</Email>
</rdf:Description>
</rdf:RDF>
```

The triples-based serialization for the same example is as follows.

```
<http://www.hospital.org/Sickpersons/sickperson1> <treated-by> <http://www.hospital.org/Physicians/physican1>.
<http://www.hospital.org/Physicians/physician1> <Name> “Kishore”.
<http://www.hospital.org/Physicians/physician1> <Email> “kishore@hospital.org”.
```

## 2.2 *Core Elements of RDFS Schema*

The section discusses the constructs of RDF schema (RDFS). RDFS vocabulary contains terms that are used to define properties and classes for a domain-specific application. Similar to RDF constructs, all the RDFS terms are also identified by pre-defined URIs with the following leading string:

http://www.w3.org/2000/01/rdf-schema#

The above URI is the namespace prefix for RDF/XML format with the prefix string RDFS. The terms are grouped as classes, properties, and utilities.

- classes

The class constructs of RDFS that are used to define classes are rdfs:Resource, rdfs: Class, rdfs:Literal, rdfs:Datatype.

- properties

The property constructs of RDFS that can be used to define properties prefixed with rdfs are range, domain, subClassOf, subPropertyOf, label, and comment.

- utilities

The miscellaneous terms are rdfs:seeAlso and rdfs:isDefinedBy.
The core schema terms are illustrated in Fig. 1 based on their purposes.
A portion of sample rdfs HOSPITAL ontology is given below for better understanding of RDF and RDFS terms usage.

```
<rdfs:Class rdf:about="http://www.hospital.org/Hospital.owl#Cancer">
  <rdfs:subClassOf
rdf:resource="http://www.hospital.org/Hospital.owl#Disease"/>
 </rdfs:Class>
<rdfs:Class
rdf:about="http://www.hospital.org/Hospital.owl#Dermatologist">
  <rdfs:subClassOf
rdf:resource="http://www.hospital.org/Hospital.owl#Doctor"/>
 </rdfs:Class>
```

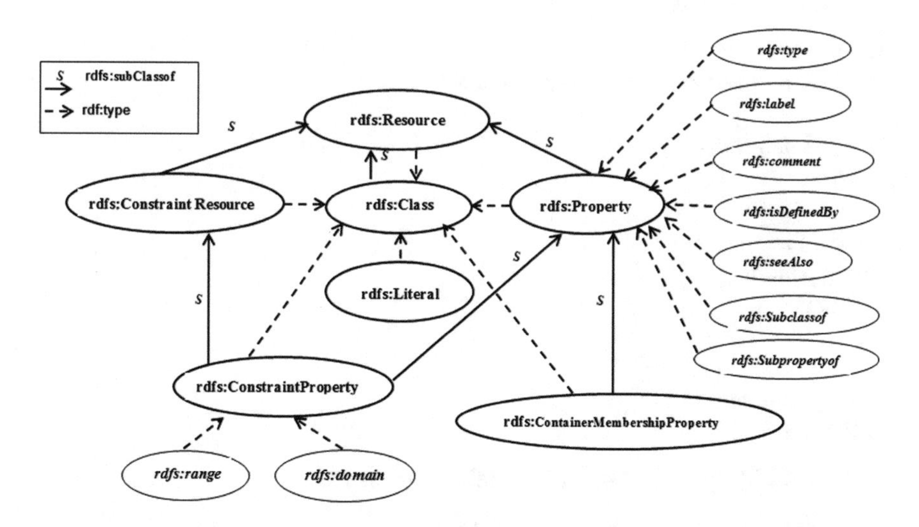


**Fig. 1** RDFS vocabulary

```
<rdfs:Class rdf:about="http://www.hospital.org/Hospital.owl#Oral">
 <rdfs:subClassOf
rdf:resource="http://www.hospital.org/Hospital.owl#Treatment"/>
</rdfs:Class>
<rdfs:Class rdf:about="http://www.hospital.org/Hospital.
owl#Sickperson">
 <rdfs:subClassOf
rdf:resource="http://www.hospital.org/Hospital.owl#Person"/>
 </rdfs:Class>
<rdfs:Class rdf:about="http://www.hospital.org/Hospital.owl#Person">
   <rdfs:subClassOf rdf:resource="http://www.w3.org/2002/07/owl#Thing"/>
 </rdfs:Class>
<rdfs:Class rdf:about="http://www.hospital.org/Hospital.owl#Surgeon">
  <rdfs:subClassOf
rdf:resource="http://www.hospital.org/Hospital.owl#Doctor"/>
</rdfs:Class>
```

## 2.3 The SPARQL Query Language for RDF Data

SPARQL is a W3C recommended query language and a protocol for accessing RDF datasets. SPARQL is "data-oriented" in that it only queries the information held in the RDF models; there is no inference in the query language itself. An RDF graph is a set of triples, and the serialization is just a way to write the triples down as there are multiple ways to encode the same RDF graph. The structure of a SPARQL query is followed as below:

```
#prefix declarations
PREFIX foo: http://example.com/resources/
...
#dataset definition
FROM ...
#result clause
SELECT...
Query pattern
WHERE {
....
}
#Query modifiers
ORDER BY....
```

A SPARQL query consists of the following, in order:

- Declarations of Prefixes to abbreviate URIs.
- Definitions of datasets to specify the RDF graph(s) that are being queried.

- A result clause to identify the information returned from the query.
- The query pattern that specifies what to query from the mentioned dataset.
- Query modifiers, slicing, ordering, and otherwise rearranging query results.

Figure 2 illustrates the working of SPARQL query language.

The below query example returns the blood test results for a sick person named "Alex" using a simple graph pattern.

```
Prefix foo:<http://www.hospital.org/Hospital.owl>
SELECT? Result
WHERE {
?sickperson name "Alex".
?sickperson hasbloodTestResult ?Result
}
```

Below query uses value constraints to list only male patients from the whole sick people list.

```
SELECT ?name ?sex
WHERE {
?sickperson sex ?sex.
FILTER (?sex="male").
?sickperson name ?name.
}
```

| ID | Subject | Predicate | Object |
|---|---|---|---|
| 1 | <Gene> | <hasName> | TrpA |
| 2 | <Gene> | <expresses> | <Protein> |
| 3 | <Protein> | <hasName> | Tryptophan Synthetase |
| 4 | <Protein> | <hasSubstrate> | <Chemical> |

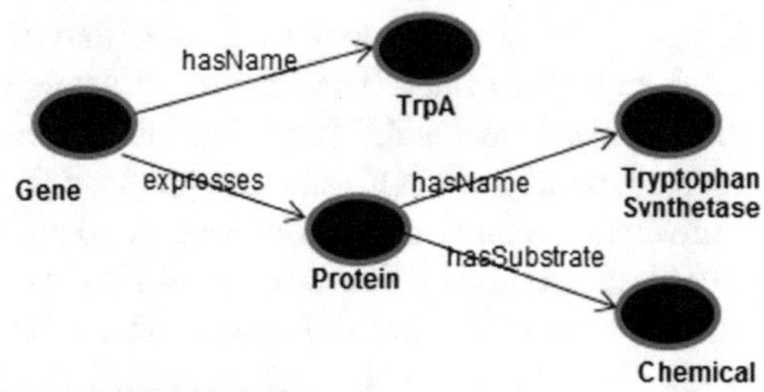


**Example SPARQL Query**

```
SELECT ?x WHERE
{
 ?x<hasName>"Tryptophan Synthetase"
 ?x <hasSubstrate> <Chemical>
```

**Answer: <Protein>**

**Fig. 2** SPARQL query execution

# 3 OWL-Based Metadata Framework

Web ontology language (OWL) [1] is a Semantic Web language which was developed in 2004. It is the formal way of representation of knowledge about concepts/entities, and relationship between the concepts. Representing the concepts and their interdependencies is called ontology. OWL is an extension of RDF syntax with additional features. The knowledge captured in OWL can be easily extracted by programs written in any language as it is a logic-based language.

OWL has a greater number of features for capturing and communicating the meaning and semantics when compared with XML, RDF, and RDFS, and hence, OWL has the ability to represent machine-understandable content on the Web. OWL is a modification of the DAML+OIL Web ontology language consolidating lessons gained from the plan and utilization of DAML+OIL. OWL includes rich vocabulary for portraying attributes and classes among others, relations between classes (e.g., disjointness), cardinality (e.g., "precisely one"), fairness, richer typing of properties and characteristics of properties (e.g., symmetry), and enumerated classes.

## *3.1 OWL Semantics*

OWL leverages upon the strength of Description Logics (DLs), a family of logic-based formal knowledge representation that enables us to capture and represent the concepts in any application domain. The DL models the concepts aka "classes" in RDF, roles aka "properties" in RDF and individuals. Individuals are the atomic elements of the domain; concepts (such as OWL) describe sets of individuals sharing the same characteristics; and describing the nature of relationships between pairs of individuals is done by roles.

Constructors are building blocks for any given DL that describes complex concepts and roles derived from simpler ones. The class constructors accessible in OWL incorporate the Booleans such as "and," "or," and "not," which in OWL parlance are called as intersectionOf, unionOf, and complementOf. The limited types of existential and universal evaluation and the corresponding OWL representation are called someValuesFrom and allValuesFrom restrictions.

OWL also supports transitive properties; in OWL, someValuesFrom limitations are utilized to depict classes, the cases of which are connected by means of an offered property to examples of some different class. Conversely, allValuesFrom limitations compel the conceivable objects of a given property and are ordinarily utilized as a sort of confined range limitation. OWL likewise takes into account property progressions, extensionally characterized classes utilizing the oneOf constructor, backwards properties utilizing the inverseOf property constructor, cardinality limitations utilizing the minCardinality, maxCardinality, and cardinality constructors.

OWL provides three increasingly expressive sublanguages evolved over time addressing specific needs by appropriate communities of implementers and adopters.

- *OWL Lite:* For those users whose primary needs are around classification hierarchy and other simple constraints. For example, restricted permitted value in cardinality constraints values is 0 or 1.
- *OWL DL* supports for advanced users who want to leverage the maximum expressiveness offered in OWL yet wanting to retain the completeness in computation.
- *OWL Full* is meant for the extreme power users who want the maximum expressiveness and the freedom associated with syntactic of RDF with no computational guarantees whatsoever.

Each of these sublanguages encompasses the simpler predecessor, both in what can be legally expressed and holding the validity of the conclusion. The following set of forward progressive relations hold and not their inverses.

- All legal OWL Lite ontology is a subset of OWL DL ontology.
- All legal OWL DL ontology is a subset of OWL Full ontology.

The list of OWL language constructs is given in Table 2.

## 3.2 Few Examples of OWL Constructs in Healthcare Domain

An important feature of OWL is the property restrictions, which can be used to set the constraints on values and specify the cardinalities.

- The Value constraints are allValuesFrom, hasValue, and someValuesFrom,
- The cardinality constraints are cardinality, maxCardinality, and minCardinality.

These are called "qualified cardinality restrictions" (QCRs), is constraining the number of values a particular property type can take on, since it is easy to express constraints like "Each individual has at most one SSN."

(a) **A Surgery Team consists of any person qualified as a doctor.**

```
:Surgery Team
 a OWL: Class ;
 OWL: equivalentClass
 [ a OWL: Restriction ;
  OWL: onProperty : Doctor
  OWL: someValuesFrom : Person
 ]
```

**Table 2** Details of OWL constructs

| RDF schema | | |
|---|---|---|
| Name | Syntax | Remark |
| Class | rdfs:Class | Kinds of things/generic concept of a type or category |
| SubClass | *rdfs:subClassOf* | Specialization of another general class |
| Property | *rdf:Property* | Characterizes those classes of things |
| SubProperty | *rdfs:subPropertyOf* | Relates a property to one of its super properties |
| Domain | *rdfs:domain* | Specifies object to which properties are applied |
| Range | *rdfs:range* | Specifies the range of a property P |
| Individual | *Individual* | Facts about class, property values, and identity |
| **(In)Equality** | | |
| *EquivalentClass* | OWL:equivalentClass | Properties applied to two or more classes will have same values but different meaning |
| *EquivalentProperty* | OWL:equivalentProperty | To show two properties have the same property span |
| *SameAs* | OWL:sameAs | Links an individual to an individual. Defining mappings between ontologies. Defines class equality, same values and same meaning |
| *DifferentFrom* | OWL:differentFrom | Links an individual to an individual, refers to different individuals |
| *AllDifferent* | OWL:AllDifferent | All individuals in the given list are different from each other |
| *DistinctMembers* | OWL:distinctMembers | Links an individual of OWL:AllDifferent to a list of individuals |
| **Property characteristics** | | |
| *ObjectProperty* | OWL:ObjectProperty | To show the link between individuals |
| *DatatypeProperty* | OWL:DatatypeProperty | Individuals are linked to data values |

(continued)

**Table 2** (continued)

| RDF schema | | |
|---|---|---|
| Name | Syntax | Remark |
| *InverseOf* | OWL:inverseOf | Defines inverse relation between properties |
| *TransitiveProperty* | OWL:TransitiveProperty | If (a,b) and (b,c) are an instances of X, this infers that (x,z) is also an instance of X |
| *SymmetricProperty* | OWL:SymmetricProperty | If (a,b) is an instance of X, then (b,a) is also an instance of X |
| *FunctionalProperty* | OWL:FunctionalProperty | Property that defines at most one unique value for each object |
| *InverseFunctionalProperty* | OWL: InverseFunctionalProperty | Property that defines two different objects will not have same value |
| **Property restrictions** | | |
| *Restriction* | OWL:Restriction | Describes constraints for the classes to satisfy |
| *OnProperty* | OWL:onProperty | Triplet for linking restriction to particular property |
| *AllValuesFrom* | OWL:allValuesFrom | Describes all possible values for property |
| *SomeValuesFrom* | OWL:someValuesFrom | Describes at least one possible value for property |
| **Restricted cardinality** | | |
| *minCardinality* | OWL:minCardinality | All objects of class must have *at least* N semantically unique values for property |
| *maxCardinality* | OWL:maxCardinality | All objects of class must have a *at most* N semantically unique values for property |
| *Cardinality* | OWL:cardinality | Restriction applied to a data value belonging to the range of the XML schema datatype |
| **Datatypes** | | |
| *xsd datatypes* | rdf:datatype = "&xsd; < type>" | Includes Boolean, numerical, string, time-related, URI, etc. |

(b) **A Team for a Cyst prostatectomy should have one male doctor.**

```
:Surgery Team
 a OWL: Class ;
 rdfs: subClassOf
   [ a OWL: Restriction ;
    OWL:onProperty :Doctor ;
    OWL:allValuesFrom :  Person
   ] .
 rdfs: subClassOf
   [ a OWL: Restriction ;
    OWL:onProperty : Doctor ;
    OWL:someValuesFrom : MalePerson
   ]
```

(c) **A Team for a C-Section Surgery should have at least three female doctors**

```
:Surgery Team
 a OWL: Class ;
 rdfs: subClassOf
   [ a OWL: Restriction ;
      OWL: onProperty : Doctor ;
      OWL: allValuesFrom : Person
   ] .
 rdfs:subClassOf
   [ a OWL: Restriction ;
    OWL: onProperty : Doctor ;
    OWL: someValuesFrom : FemalePerson
    OWL: minCardinality "3"^^xsd:integer
   ]
```

## *3.3 Reasoning Using OWL*

The OWL ontology requires an inference engine to reason over the knowledge base to discover the hidden relationships from the ontologies. Some of the well-known reasoners include Pellet, Hermit, Fact++, and RacerPro [2]. The availability of such reasoners was the key motivation of W3C to base OWL as a DL. Ontology IDE tools (like SWOOP, code, Protégé 4, and TopBraid Composer) provide feedback to developers on the logical implication of their designs using DL reasoners.

Reasoners can help us identify any missing relationships that of type subclass. For example, When Fact++ was used against SNOMED, it helped discovering 180 missing relationships that were of type subclass.

For instance, researchers utilize ontologies, (for example, the gene ontology and the biological pathways exchange ontology) for annotating information from gene sequencing tests, enabling them to answer complex queries, (for example, "What DNA binding products interact with insulin receptors?"). For the reasoner to answer to not just recognize people that are (maybe just verifiably) cases of DNA-restricting items and of insulin receptors yet to distinguish which sets of people are connected. Rules have to be written using ontology development tools to do the appropriate reasoning. One such language proposed for defining rules for Semantic Web is discussed in Sect. 4.1.

### 3.3.1 The Semantic Web Rule Language (SWRL)

SWRL is the proposed language to express rules and logics in Semantic Web. Rules help us to infer additional information from the dataset. SWRL allows users to write rules expressed in terms of OWL concepts to reason about OWL individuals. Using the rules, new knowledge can be inferred from existing knowledge bases. There are many built-ins which will provide an extension mechanism whereby the modeling language can be enhanced with domain-specific built-ins. The predicates can be class expressions, property expressions, data range restrictions, sameIndividual, differentIndividuals, core SWRL built-ins, and user-defined SWRL built-ins. Table 3 lists the available built-ins which are defined for various comparisons.

Consider the following indications recorded in the database by patients.

indication(fever), indication(running_nose), indication(headache), indication (body_ache), indication(sore_throat), indication(cough), indication(chills), indication(conjunctivitis)

**Table 3** SWRL built-ins for comparisons

| Built-ins | Syntax with comment |
|---|---|
| swrlb:equal | swrlb:equal(?x,?y)<br>If the argument1 and the argument2 are same |
| swrlb:notEqual | swrlb:notEqual(?x,?y)<br>If argument1 and the argument2 are not same.<br>It is the negation of swrlb:equal |
| swrlb:lessThan | swrlb:lessThan(?x,?y)<br>If the argument1 is less than the argument2 |
| swrlb:lessThanOrEqual | swrlb:lessThanOrEqual(?x,?y)<br>If the argument1 is less than or equal to the argument2 |
| swrlb:greaterThan | swrlb:greaterThanOrEqual(?x,?y)<br>If the argument1 is greater than the argument2 |
| swrlb:greaterThanOrEqual | swrlb:greaterThanOrEqual(?x,?z)<br>if the argument1 is greater than or equal to the argument2 |

Suppose doctor gives a rule for Viral_Fever as "If indications are fever, cough, running nose and chills then patient probably have Viral Fever." The rule can be written as

**Diagnosis (Viral_Fever) :- indication(fever), indication(cough), indication (running_nose), indication(chills)**

The rules written in SWRL can be integrated through a Rule Editor (protégé), and the additional information can be inferred from the patient database.

**Indication(?x) ^Indication(?y) ^Indication(?z) ^Indication(?v)←Diagnosis(?X)**

Another way of adding rules to the ontology is through JENA. JENA can not only read ontologies, it can also reason out information from the ontology, like other reasoners RacerPro, Pellet.

## 4 Clinical Decision Support System (CDSS)

Clinical decision support system majorly helps in improving clinical practices. It is a computer-based information system which helps in clinical decision making. The diagnostic features of a patient are mapped to a computerized clinical database using which patient-specific assessment or suggestions are disclosed to the clinician or the patient for decision making [3].

A large quantity of data from medical devices, diagnosing patients, and medical imaging has already been produced periodically in health clinic and health centers [4]. This will continue to happen which creates the need for digitalization of health through high-tech devices and information systems. Due to this explosive growth of data, there is a need to discover or uncover valuable information from that database in order to transform such data into knowledgebase that could help in improving healthcare practice and develop better biomedical products [5]. Due to many challenges, this ultimate goal persists to be a tedious task.

Interoperability is one of the major challenges in managing CDSS [6]. Generally, the distributed data repositories store the generated data from different sources. This database will have inconsistency in naming, structure and format. This distributed data should be transformed into a common format which is accurate, manageable, and efficient in processing the data to integrate with other systems [7].

Since the cost of data integration is high, many hospitals and medical centers are now trying to use Semantic Web Technologies to integrate the data. The advantage of Semantic Web includes the integration of heterogeneous data using explicit semantics, simplified annotation, and sharing of findings. In order to help the organization to adopt Semantic Web, the World Wide Web Consortium (W3C) has formed the Semantic Web for Health Care and Life Sciences Interest Group (HCLS IG). The HCLS IG was established to increase the use of Semantic Web Technologies to support collaboration, innovation, and adoption of Semantic Web in the domains of Health Care and Life Sciences [11].

## *4.1 Medical Ontologies*

BioPortal is a comprehensive repository of biomedical Ontologies [12]. Few of them are discussed below:

### 4.1.1 SNOMED CT

To develop a comprehensive high-quality clinical/health information in a consistent and reliable manner, SNOMED CT places an integral part [8]. It helps in automatic interpretation of clinical phrases selected by the clinician in a more standardized way. SNOMED CT supports evidence-based care as it is clinically validated and semantically rich, thus benefitting individual patients and clinicians in any location. Processing and presenting the same clinical information to serve different purposes is one of the major advantages of SNOMED CT.

### 4.1.2 LOINC

Generally, clinical information consists of health measurement, observations, test results, document, etc. LOINC as a common language helps in creating different codes for measurements, observations, test, etc., that has a clinically different vocabulary. To do that, LOINC codes distinguish a given observation across six dimensions called component, property, time, system, scale, and method. It consists of 192,372 classes and 152 properties. It is available only in the Unified Medical Language System (UMLS) format.

### 4.1.3 MedDRA

Some clinicians are convenient with using and understanding clinical information in their native languages. MedDRA supports in successfully achieving this purpose. It is a multilingual terminology allowing the clinicians and patients to use their native languages [9]. MedDRA uses a hierarchical structure to analyze individual medical events or issues involving a system, organ, or etiology. It uses its multiaxial hierarchy and a set of standardized queries (MedDRA queries) to detect and monitor clinical syndromes whose symptoms consist of numerous systems or organs. A list of few more ontologies which are currently available in BioPortal and number of classes in the ontology, properties, and ontology format are summarized in Table 4.

**Table 4** List of ontologies in BioPortal

| Name of the ontology | Number of classes | Number of properties | Format |
|---|---|---|---|
| LOINC | 1,92,372 | 152 | UMLS |
| SNOMEDCT | 3,27,128 | 152 | UMLS |
| MedDRA | 69,107 | 16 | UMLS |
| FMA | 1,00,080 | 188 | OWL |
| ICD10 | 12,445 | 1 | UMLS |
| RADLEX | 46,433 | 91 | OWL |
| NBO | 1,91,799 | 195 | OWL |
| DRON | 4,34,663 | 20 | OWL |
| MFOEM | 899 | 29 | OWL |
| VO | 6211 | 137 | OWL |

## *4.2 Architecture of Clinical Decision Support System (CDSS)*

Zenuni et al. [10] proposed the system architecture for CDSS depicted in Fig. 3 that includes four modules: the inference engine, the graphical user interface, the SWRL rule, and the ontologies.

In the CDSS architecture, the user can interact through a graphical user interface and place request for any test or diagnostic for a specific illness in the system. The inference engine uses the rules in the knowledge base as the reasoning component through ontologies and ultimately infers a diagnostic for the specific illness tested in

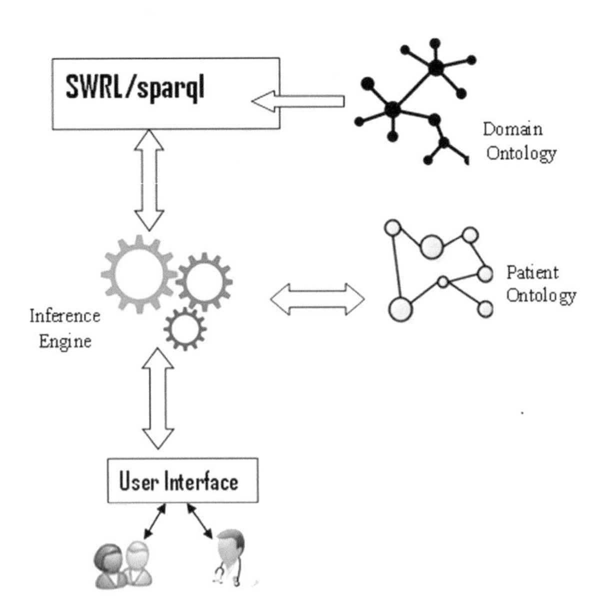


**Fig. 3** Architecture of clinical decision support system (CDSS)

the system. The rules in the knowledge base are constructed from the expertise in the medical domain, and these rules are expressed in SWRL format after mapping the domain ontology concepts. During this process, the ontologies feed the reasoning component(s) with the necessary concepts and match their relationships which help the inference engine to combine the rules with concept instances while inferring the diagnostic.

### 4.2.1 Domain Ontology

A portion of domain ontology which is used for translation medicine during the clinical use case scenario is illustrated below. List of classes and relationships modeled in this ontology are: LaboratoryTestOrder class which denotes the order of the laboratory test for a patient. The class Panel contains one or more tests represented in the class Test and the order may be for a Panel of test represented in the class Panel. The class INDAddress represents the receipientAddress and a payorAddress in the order for a laboratory test. INDAddress represent the set of addresses in the country India. The class Patient is a major concept that features the information about the patient such as the results of diagnostic tests and his family history. It is a subclass of Person. Details related to a patient's family and his/her relatives are denoted using the is_related relationship.

If the family of the patient being evaluated has similar symptoms of disease or ailments, it is captured under the class FamilyHistory and this information related to the Patient class through the relationship called hasFamilyHistory. The information about the results of laboratory tests of the patient is captured under the StructuredTestResult class which is connected to the Patient class through the relationship called hasStructuredTest and the Test class through the relationship called associatedResult.

The MolecularDiagnosticTestResult class represents the results of a molecular diagnostic test result. Molecular diagnostics identify mutations represented using the identifiesMutation relationship and indicate diseases represented using the indicatesDisease relationship in a patient.

Gene class represents information about genes and is associated with the class Patient through the relationship called hasGene. Mutation class represents the genetic variants or mutations of a given gene linked to the class Patient through the relationship called hasMutation. The relationship isMutationOf represents the relationship between gene and mutation.

The class Disease represents the conditions of the disease prevailing in the patient who is diagnosed. It is associated to the class Patient through the relationship suffersFrom and is also associated with molecular diagnostic test class through the relationship indicatesDisease.

OWL representation of the above domain ontology in Turtle format is as follows:

```
@prefix OWL: <http://www.w3.org/2002/07/OWL#> .
@prefix rdfs: <http://www.w3.org/2000/01/rdf-schema#> .
@prefix xsd: <http://www.w3.org/2001/XMLSchema#> .
<http://medical/#Person> a OWL:Class .
<http://medical/#Patient> a OWL:Class ;
 rdfs:subClassOf <http://medical/#Person>, [
  a OWL:Restriction ;
  OWL:onProperty <http://medical/#isRelatedTo> ;
  OWL:allValuesFrom <http://medical/#Relative>
 ] <http://medical/#Relative> a OWL:Class ;
 rdfs:subClassOf <http://medical/#Person> .
<http://medical/#StructuredTestResult> a OWL:Class ;
 rdfs:subClassOf [
  a OWL:Restriction ;
  OWL:onProperty <http://medical/#hasPatient> ;
  OWL:cardinality ”““
1 ”““^^xsd:nonNegativeInteger ] .
<http://medical/#MolecularDiagnosticTestResult> a OWL:Class .
<http://medical/#FamilyHistory> a OWL:Class .
<http://medical/#Disease> a OWL:Class .
<http://medical/#Gene> a OWL:Class .
<http://medical/#Mutation> a OWL:Class ;
 rdfs:subClassOf [
  a OWL:Restriction ;
  OWL:onProperty <http://medical/#isMutationOf> ;
  OWL:someValuesFrom <http://medical/#Gene>
 ] .
<http://medical/#LaboratoryTestOrder> a OWL:Class .
<http://medical/#Panel> a OWL:Class .
<http://medical/#Test> a OWL:Class .
<http://medical/#INDAddress> a OWL:Class .
<http://medical/#isRelatedTo>
 a OWL:ObjectProperty, OWL:TransitiveProperty ;
 rdfs:domain <http://medical/#Patient> ;
 rdfs:range <http://medical/#Relative> .
<http://medical/#hasFamilyHistory> a OWL:ObjectProperty ;
 rdfs:domain <http://medical/#Patient> ;
 rdfs:range <http://medical/#FamilyHistory> .
<http://medical/#associatedRelative> a OWL:ObjectProperty ;
 rdfs:domain <http://medical/#FamilyHistory> ;
 rdfs:range <http://medical/#Relative> .
```

```
<http://medical/#hasStructuredTestResult> a OWL:ObjectProperty ;
 rdfs:domain <http://medical/#Patient> ;
 rdfs:range <http://medical/#StructuredTestResult> ;
 OWL:inverseOf <http://medical/#hasPatient> .
<http://medical/#hasMolecularDiagnosticTestResult> a OWL:
ObjectProperty ;
 rdfs:subPropertyOf <http://medical/#hasStructuredTestResult> ;
 rdfs:range <http://medical/#MolecularDiagnosticTestResult> .
<http://medical/#identifiesMutation> a OWL:ObjectProperty ;
 rdfs:domain <http://medical/#MolecularDiagnosticTestResult> ;
 rdfs:range <http://medical/#Mutation> .
<http://medical/#indicatesDisease> a OWL:ObjectProperty ;
 rdfs:domain <http://medical/#MolecularDiagnosticTestResult> ;
 rdfs:range <http://medical/#Disease> .
<http://medical/#suffersFrom> a OWL:ObjectProperty ;
 rdfs:domain <http://medical/#Patient> ;
 rdfs:range <http://medical/#Disease> .
<http://medical/#hasMutation> a OWL:ObjectProperty ;
 rdfs:domain <http://medical/#Patient> ;
 rdfs:range <http://medical/#Mutation> .
<http://medical/#hasGene> a OWL:ObjectProperty ;
 rdfs:domain <http://medical/#Patient> ;
 rdfs:range <http://medical/#Gene> .
<http://medical/#isMutationOf> a OWL:ObjectProperty ;
 rdfs:domain <http://medical/#Mutation> ;
 rdfs:range <http://medical/#Gene> .
<http://medical/#recipientAddress>
 a OWL:ObjectProperty ;
 rdfs:domain <http://medical/#LaboratoryTestOrder> ;
 rdfs:range <http://medical/#INDAddress> .
<http://medical/#payorAddress>
 a OWL:ObjectProperty ;
 rdfs:domain <http://medical/#LaboratoryTestOrder> ;
 rdfs:range <http://medical/#INDAddress> .
<http://medical/#testPanel>  a OWL:ObjectProperty ;
 rdfs:domain <http://medical/#LaboratoryTestOrder> ;
 rdfs:range <http://medical/#Panel> .
<http://medical/#test>  a OWL:ObjectProperty ;
 rdfs:domain <http://medical/#Panel> ;
 rdfs:range <http://medical/#Test> .
```

```
<http://medical/#associatedResult> a OWL:ObjectProperty ;
 rdfs:domain <http://medical/#Test> ;
 rdfs:range <http://medical/#StructuredTestResult> .
<http://medical/#orderDateTime> a OWL:DataTypeProperty, OWL:
FunctionalProperty ;
 rdfs:domain <http://medical/#LaboratoryTestOrder> ;
 rdfs:range xsd:datetime .
<http://medical/#PatientWithMYH7Gene> a OWL:Class ;
 rdfs:subClassOf [
  a OWL:Restriction ;
  OWL:onProperty <http://medical/#hasGene> ;
  OWL:hasValue <http://medical/#MYH7>
 ] .
<http://medical/#NormalStructuredTestResult> a OWL:Class ;
 rdfs:subClassOf <http://medical/#StructuredTestResult> ;
 OWL:disjointWith <http://medical/#AbnormalStructuredTestResult> .
<http://medical/#AbnormalStructuredTestResult> a OWL:Class ;
 rdfs:subClassOf <http://medical/#StructuredTestResult> ;
 OWL:disjointWith <http://medical/#NormalStructuredTestResult> .
```

### 4.2.2 Rules for Clinical Decision Support and Integration of Ontologies

In this section, the rules to integrate with the domain ontology to infer new knowledge as recommendation will be discussed. In some cases, all these rules cannot be written in ontology using OWL and hence such rules are expressed using SWRL. Consider the example: if a patient has a structured test result which is indicative of a particular disease, then the patient suffers from that disease. Properties involved in these rules are hasStructuredTestResult, indicatesDisease, and suffersFrom. These types of rules cannot be expressed using OWL axiom, but it can be expressed using SWRL representation as follows.

hasStructuredTestResult(?x,?y) ^ indicatesDisease(?y,?z) ->suffersFrom(?x, ?z)

Consider another example: if the patient has an allergy to fibric acid or has an abnormal liver panel, then the patient is recommended for ZetiaLipdManagemetn therapy. This rule can be written in SWRL as follows:

hasALPValue (?x,?y) ^ swrl:greaterThan(?y, NormalALPValue) ^ isAllergicTo (?x, FibricAcid)->recommendedTherapy(?x, ZetiaLipidManagementTherapy)

With the help of a reasoner, the above rule can be executed with domain ontology in order to infer new knowledge. Once the doctor enters all the details about patient, the system will recommend decision based on the rules written in the knowledgebase and domain ontology.

## 5 Summary

In this chapter, we tend to confer an in-depth discussion of two major information frameworks based on RDF and OWL specifications for developing ontologies. The information models and query languages of those specifications were given. There are two main kinds of clinical decision support systems. One form of CDSS, which uses a knowledge domain, applies rules to patient information using a reasoning engine and displays the results to the end user. Systems without a knowledge domain, on the other hand, accept machine learning to investigate clinical information. We have given here a DSS solution based on Semantic Web specifications to a clinical use case situation that takes the benefits of using ontologies as its knowledge domain.

## References


1. Patel-Schneider, P. F., Hayes, P., & Horrocks, I. (2004). OWL web ontology language semantics and abstract syntax. W3C recommendation.
2. Carroll, J. J., & Klyne, G. (2004). Resource description framework (RDF): Concepts and abstract syntax. W3C recommendation.
3. Castaneda, C., et al. (2015). Clinical decision support systems for improving diagnostic accuracy and achieving precision medicine. *Journal of Clinical Bioinformatics, 5*(1), 4.
4. Gorman, S. P., Greenes, R. A., Haynes, R. B., Kaplan, B., Lehmann, H., & Tang, P. C. (2001). Clinical decision support systems for the practice of evidence-based medicine. *Journal of the American Medical Informatics Association, 8*(6), 527–534.
5. Riañoa, D., Reala, F., López-Vallverdúa, J. A., Campanab, F., Ercolanic, S., Mecoccic, P., et al. (2012). An ontology-based personalization of health-care knowledge to support clinical decisions for chronically ill patients. *Journal of Biomedical Informatics, 45*(3), 429–446.
6. Jaspers, M. W., Smeulers, M., Vermeulen, H., & Peute, L. W. (2011). Effects of clinical decision-support systems on practitioner performance and patient outcomes: A synthesis of high-quality systematic review finding. *Journal of the American Medical Informatics Association, 18*(3), 327–334.
7. Garg, A. X., Adhikari, N. K., McDonald, H., Rosas-Arellano, M. P., Devereaux, P., Beyene, J., et al. (2005). Effects of computerized clinical decision support systems on practitioner performance and patient outcomes: A systematic review. *JAMA, 293*(10), 1223–1238.
8. SNOMEDCT. (May, 2017). Systematized nomenclature of medicine—clinical terms. Retrieved from http://www.ihtsdo.org/snomed-ct.
9. MedDRA (May, 2017). Medical dictionary for regulatory activities. Retrieved from http://www.meddra.org/.
10. Zenuni, X., Raufi, B., Ismaili, F., & Ajdari, J. (2015). State of the art of semantic web for healthcare. *Procedia, Social and Behavioral Sciences, 195,* 1990–1998.

11. HCLSIG. (May, 2017). Semantic web health care and life sciences interest group. Retrieved from http://www.w3.org/blog/hcls/.
12. BioPortal. (May, 2017). The world's most comprehensive repository of biomedical ontologies. Retrieved from https://bioportal.bioontology.org/.
13. LOINC. (May, 2017). LOINC logical observation identifier names and codes. Retrieved from http://loinc.org/.
14. WHO. World Health Organization. (May, 2017). The ICD-10 classification of mental and behavioural disorders: clinical descriptions and diagnostic guidelines. Geneva: World Health Organization. http://www.who.int/classifications/icd/en/.
15. RadLex. Radiology Lexicon. (May, 2017). http://www.rsna.org/RadLex.aspx.

# Prediction of Land Cover Changes in Vellore District of Tamil Nadu by Using Satellite Image Processing

**M. Prabu and S. Margret Anouncia**

**Abstract** Prediction of land cover changes is important to evaluate the land use or land cover changes to monitor the land use changing aspects for the Vellore district. Due to land use and land cover change, most of the rural areas around the Vellore district become unable to cope with environmental risk and agriculture. Population is one of the main issues in increasing the land cover changes in Vellore district. From the satellite, data can easily find out the changes in Vellore district. Result is compared with real time to show the extreme changes in the study area. Vegetation cover decreased, and settlement and built up areas increased due to increasing population. The Objective is to find out the land cover changes and predict how the Vellore district in future. And also, this study suggests some remedial measures to protect agriculture of Vellore district.

**Keywords** Land cover changes · Satellite image processing · Vegetation index GIS · Remote sensing

## 1 Introduction

Land use classification technique could be used to yield data-relating changes in land uses. Landsat data have been used to find out land cover changes. The classification system was planned to mostly rely on remote sensing. Preprocessed spectral band image was classified using ISODATA classification technique. Previous hybrid classification was used to find out land cover area changes using unsupervised training data followed by supervised training data. The hybrid classification method used the signs that were produced by the ISODATA sorting. Hybrid supervised–unsupervised classification produced land cover mapping [1].

M. Prabu (✉) · S. Margret Anouncia
School of Computer Science and Engineering, VIT University,
Vellore, Tamil Nadu, India
e-mail: Prabu7mca@gmail.com





S. Margret Anouncia and U. K. Wiil (eds.), *Knowledge Computing and its Applications*, https://doi.org/10.1007/978-981-10-8258-0_5

In order to improve the satellite image features, a PCA fusion method was applied by using the panchromatic bands and multispectral bands of Aster data and Landsat data. Supervised maximum likelihood classification is applied to Landsat images to find land cover changes. The Multivariate Alteration Detection (MAD) transformation was based on a traditional statistical transformation referred to as canonical correlation investigation to enhance the information. Results of supervised classification of Aster and ETM data were evaluated for the particular area; this was pointed at using optical satellite remotely sensed images used in change detection of vegetation in protected areas [2].

Sub-pixel impervious surface mapping and related change detection include initial regression tree modeling and assessment, imperviousness change detection and interpretations. Machine Learning algorithm and Regression algorithm conducts a binary-repeated partitioning to find out changes in land cover. The comparative status of predictive variables was evaluated based on occurrence of a variable used and the location of the variable within the multivariate linear regression equations. This approach is used in urban land cover changes by measuring sub-pixel percent unyieldingness using Landsat data and high-resolution imagery [3].

In one method, Pixel-based Automated Classification Chain methodology and TWinned Object were used to map the changes. This TWOPAC classification methodology is a supervised decision tree classification; in this, the tree is derived using illustrative areas. But the feature generation and segmentation were not included in TWOPAC environment. Supervised classification is achieved in the TWOPAC environment with a positive user-defined sample. The implementation of this method features the decision tree classification. This algorithm examines through given set of training cases to discover the finest quality for segregation into pre-defined classes. Many datasets from different spatial resolutions and sensors were treated with the framework of TWOPAC [4].

In another method, unsupervised classification was performed as a result of which ISO clusters were generated. Using the clumped image, mean values of variables like NDVI, mean water mask were generated and used as an input for expert classification. Land cover change analysis was conducted using post-classification comparison method. The classified lands cover information comprising of eight classes and combine into a single map. It demonstrated knowledge-based image classifications using Landsat ETM+ and MSS imagery as a useful technique for classification [5].

A mixture of the reflective spectral bands of different time series images was used for image classification in another method. A hybrid supervised and unsupervised training method is referred to as "guided clustering," classes are clustered into subclasses, and classifier training was introduced with maximum likelihood classification. The outcomes enumerate the change patterns of land cover in the urban area and validate the potential of multitemporal satellite data to provide an exact, inexpensive way to map and examine changes in land cover; it can be used as inputs for land management [6].

Land use and land cover change information has a significant role to play at local and provincial as well as at macro-level planning [7]. The land cover changes happen naturally in a continuous improvement and stable way; however, sometimes it may be fast and unexpected due to activities of anthropogenic. Land use and land cover changes have a significant role in regional planning at different temporal and spatial scales. This along with the temporal and spatial analysis technologies like GIS and GPS helps for good planning and cost-effective decision.

Indecision is essential in land cover change and land use predictive studies and its assessment [8]. The existence of fields with little vegetation or no vegetation during the winter period increases impurity fluxes to the rivers. The land cover change is also used to find out the changes in trace soil fertility and carbon stocks in the study areas; meanwhile, it is used to find the global climate change and atmospheric change conditions. Many modules of land cover change are of awareness. The most important need is the change of primary forest to crop land areas, field other non-forested land-cover types [9].

This work presents a comprehensive study of land cover changes in study area. Landsat TM is acquired for the study areas between 2009 and 2017. It will define a multistage process of consistent protocols; it is applied for large volumes of images to map the land cover changes using set of defined system.

## 2 Study Area

Vellore district is situated in Tamil Nadu state of India and located in the border of Andhra Pradesh state. The maximum temperature in summer is up to 45 °C. From the local census calculation, the town has approximately 210,000 people at the time of 2012. This study area covers 1055 and 595 ha which are already covered by people. Geographical location of study is shown in Fig. 1.

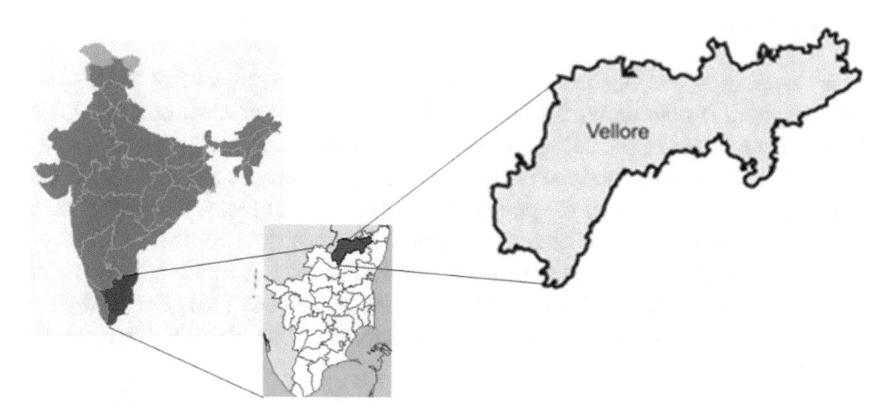


**Fig. 1** Geographical location of study area—Vellore district, India

# 3 Materials and Methodology

## *3.1 Materials Used*

The Landsat TM data for the years 1999 and 2017 were used for the land use mapping coupled with intensive ground truth verifications. Here, two sets of data were taken. Those images are years of 2009 and 2017.

(1) Different land use types and its changes were calculated.
(2) Land use modeling for the years 2009–2017 was created.
(3) Future land use was predicted.

The model is particularly valuable for measuring changes in spatial patterns of ever-changing land uses because of the clear attention given to connections [10]. Progressions will affect future land use possibilities in the study area [11].

## *3.2 Methodology*

### 3.2.1 High-Resolution Satellite Image

The high-resolution satellite image data of the earth are downloaded from NASA Web site. They provide free Landsat7 data of the earth. Landsat7 provides 1000 and 250 m resolution data of earth images at free cost. Figure 2 shows framework of methodology followed for prediction of changes.

### 3.2.2 Preprocessing Techniques

Color Mapping

Color mapping is one of the main processes in image preprocessing techniques. Color mapping is nothing but a function, which helps map or transform the colors of one source image to the colors of another target image. It is also called as color transfer [12]. There are two types of color mapping techniques available. First one is that utilize the statistics of the colors of two different images, and another one is that rely on already given pixel communication between the two different images.

In this study, first technique is used for color, in which two different time period images have been classified with color mapping techniques [13]. For that color statistics has been taken for those images and develops it into diagrams for mapping.

**Fig. 2** Framework of change detection

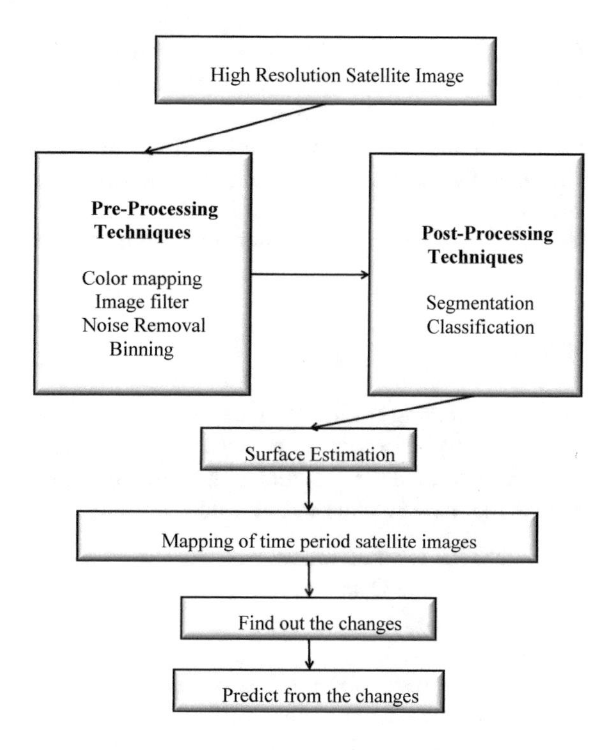


Image Filter

Image filtering is used to apply various effects on satellite images. The image filtering techniques are based on type of an image. In this study, median filtering is used for color filter. This filter is to reduce noise in a satellite image, and also it is similar to mean filter [14]. One of the main advantages of median filter is it preserves the useful information and details in the satellite image. Median filter preserves edges of the images the process of removing noise from the image. Median filter algorithm is given below.

Step 1: Two-dimensional windows of size $3 \times 3$ are designated and placed throughout the processed pixel of $p(x, y)$ in the place of noisy in an image.
Step 2: Categorize the pixels as per rising order, and determine the pixel value of median and maximum of the sorted vector.
Step 3: Processed pixel should be greater than minimum pixel value and maximum pixel value should be less than 255 if the processed pixel presented in within the range. So, it is classified as uncorrupted pixel and it is left unchanged.
Step 4: If processed pixel is found as corrupted pixel, then the image has the following two cases:

Case 1: If minimum pixel value less than median pixel value is less than maximum pixel value and median pixel value is less than 255, then exchange the corrupted pixel with median pixel value.
Case 2: If first case is not fulfilled, then median pixel value is a noisy pixel. Calculate the difference between pair of adjacent pixels, and obtain the difference vector.
Step 5: Steps one to four are repetitive until the processing is completed.

Binning

Binning is nothing but the combination of two or more image sensor pixels to figure a new super-pixel earlier to readout and digitizing. The result of the binning produces charge carrier's accumulation from the single pixel [15]. Binning is used to reduce the pixel resolution, but it does not hide the information on the image.

Algorithm for Binning image:
Read the Satellite Data
Check resolution
If resolution i $\geq$ Large
    Resize the resolution i as low resolution
    Check image quality after resizing
    If Image quality i% < original image quality
        Enlarge the image
    Else
        Store the image
    Else
        Show the original image

### 3.2.3 SVM Classification

Nowadays, SVM classification techniques are involved in many applications like image classification, handwriting recognition. Support Vector Machine classification technique is very effective in solving bioinformatics problems, particularly analyzing microarray expression data, and detecting remote protein homologies. SVM performs well from other classification techniques. Decision planes, that splits between a set of things having dissimilar class associations. Illustration of SVM classification method is shown in Fig. 3.

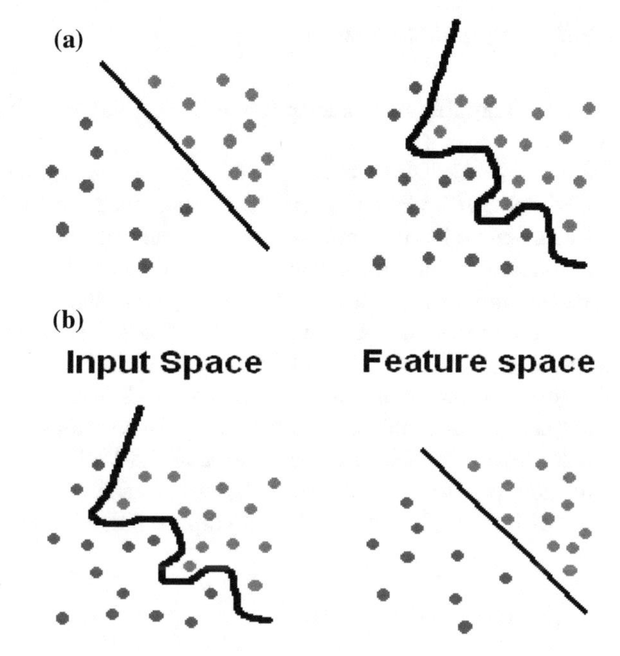


**Fig. 3** **a** SVM classification methodology; **b** Support Vector Machine linear classification method

From this example, the objects fit either to class green or red. The separation line describes a border on the right side of objects in green and objects in the left in red [16]. So, the classifier splits a set of objects into their separate colors with a line and it is known as hyperplane classifiers.

Figure 3b illustrates the basic concept of SVM. Left side of the schematic shows the original objects, which means rearranged objects. The process of rearranging the things is called as mapping. Right-side schematic shows the mapped objects. Left schematic shows the mapped object which is linearly separated.

Linear SVM Mathematical Model

$$M = \frac{(x^{+} - x^{-}) \cdot w}{|w|} = \frac{2}{|w|}$$

Goal of the linear SVM model is to correctly classify all training data

$$\begin{gathered} \text{if } y_i = +1 \\ \text{if } y_i = -1 \\ \text{for all } i \end{gathered}$$

and maximize the margin $M = \frac{2}{|w|}$ same as minimize $\frac{1}{2} w^t w$.

## 3.3 *Surface Estimation*

### 3.3.1 Mapping of Different Time Period Satellite Images

Figures 4 and 5 show the original images of Vellore district at the time period of 2009 and 2017. These original images show their properties, like where the greenery lands, people living areas, sands, wastelands situated. Even though proper classification is must for research study. As we discussed before in SVM classification, the satellite images are undergoing with linear classification method [17, 18].

From the classification image, Figs. 6 and 7 show many changes happened in between 2009 and 2017 in Vellore district. From the figure, blue represents wastelands that means unused lands for both agriculture and resident. Green represents dry lands that can be used only for residential purpose; yellow with cyan color represents residential areas around the Vellore district [19]. Here the classification process proved that many wastelands, which means that unused lands would be occupied for residential purpose (Blue color).

### 3.3.2 Finding the Changes

Figures 8 and 9 show the changes occurred during 2009–2017. These figures tell what happened between the two different time periods. And it shows that the agricultural areas became residential areas between these two different time periods.

Figures 8 and 9 easily conclude that green and blue percentage has been decreased due to population and new residential areas arrived in unused lands.

**Fig. 4** Satellite image of 2009 Vellore district

**Fig. 5** Satellite image of 2017 Vellore district

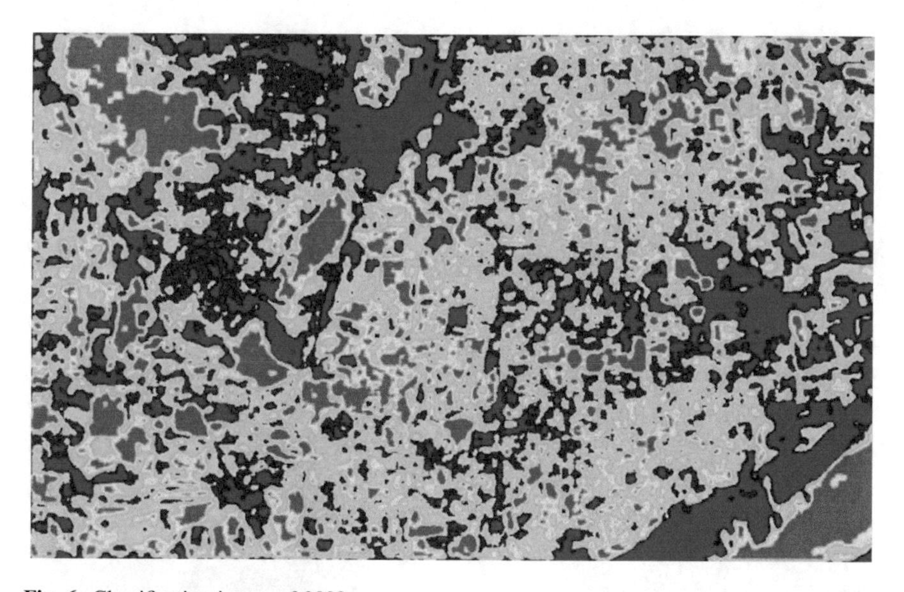

**Fig. 6** Classification image of 2009

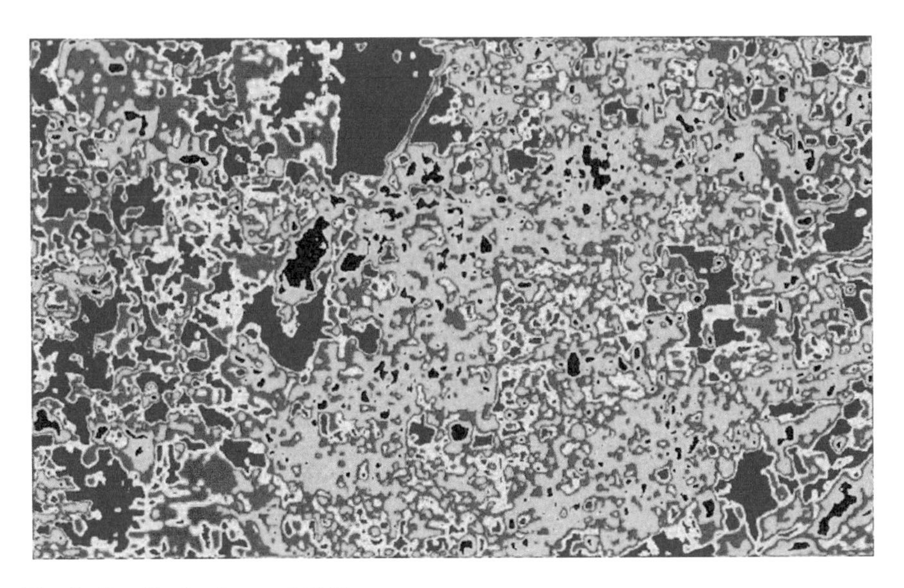

**Fig. 7** Classification image of 2017

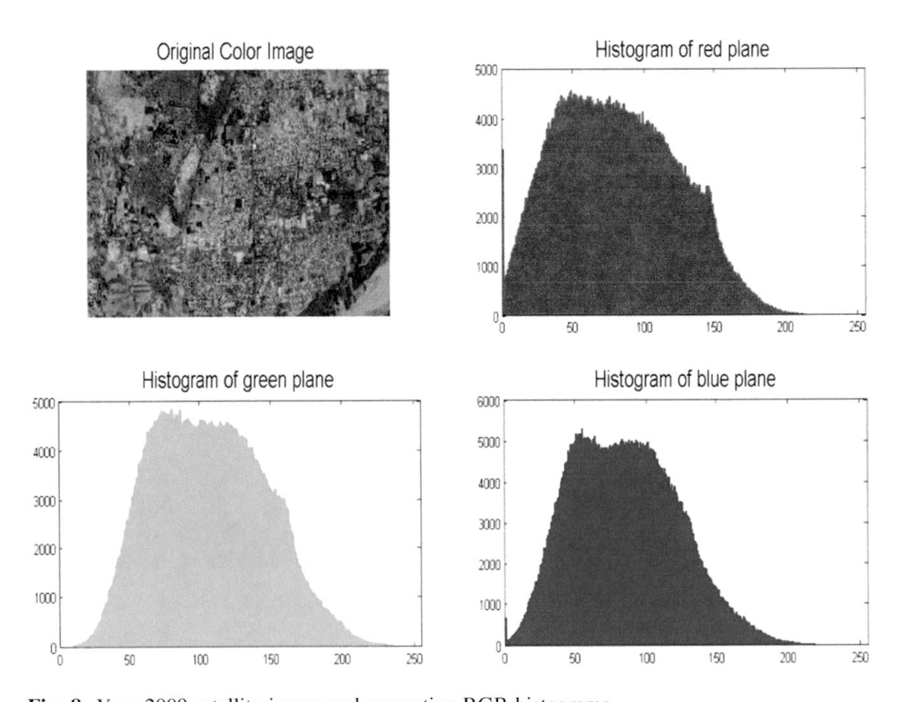


**Fig. 8** Year 2009 satellite image and respective RGB histograms

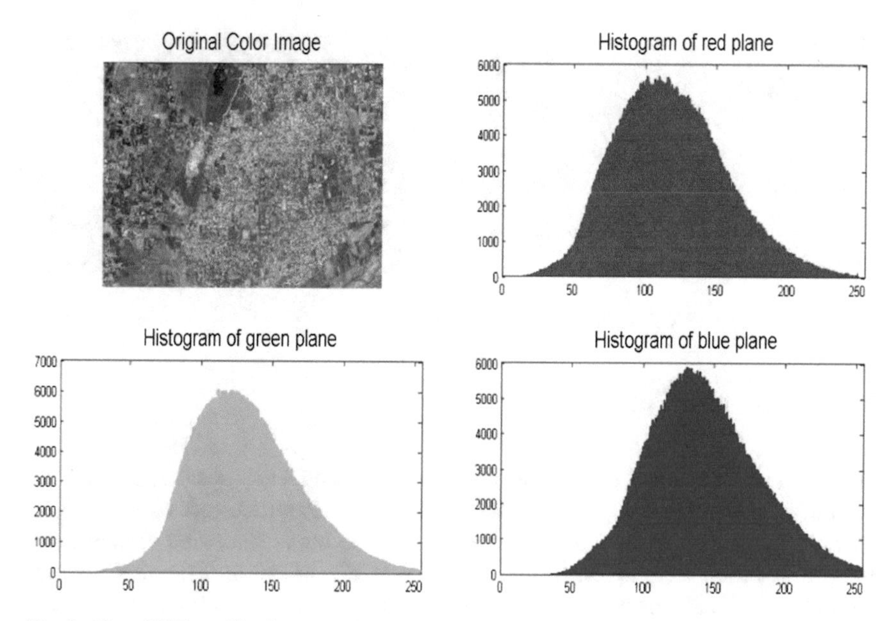


**Fig. 9** Year 2017 satellite image and respective RGB histograms

## *3.4 Predict from the Changes*

From the above study of changes in Vellore district during the time period of 2009 to 2017, research can easily predict in future very vast development will be there in Vellore district due to population, immigration of people from other areas, and very large industrial area arrival around the Vellore district.

Figure 10 shows the prediction of Vellore district from 2009 to 2025. In year 2025, the Vellore district may change as shown in Fig. 10. Figure 10 is an edited image with respect to changes in image classification of 2009 and 2017, but in real this will happen due to population. Apart from that, the agricultural areas will be used for residential purpose. There are some remedial measures to stop this. Those suggestions are given below:

- Sampling trees.
- Save trees and preserve water for future generation.
- Government should give some offers to farmers for agricultural development.
- Encouraging future generation for agricultural education.
- Local government restricts building approvals with some conditions.
- Continuous survey.
- Continuous monitoring of forest and agriculture changes, either by remote sensing or direct visit.

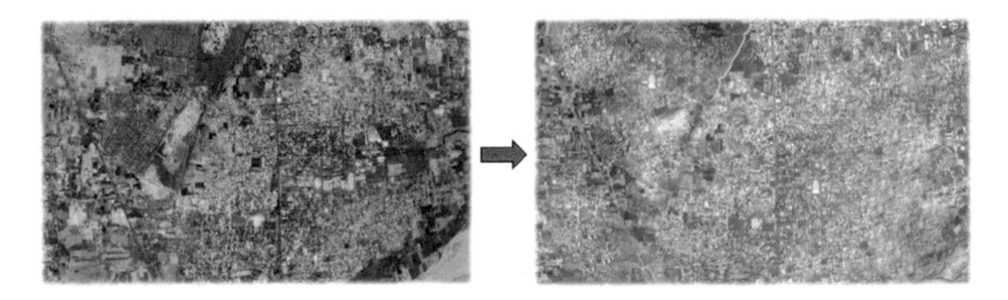

**Fig. 10** Prediction of Vellore district from 2009 to 2025

# 4 Results and Discussion

The Landsat data were undergone with preprocessing techniques such as color mapping, image filtering, noise removal, binning. Preprocessed data can give accurate result than the raw data. These images post-processed using segmentation and classification techniques. In this study, SVM classification method is used for the better and efficient classification. Comparing with other classification techniques, SVM has better place for classification of images. As we discussed before in this study, different time periods of Landsat study area images are classified to get the accurate result for mapping of land cover changes. The classification methods accuracy was assessed. It was found that Support Vector Machine classification produces more exact results than other classification methods.

This study refers this classification method for future studies to examine and get accuracy with added values. Main objective of this work is to show calculation of land cover change, and land use is to improve effective urban planning.

# 5 Conclusion

Approaches to monitoring and prediction of land use change are presented in this paper. The results show that there is a difference between 2009 land cover and 2017 land cover images. From those images, we can easily conclude that the population has been raised from 2009 to till now. Due to land cover area changing, the agricultural areas were vanished and it became unused lands. From this study, we know that the vegetation land and farmland will decrease sharply. This change detection technique is used with remote sensing data to provide accurate information on land cover and land uses in the study area. Resulted information may be useful in ecological and reasonable town planning of the Vellore district.

## References


1. Rozenstein, O., & Karnieli, A. (2011). Comparison of methods for land-use classification incorporating remote sensing and GIS inputs. *Applied Geography, 31*(2), 533–544. https://doi.org/10.1016/j.apgeog.2010.11.006.
2. Nori, W., & Elsiddig, E. N. (2008). Detection of land cover changes using multi-temporal satellite imagery. *Remote Sensing and Spatial Information Sciences*, (2004), 2004–2009. Retrieved from http://www.isprs.org/proceedings/XXXVII/congress/7_pdf/5_WG-VII-5/36.pdf.
3. Yang, L., Xian, G., Klaver, J. M., & Deal, B. (2003). Urban land-cover change detection through sub-pixel imperviousness mapping using remotely sensed data. *Photogrammetric Engineering Remote Sensing*, *69*(9), 1003–1010. https://doi.org/10.14358/PERS.69.9.1003.
4. Huth, J., Kuenzer, C., Wehrmann, T., Gebhardt, S., Tuan, V. Q., & Dech, S. (2012). Land cover and land use classification with TWOPAC: Towards automated processing for pixel- and object-based image classification. *Remote Sensing, 4*(9), 2530–2553. https://doi.org/10.3390/rs4092530.
5. Uddin, K., & Gurung, D. R. (2010). Land cover change in Bangladesh—A knowledge based classification approach. *Area, 1977,* 41–46.
6. Yuan, F., Sawaya, K. E., Loeffelholz, B. C., & Bauer, M. E. (2005). Land cover classification and change analysis of the Twin Cities (Minnesota) metropolitan area by multitemporal Landsat remote sensing. *Remote Sensing of Environment, 98*(2–3), 317–328. https://doi.org/10.1016/j.rse.2005.08.006.
7. Zhou, R., Hu, Y., Li, Y., & He, H. (2010). Land use change modeling and predicting of Xinzhuang town based on CLUE-S model. In *2010 International Conference on Intelligent Computation Technology and Automation, ICICTA 2010*, (Vol. 2, pp. 581–584). https://doi.org/10.1109/ICICTA.2010.525.
8. Xiongwei, H. (2008). The dynamic monitoring and prediction of the regional land use change based on RS and GIS: A case study of Changsha County. In *2008 International Workshop on Education Technology and Training & 2008 International Workshop on Geoscience and Remote Sensing*, (pp. 596–599). https://doi.org/10.1109/ETTandGRS.2008.276.
9. Roberts, D. A., Numata, I., Holmes, K., Batista, G., Krug, T., Monteiro, A., ... Chadwick, O. A. (2002). Large area mapping of land-cover change in Rondônia using multitemporal spectral mixture analysis and decision tree classifiers. *Journal of Geophysical Research D: Atmospheres*, *107*(20). https://doi.org/10.1029/2001JD000374.
10. Selvaraj, R. S. (2011). Prediction of monthly rainfall in Chennai using back propagation neural. *Network, 3*(1), 1994–1996.
11. Liu, W., Seto, K., Sun, Z., & Tian, Y. (2006). Urban land use prediction model with spatiotemporal data mining and GIS. *Urban Remote Sensing*, (January), 165–178. https://doi.org/10.1201/b15917-12.
12. Procházka, A., Kolinova, M., Fiala, J., Hampl, P., & Hlavaty, K. (2000). Satellite image processing and air pollution detection. 2282–2285.
13. Starks, S. A. (n.d). Interval and fuzzy methods in remote sensing and satellite image processing. In *Proceedings Joint 9th IFSA World Congress and 20th NAFIPS International Conference* (*Cat. No. 01TH8569*), *4*, 2019–2022. https://doi.org/10.1109/NAFIPS.2001.944378.
14. Kumar, N. S., Anouncia, S. M., & Prabu, M. (2013). Application of satellite remote sensing to find soil fertilization by using soil colour. *International Journal of Online Engineering*, *9*(2), 44–49. Retrieved from http://dx.doi.org/10.3991/ijoe.v9i2.2530.
15. Prabu, M., & Anouncia, S. M. (2016). NDVI generation of Chlorophyll from OCM data for the Indian Ocean region using multispectral images. *Research Journal of Pharmaceutical, Biological and Chemical Sciences, 7*(5), 2855.
16. Zhu, W., Pan, Y., Hu, H., Li, J., & Gong, P. (2004). Estimating net primary productivity of terrestrial vegetation based on remote sensing : A case study. *Sciences-New York 0*(C), 528–531.

17. Huang, J., Wan, Y., & Shen, S. (2009). An object-based approach for forest-cover change detection using multi-temporal high-resolution remote sensing data. In *Proceedings—2009 International Conference on Environmental Science and Information Application Technology, ESIAT 2009*, (Vol. 1, pp. 481–484). https://doi.org/10.1109/ESIAT.2009.163.
18. Ramachandra, T., & Kumar, U. (2004). Geographic resources decision support system for land use, land cover dynamics analysis. In *Proceedings of the FOSS/GRASS* ..., (September), (pp. 12–14).
19. Kumar, N. S., Anouncia, S. M., & Prabu, M. (2014). Fuzzy-based satellite image classification to find Greenery and used Land. *International Journal of Tomography and Simulation*, *25*(1), 1–2.

# An Integrated Approach for Optimal Feature Selection and Anatomic Location Identification on Pediatric Foreign Body Aspired Radiographic Images

**M. Vasumathy and Mythili Thirugnanam**

**Abstract** Foreign body aspiration is a frequent pediatric emergency, with incidence peaking at two years of age. Foreign body (FB) can be described as the intrude object which is not belong to the human body. The localization of FB needs radiography X-ray, CT, MRI assessment. Foreign bodies such as coin and metallic items are easily seen on radiographs, but it is difficult to identify food and plastic objects on foreign body aspired radiography images. The process of location identification takes more time which leads more complication, even it leads to fatal (Lecron and Benjelloun in Med Imaging SPIE Proc 8314:1–8, 2012 [8]). Therefore, the proposed work aims to develop an approach for identifying the anatomic location in which the complications of diagnosis process will be reduced. Image processing plays vital role in this scenario, especially in automating the process of determining the anatomic location of the foreign body on pediatric radiographic images. This chapter mainly focuses on identifying the relative advantages of using specific combination of image enhancement, segmentation, feature extraction for optimal feature selection and proposing an approach for automatic anatomic location identification process on pediatric foreign body aspired images. This process includes the radiographic image acquisition of the foreign body aspired pediatric patients, image enhancement, and segmentation methods. The identification of suitable segmentation method for extracting the optimal features is related to a range of research studies published on image segmentation, feature extraction, feature selection methods. The observation of the existing work helps to understand the importance of various segmentation methods and also supports to develop improved segmentation methods such as constraint-based median filtering, constraint-based iterative thresholding, constraint-based Sobel boundary detection, and *K*-means clustering. The ability of the enhanced segmentation techniques are

M. Vasumathy (✉) · M. Thirugnanam
School of Computer Science and Engineering, VIT University, Vellore, India
e-mail: vasumathy.mano@vit.ac.in

M. Thirugnanam
e-mail: tmythili@vit.ac.in



S. Margret Anouncia and U. K. Wiil (eds.), *Knowledge Computing and its Applications*, https://doi.org/10.1007/978-981-10-8258-0_6

determined by the performance comparison with the existing segmentation techniques which is done by the quality metrics evaluation. The feature extraction is used for describing the true region of interest based on shape-, edge-, and texture-based descriptors. The most influenced features are identified by applying hybrid feature selection method which is a combination of filter and wrapper methods in predicting the location and shape of the foreign body. A novel, automatic anatomic location identification approach (AALIA) using 8-connected block searching algorithm and corner identification methods are applied to identification and classification of the anatomic location of foreign body. To evaluate performance of the proposed approach, the accuracy measure precision, recall, *F*-Measure, and receiver operator characteristic (ROC) with respect to sensitivity, specificity, positive predicted rate, and negative predicted rate are considered. The results obtained from the developed approach are comparatively better than the existing works.

**Keywords** Foreign body aspiration · Image segmentation · Object recognition
Location identification

## 1 Introduction

Generally, in computer vision and medical image processing, the researchers aim to develop new technologies to gain a better semantic understanding of images or improved performance that gives different perspectives on the same image to understand not only the content, but also the significance and internal meaning. Especially foreign body recognition on pediatric radiographic images involves various image processing key stages for better semantic understanding of the foreign body aspiration. The recognition of foreign body objects is a trivial task, and a number of researchers have addressed this foreign body recognition with manually or semi-automated. Foreign body can get into the human body by accidental aspiration or by ingestion with the food during eating or without food. From the various study reports, it is reported that in every hour, eight children die from FBA. Identifying the relative advantages using specific combinations of image enhancement and segmentation algorithms for feature extraction on pediatric foreign body aspired images and developing a robust methodology for foreign body object location recognition are elaborated in this chapter. In order to select good feature extraction methods, radiologic benchmark is applied to demonstrate their ability. The good results on these benchmark show that these methods are capable of recognizing different foreign body object. The anatomic location of the aspired foreign body is automatically identified with 8-connected block searching algorithm and corner identification methods to identify the anatomic location of the aspired foreign body.

**Table 1** Survey on pediatric FBA for assessing the significance of identifying aspired foreign body

| Author(s) | Observation for assessing the significance of identifying aspired foreign body |
|---|---|
| Vasumathy and Thirugnanam et al. [1] | The suitable segmentation techniques are identified by evaluating quality metrics and the significance of assessing aspired foreign body characteristics |
| Khan et al. [2] | The author proposed algorithm with image fusion using a discrete wavelet transform on pediatric X-ray image and produce significant enhancement in the degraded pediatric X-ray images |
| Firoz et al. [3] | The enhancement method results indicate that the contrast is improved by morphological transform method that helps in better diagnosis |
| Kramer et al. [4] | The review concludes to determine treatment procedure of FBA that requires assessment of foreign body size, type, location |
| Kaviani et al. [5] | The work concludes that the detection of foreign bodies is dependent on the imaging and the characteristics of the foreign body such as the material, size, and its location |
| SeikholetKuki et al. [6] | The study results that the size and thickness of foreign body determine the presence of objects in any X-ray image |

## 2 Literature Review

Some of the current research works on pediatric FBA for assessing the significance of identifying aspired foreign body and the significance of assessing foreign body size and shape are discussed in Table 1. Image segmentation is one of the key stages in medical image processing which aims to recognize distinct homogeneity in gray levels within the regions of the same image. The existing surveys on various image segmentation techniques used for anatomic localization on radiographic images are shown in Table 2.

### *2.1 Survey on Current Research Works on Pediatric Foreign Body Aspiration*

Based on review, it is concluded that the detection of foreign bodies depends on the imaging characteristics such as location, shape, and size. Most of the research work focused on risk factors, types of foreign body, significance of early diagnosis of aspired foreign body and system was implemented as semi-automated detection of foreign body imaging features with expert knowledge. Hence, there is a need for developing an automatic anatomic location and shape identification of foreign body presented in pediatric foreign body aspired radiography images without manual intervention.

**Table 2** Review on various image processing techniques for anatomic localization on radiographic images

| Author(s) | Technique used for anatomic location identification | Considered anatomic location for the work | Observation of the proposed work |
|---|---|---|---|
| Lalendra upreti et al. [7] | Histogram approach | Airway, trachea, bronchial | The histogram tool helps to identify foreign body lodgment and diagnosis on digital radiographs of pediatric chest |
| Lecron et al. [8] | Model-based vertebra detection, identification. | Vertebra detection | Generalized Hough transform (GHT) is used for automatic vertebra detection with accuracy rate of 92% |
| Juhasz et al. [9] | KNN and curve fitting methods. | Chest localization | Describe a method for segmenting anatomic structures on chest radiographs. The results are analyzed with an efficient GPU-based real-time evaluation |
| Klinder [10] | Multi-class SVM | Vertebra detection | To automate the vertebra detection in X-ray, contrast-limited adaptive histogram equalization, canny edge detection, and an edge polygonal approximation are used |
| Benjelloun [11] | Interest point detection | Spine localization | The point of interest is detected to locate a vertebra on X-ray images and support vector machine is used to classification |
| Upadhyaya et al. [12] | Prediction based on expert advice | Esophagus identification | Discussed about multiple coins swallowing with its diagnosis and removal technique |

## 2.2 Survey on Anatomic Localization Identification

Identifying the anatomic location in pediatric X-ray image remains challenging task in medical image processing. Various image processing techniques used for anatomic localization described are shown in Table 2.

Based on the above review, it is revealed that the process of identifying region of interest in brain tumor and heart-related diagnosing is done by using image processing techniques such as histogram equalization and interest point detection method with manual intervention, but no work focuses on automatic segmentation and feature extraction to identify object size, anatomic location of the aspired foreign body in pediatric radiography images. Hence, this work analyzes various image segmentation techniques on pediatric radiography images to identify the suitable segmentation technique which shall be applied to extract the region of interest and features on the foreign body aspired pediatric radiography images.

## 2.3 Survey on Influenced Feature Selection

The influenced feature selection method helps to reduce the feature set which minimizes the computation time and improves the prediction accuracy during the process of classification. Widely used feature selection methods on medical image dataset are elaborated in Table 3.

From the above literature review, it was observed that there are three general classes of feature selection algorithms such as filter methods, wrapper methods, and embedded methods that are used for medical data mining application. From the analysis, it is understood that each method has their own limitations in selection of optimal features and the combination of feature selection yields better accuracy in terms of ROC measure [15–18, 21]. With this motivation, this work developed hybrid feature selection method for selecting optimal features to determine the location of the aspired foreign body. Review of analysis concluded that introducing the hybrid approach in feature selection achieved better accuracy rate in predicting the various medical data. Most of the researchers used ROC measure as a metric to evaluate their predicting results of the applied dataset. So with this context, the developed approach adopted the concept of hybrid approach in feature selection, and ROC measure is used as a metric to assess the classification accuracy.

**Table 3** Review on various feature selection methods for identifying influenced features

| Author(s) | Applied feature selection method | Observation of the work |
|---|---|---|
| Pohjalainen et al. [13] | Filter, wrapper, and embedded feature selection methods with k-nearest neighbors (KNN) for classification | The proposed automatic feature selection reduces the overall features to feature set size with comparable or even better performance than other selection methods |
| Blum et al. [14] | Filter, wrapper, and embedded and feature weighting methods | Selecting relevant features remains challenging task in machine learning |
| Choras et al. [15] | Embedded feature selection with KNN | Investigated the use of feature extraction and selection for image retrieval in CBIR and biometrics systems |
| Guyon et al. [16] | Introduction and use of filter, wrapper, and embedded methods | The author provides a better definition of the efficient search methods objective feature construction, feature validity assessment, and feature ranking methods |
| Ladha et al. [17] | Filter, wrapper, embedded | Gives detailed taxonomy of feature selection techniques |
| Janecek et al. [18] | Filter, wrapper, embedded with PCA, WEKA toolkit | The wrapper method produces the smallest feature subsets with good classification accuracy |

# 3 Proposed Work

The proposed work mainly focused on identifying foreign body on pediatric radiography images with it is shape and location. Performance evaluation of the existing segmentation technique is observed to identify the location of the true region of interest. The proposed approach is developed with different phases which include initial phase, segmentation phase, feature extraction phase, feature selection phase, and anatomic location identification phase. Initial phase is concentrated to improve the quality of the image by applying image preprocessing technique such as scaling transformation and histogram equalization. Various image segmentation techniques such as edge- and region-based segmentation techniques are identified from the existing work in segmentation phase. These segmentation techniques are implemented on foreign body aspired pediatric X-ray images to analyze the segmentation performance in order to identify the suitable segmentation technique. The segmentation phase applies the identified suitable segmentation techniques to identify the true ROI. Feature extraction plays a major role in describing data presented in the input image. In feature extraction phase, the edge-, shape-, and texture-based feature extraction techniques are applied to extract the features and stored in feature database. The feature selection methods such as filter, wrapper, and embedded methods are applied to improve the shape and location prediction accuracy in feature selection phase. The location identification phase applies automatic anatomic location identification approach (AALIA) for identification and classification of the anatomic location of foreign body. The results are presented with the performance evaluation carried out with respect to benchmarking standard metrics which includes precision, recall, F-Measure, and ROC. The overall schematic view of the developed approach is shown in Fig. 1.

## *3.1 Initial Phase*

The initial phase mainly focused on the performance evaluation of existing segmentation technique. To create visual representations of the interior organs of the body, medical experts prefer number of imaging techniques such as X-ray, CT, MRI, and ultrasound to assist medical experts. Especially the treatment process of foreign body aspiration incident needs radiographic assessment for localization of foreign body.

### 3.1.1 Performance Evaluation of Existing Image Processing Techniques

Segmentation of medical images is one of the key tasks in image processing. Segmentation subdivides an image into meaningful regions or objects. Segmentation

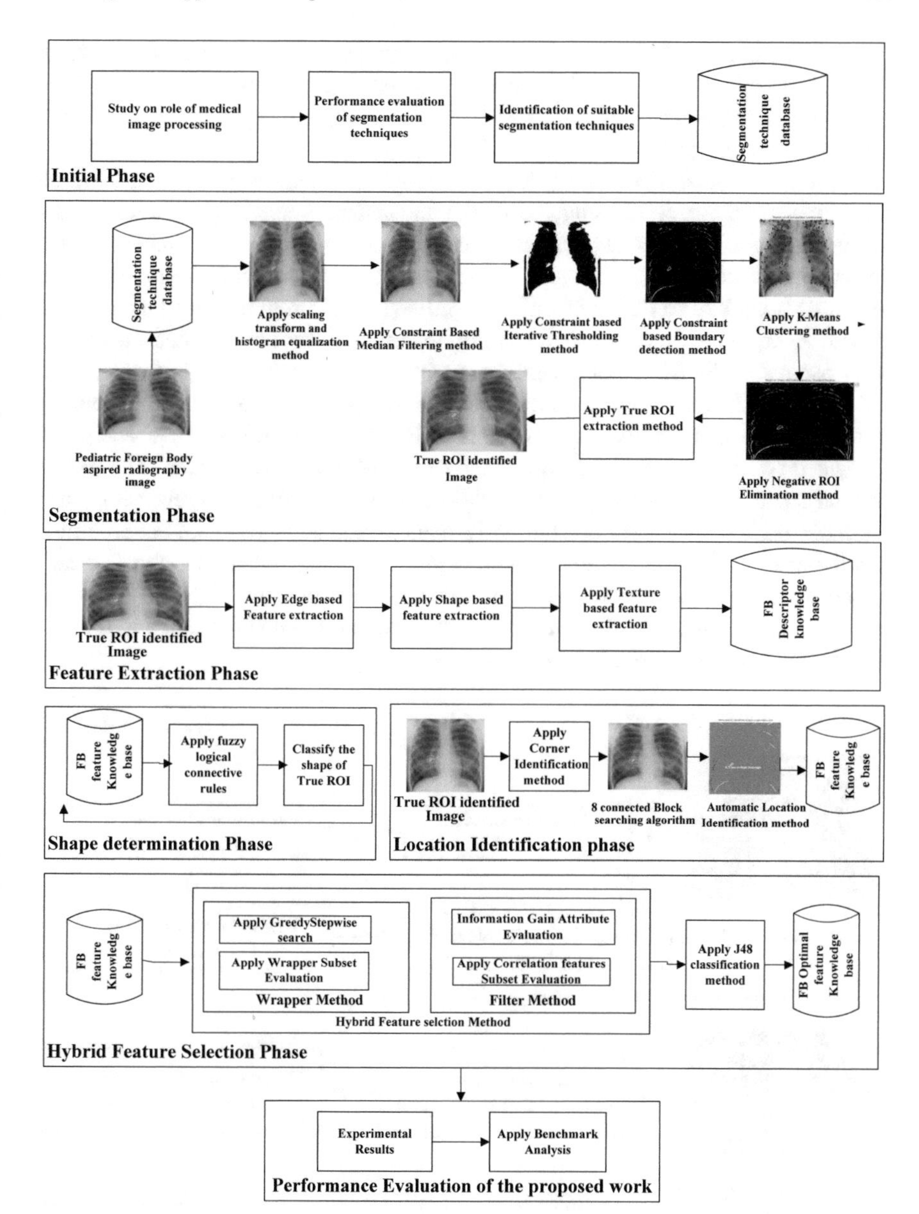


**Fig. 1** Overall schematic view of the developed approach

techniques involve thresholding, edge detection, and clustering for segmenting the boundary or detecting the edges from an image. Assessing the performance of various segmentation techniques by the quality metrics helps to prove their ability in object identification. There are various quality analysis metrics available for examining the image quality, such as signal-to-noise ratio (SNR), peak signal-to-noise ratio (PSNR), mean absolute error (MAE), root mean square error (RMSE). These metrics are validated based on the fundamental benchmark of the quality analysis metrics. The steps involved in evaluating the various segmentation techniques are illustrated in Fig. 2.

The related works help to identify various image processing techniques which are widely used for performing the image enhancement and segmentation of the medical images. Segmentation phase contains noise filtering, thresholding, edge detection, and clustering techniques, texture-based segmentation for segmenting the region or boundary to detect the meaningful segmented region from pediatric FB aspired X-ray images. These segmentation techniques are implemented on foreign body aspired pediatric X-ray images to analyze the segmentation performance in order to identify the suitable segmentation technique as discussed in Vasumathy et al. [1]. The quantitative metrics such as SNR, PSNR, RMSE, and MAE are computed for evaluating the performance of segmentation techniques. The experimental results obtained MIK segmentation technique is comparatively better than the other segmentation. As per the performance evaluation results, it is identified that the median filtering, iterative thresholding, Sobel edge detection, and *K*-means clustering techniques are found to be suitable, and the improved hybrid version of these techniques called MIK segmentation technique is applied for isolating the true ROI in foreign body aspired radiography images.

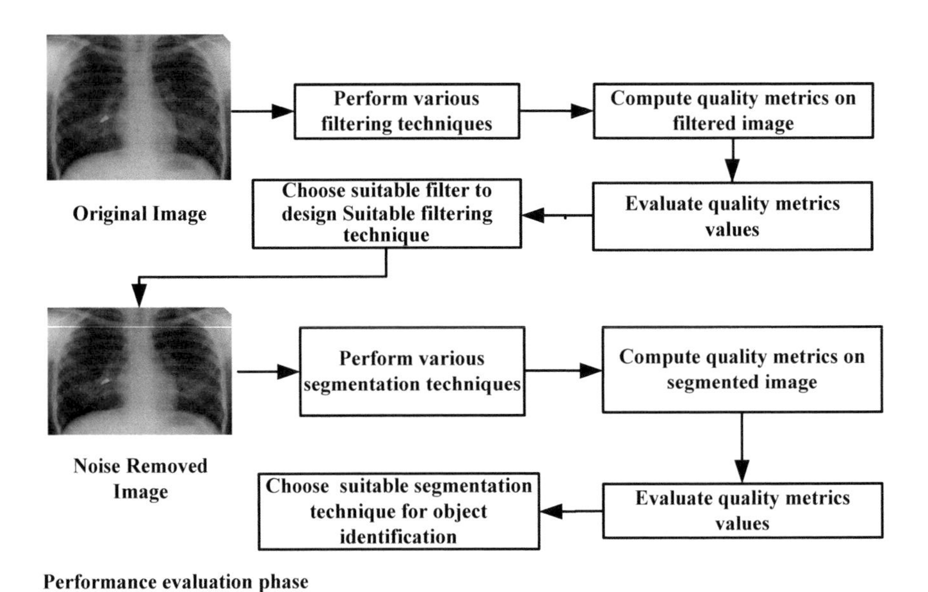


**Fig. 2** Process of performance evaluation of various segmentation techniques

## 3.2 *Image Segmentation Phase*

Image enhancement is performed as a preprocessing step before image segmentation to improve quality of the image. Feature extraction methods are applied to segmented structure to extract the required data such as shape and texture to describe a set of relevant information from the input image for object identification. The negative ROI elimination method reduces the smaller region as well as the very large region based on the standard deviation which gives the true ROI. True ROI features are extracted and stored for further object determination process. The segmentation process of FB aspired pediatric radiography image is shown in Fig. 3.

### 3.2.1 Scaling Transformation and Histogram Equalization

Intensity scaling transformation allows the observer to focus on specific intensity band of interest. If *f*1 and *f*2 are known to define the intensity band of interest, a scaling transformation may be defined as

$$e = \begin{cases} f, & f1 \leq f \leq f2 \\ 0, & \text{otherwise} \end{cases} \tag{1}$$

$$g = \left\{ \frac{e - f1}{f2 - f1}, (f\max), \right. \tag{2}$$

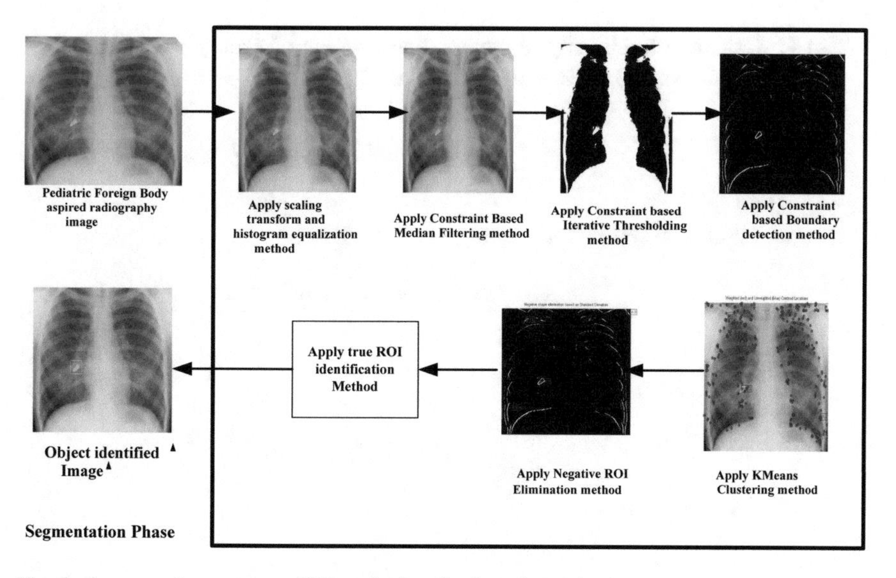


**Fig. 3** Segmentation process of FB aspired pediatric radiography image

where $e$ is an intermediate image, $g$ is the output image, and $f$max is the maximum intensity of the display.

### 3.2.2 Constraint-Based Median Filtering

In image filtering, the input image is converted into a two-dimensional array of numbers to determine which pixels in an image have been affected by noise. A pixel that is differ from a majority of its neighbors, as well as not structurally aligned with those pixel to which it is similar, is designated as noise. These noisy pixels are then replaced by the median value pixels in the neighborhood that have passed the noise detection test.

### 3.2.3 Constraint-Based Iterative Thresholding

Iterative thresholding is adopted in this work to replace the noisy pixels and to enhance the contrast.

Steps involved in iterative thresholding algorithm

Step 1: Load all the pixel value from the input image.
Step 2: Decompose color code in RGB components using pixel grabber.
Step 3: Compute grayscale value for each pixel and store.
Gray value = (0.3 * red + 0.59 * green + 0.11 * blue);

$$G_{\mathrm{pv}} = \sum_{i=0}^{n} \left(0.3 * p_{\mathrm{r}}, 0.59 * p_{\mathrm{g}}, 0.11 * p_{\mathrm{b}}\right) \tag{3}$$

Step 4: Calculate luminosity, lightness, and average of each pixel
Luminosity: 0.21 R + 0.72 G + 0.07 B.

$$P_{\mathrm{lu}} = \sum_{i=0}^{n} \left(0.21 * p_{\mathrm{r}}, 0.72 * p_{\mathrm{g}}, 0.07 * p_{\mathrm{b}}\right) \tag{4}$$

Lightness: (max(R, G, B) + min(R, G, B))/2.

$$P_{\mathrm{lig}} = \sum_{i=0}^{n} \left(\frac{\max(p_{\mathrm{r}}, p_{\mathrm{g}}, p_{\mathrm{b}}) + \min(p_{\mathrm{r}}, p_{\mathrm{g}}, p_{\mathrm{b}})}{2}\right) \tag{5}$$

Average: (R + G + B)/3.

$$P_{\mathrm{avg}} = \sum_{i=0}^{n} \left(\frac{\sum (p_{\mathrm{r}}, p_{\mathrm{g}}, p_{\mathrm{b}})}{3}\right) \tag{6}$$

where $n$ is total number of pixels in region, $p_r$ is red value, $p_g$ is green value, $p_b$ is blue value, lu is luminosity, lig is lightness, and avg is average of each pixel in an image

Step 5: Calculate the threshold value by adding three constraints as follows:
Threshold: (Luminosity + Lightness + Average)
Mean Threshold: Mean of (Luminosity + Lightness + Average)

Step 6: Create gray scale chart of all the pixels of the image with $x$ and $y$ coordinates and threshold value.

$$\mathrm{ITC_{gr}} = \frac{\sum_{i=0}^{n} (x_i, y_j)}{n} \tag{7}$$

where n is total number of pixels in region and $(x_i,y_i)$ is the pixel position.

Step 7: Compare the mean threshold value of each pixel with the grayscale chart.

Step 8: Construct the output image by applying mean threshold value to the pixels of input image.

### 3.2.4 Constraint-Based Sobel Edge Detection Method

Boundary-based segmentation is performed based on information about edges in the image. Edge is a boundary between two homogeneous regions. Boundary detection refers to the process of identifying and locating sharp discontinuities in an image. Detected edges are used to identify the objects present in an image. Based on the existing survey, the proposed work applies Sobel edge detection algorithm for detecting the boundary. The Sobel function outputs an image with the boundaries detected.

### 3.2.5 *K*-Means Clustering Algorithm

From the performance evaluation, the *K*-means clustering algorithm found to be suited for this work which initially groups random pixels into n number of clusters. The cluster then iterated every time where the centroid value of the cluster is computed. Then Euclidian distance between centroid and each neighboring pixel is computed. The pixels with minimum distance are joined together, and the cluster undergoes alteration process. The algorithm groups normal region pixels into single cluster and abnormal region cluster into single cluster.

### 3.2.6 Negative ROI Elimination Method

The negative ROI elimination method rapidly eliminates the smaller and very larger regions based on standard deviation of the cluster regions.

Steps involved in negative ROI elimination method

1. Read the cluster image for labeled regions.
2. Each region is iterated to calculate the standard deviation based on the pixel values.
3. To eliminate, the negative region standard deviation is weighted as less than 30 for the smaller region and greater than 100 for larger regions.
4. If the region standard deviation ranges between less than 30 and greater than 100, then the region pixels are weighted as black pixel value.
5. The regions which contain black pixels are considered as background of the image and eliminated for further iteration.
6. The remaining true regions are marked with numerical label value for feature extraction.
7. Determine the true ROI based on the area of the identified regions.

## *3.3 Feature Extraction*

In this work, for feature extraction, the true ROI boundaries are detected, and from the boundary, shape and texture features are calculated.

### 3.3.1 Edge-Based Feature Extraction

The edge-based features are such as foreign body size in centimeter, area, perimeter, standard deviation, minimum intensity, maximum intensity and standard deviation. These calculated edge-based features are the base for calculating shape-based features such as convex area, compactness, elongation, rectangularity, eccentricity and dispersion.

### 3.3.2 Shape-Based Feature Extraction

Shape descriptor features are calculated from objects contour such as circularity, compactness, elongation, dispersion, aspect ratio, irregularity, length irregularity, complexity, sharpness. The shape information is derived from the object contour. The set of shape features from the contour image is extracted and stored. Various shape-based descriptors with description are described in Table 4.

**Table 4** Various shape-based descriptors

| S. No. | Descriptor | Expression |
|---|---|---|
| 1 | Circularity | $\text{Cir} = \left(\frac{4*\pi*\text{Area}}{(\text{Perimeter})^2}\right)$ |
| 2 | Compactness | $\text{Compact} = \left(\frac{2*\sqrt{\text{Area}*x}}{\text{Perimeter}}\right)$ |
| 3 | Elongation | $\text{Elong} = \left(\frac{\text{Area}}{(2*R_{\text{Max}})^2}\right)$ |
| 4 | Rectangularity | $\text{Rect} = \left(\frac{\text{Area}}{\text{perimeter}}\right)$ |
| 5 | Eccentricity | $\text{Ecc} = \sqrt{1 - \left(\frac{R_{\min}}{R_{\max}}\right)^2}$ |
| 6 | Solidity | $\text{Solidity} = \left(\frac{\text{Area}}{\text{ConvexArea}}\right)$ |
| 7 | Dispersion | $\text{Disp} = \left(\frac{R_{\text{Max}}}{\text{Area}}\right)$ |

### 3.3.3 Texture-Based Feature Extraction

Texture is one of the key characteristics used in identifying objects and region of interest in medical image analysis. Various texture-based feature extraction approaches such as gray-level histogram, co-occurrence matrices, run length matrix, gray-level difference matrices, gradient matrices, texture feature coding method, autocorrelation coefficients are implemented, and collective features are get stored. Various texture-based descriptors are described in Table 5.

**Table 5** Various texture-based descriptors

| S. No. | Descriptor | Expression |
|---|---|---|
| 1 | Energy | $\text{Energy} = \sum_{i,j=0}^{N-1} (p_{ij})^2$ |
| 2 | Entropy | $\text{Entropy} = \sum_{i,j=0}^{N-1} - \ln(p_{ij}) P_{ij}$ |
| 3 | Contrast | $\text{Contrast} = \sum_{i,j=0}^{N-1} (p_{ij}(i-j))^2$ |
| 4 | Homogeneity | $\text{Homogeneity} = \sum_{i,j=0}^{N-1} \frac{p_{ij}}{1+(i-j)^2}$ |
| 5 | Correlation | $\text{Correlation} = \sum_{i,j=0}^{N-1} p_{ij} \frac{(i-\mu)(j-\mu)}{\sigma^2}$ |
| 6 | Prominence | $\text{prominence} = \text{sgn}(B)\lvert B^{1/4}\rvert$ |
| 7 | Angular second moment | $\text{Angular Second Moment} = \sum_i \sum_j p(i,j)^2$ |
| 8 | Sum of variance | $\text{Sum of Variance} = \sum_{i=2}^{2N_g} (i - f_S)^2 p_{x+y}(i)$ |
| 9 | Sum of entropy | $\text{Sum of Entropy} = -\sum_{i=2}^{2N_g} p_{x+y}(i) \log\{p_{x+y}(i)\} = f$ |
| 10 | Info measure of correlation1 | $\text{Info. Measure of correlation1} = \frac{HXY-HXY}{\max\{HX,HY\}}$ |

## *3.4 Automatic Anatomic Location Identification Approach (AALIA)*

The automatic anatomic location identification of foreign body objects in X-ray images utilizes the existing knowledge of pediatric anatomy structures, the knowledge of predefined foreign body characteristics, and the knowledge of the various image processing operations to extract the aspired foreign body anatomic location and other characteristics from the FBA pediatric X-ray images. Novel automatic location identification method using 8-connected block searching algorithm and a novel corner identification method are used to locate the position of interested foreign body region pediatric X-ray image. Finally, a knowledge or evidence base with the identified influenced descriptors such as size, shape, and anatomic location is created to assist the medical practitioner in treatment management process of pediatric foreign body aspiration. The schematic view of automatic anatomic location identification approach of aspired foreign body in pediatric radiography images is shown in Fig. 4.

The input image is converted into a two-dimensional array of numbers that represent the real, continuous intensity distribution for further enhancement process. The enhancement techniques are applied with the aim of improving the quality of a digitized foreign body aspired X-ray image. The edge-, shape-, and texture-based features are extracted from the true ROI. The influenced features are

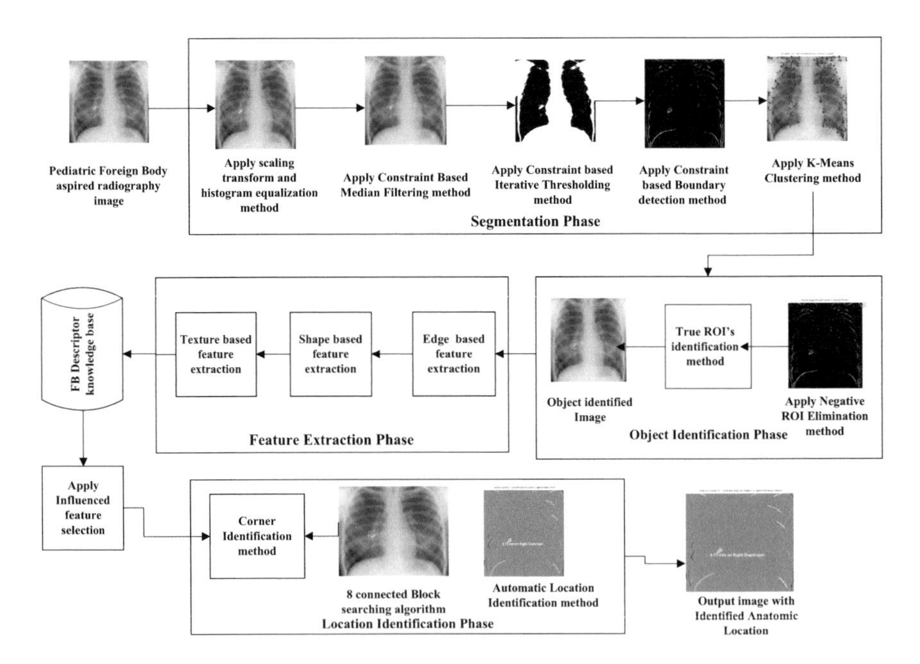


**Fig. 4** Schematic view of automatic anatomic location identification approach

identified to determine the anatomic location of the true ROI using combination of filter and wrapper feature selection method. The identified features are than applied for anatomic location identification.

### 3.4.1 Location Identification Phase

Identifying the anatomic location in pediatric X-ray image remains challenging task in medical image processing. Location identification phase involves corner identification method, 8-connected block searching method, and Automatic Anatomic Location Identification Approach (AALIA) to identify the common anatomic locations such as esophagus, airway, bronchus, trachea, gastrointestinal region in the pediatric FB aspired X-ray image.

Corner Identification Method

Corner identification method finds corners in an image based on the start, end, and center pixels' position. For any image, the starting point of the image pixel is ($x = 0$, $y = 0$) and the end point should be the length and width of the image. The half of the length and width gives the center point of the image. With these points, the algorithm iterates all the pixels of the image to find eight corners of the image. There is a uniqueness constraint in pixel, so points from one block cannot participate in another block.

Steps involved in corner identification method

1. Read all the pixel values of the input image
2. Calculate the minimum $x$ and $y$ values that correspond to staring point of the image that denotes as SP(min$x$, min$y$)
3. Calculate the maximum x and y values of the image that correspond to image length and width and also denoted as EP(max$x$, max$y$)
4. Calculate the centroid value of the image by using the image length and width

$$\mathrm{CP}(x, y) = (\mathrm{length}/2, \mathrm{width}/2)$$

5. The upper left corner of the image is calculated by finding minimum of $x$ and $y$ values, and this is denoted as ULC(min$x$, min$y$)
6. The upper right corner of the image is calculated by finding maximum of $x$ and minimum of $y$ values, and this is denoted as URC(max$x$, min$y$)
7. The bottom left corner of the image is calculated by finding minimum of $x$ and maximum of $y$ values, and this is denoted as BLC(min$x$, max$y$)
8. The bottom right corner of the image is calculated by finding maximum of $x$ and maximum of $y$ values, and this is denoted as BRC(max$x$, max$y$)

9. The right half corner of the image is calculated by finding maximum of $x$ and maximum of $y$ divided by 2 value, and this is denoted as RHC(max$x$, max$y$/2)
10. The left half corner of the image is calculated by finding maximum of $x$ divided by 2 and maximum of $y$ values, and this is denoted as LHC(max$x$/2, max$y$).

8-Connected Block Searching Method

A novel block searching method searches for a particular group of connected component pixel values in a pediatric FB aspired radiography image. The blocks are constructed using identified corners of the image. If two adjoining pixels are on, they are part of the same object, regardless of whether they are connected along the horizontal, vertical, or diagonal direction.

Steps involved in 8-connected block searching method

1. Read input image and apply corner identification method to create blocks on the image.
2. The algorithm matches the block of pixels in region $r$ to a block of pixels in region $r + 1$ by moving the block of pixels over a search region.
3. For each subdivision or block in region $r + 1$, the block matching block establishes a search region based on the value of region and centroid ($r$, $c$) parameter.
4. If the block contains the labeled pixel ($r$, $c$), then it is marked with a label name and continues the next iteration until the desired pixel is identified.
5. If the pixel block is identified, then it is added to a functional vector with corresponding block id.

### 3.4.2 Steps Involved in (AALIA) Automatic Anatomic Location Identification Approach

The Automatic Anatomic Location Identification Approach makes use of corner identification method and 8-connected block searching method to identify some of the common anatomic locations such as esophagus, airway, bronchus, trachea, gastrointestinal region in pediatric FB aspired X-ray image.

1. Read the segmented image with true ROI (output of negative ROI elimination method).
2. Apply corner identification method to obtain nine corner coordinate values of the identified true ROI region.
3. Iterate the entire true region to identify the position of centroid pixel.
4. Map the region with centroid pixel value.
5. Apply 8-connected block searching method to obtain nine blocks in the input image.

6. A location vector with anatomic location and the corresponding blocks are mapped as per the following steps.
7. The airway region falls between upper half and middle block of the X-ray image.
8. The left lung region falls between upper left corner, upper half, left half, and middle region.
9. The right lung region falls between upper half, upper right corner, middle and right left regions.
10. The gastrointestinal tract falls between bottom left and bottom right corners of the image.
11. The left main bronchus falls between left half and the middle regions.
12. The right main bronchus falls between middle and the right half corners.
13. Location vector is iterated with the region centroid value to identify the true ROI location.
14. If the block contains the centroid value, then the corresponding location is mapped for the input centroid value and matched location is stored along with the features in FB descriptor knowledge base.

## 3.5 *Hybrid Feature Selection Method*

From the detailed survey, it is understood that most of the researchers were introduced combination of feature selection methods to improve the performance of feature subset selection. With this motivation, this work introduced hybrid feature selection method to select the optimal features which helps to determine the location on pediatric foreign body aspired radiography image. The steps involved in feature selection process are illustrated in Fig. 5.

### 3.5.1 Selecting the Attributes Based on Ranking

The FB descriptor knowledge base is iterated for optimal feature selection. The optimal features are selected based on ReliefF feature evaluator method and correlation features subset evaluation method. The procedure adopted for selecting the optimal features for location identification is illustrated below.

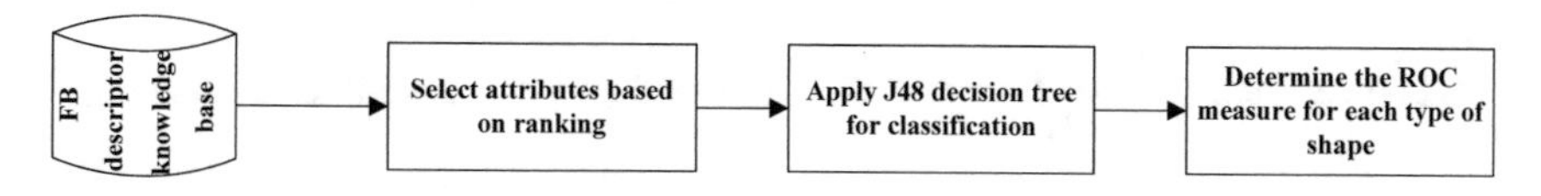


**Fig. 5** Steps involved in feature selection process

Steps involved in hybrid feature selection method

1. Load all the features from feature database and create feature vector ($X$).
2. Rank the features by evaluating the significance of each feature conjunction with ReliefF attribute evaluator.
3. The ReliefF feature evaluator iterates the feature vector ($X$) and applies euclidean distance measure to find the instance closeness between attributes of feature vector ($X$). The distance formula is given below.

$$\text{ReliefF} = F_w - (x_i - \text{nearHit}_i)^2 + (x_i - \text{nearMiss}_i)^2 \quad (8)$$

ReliefF is a ranking measure, $F_w$ is the feature instance, the closest same class instance is called near hit, and the closest different class instance is called near miss.

4. Set threshold to $-1$ by which attributes can be discarded.
5. Select the feature, if its relevance is greater than a threshold value.
6. The selected features are used to create relevance feature vector, and the relevant features are ranked between 0 and 1.
7. The correlation features subset evaluation method evaluates the worth of a subset of attributes by considering the individual predictive ability of each feature along with the degree of redundancy between them.

The correlation between each of the feature is predicted by the following formula

$$\text{CFS}_{\text{ZC}} = \frac{k\overline{r_{zi}}}{\sqrt{k + k - (k-1)\overline{r_{ii}}}} \quad (9)$$

where

$\text{CFS}_{zc}$ correlation between the summed components and the outside variable.
$k$ number of components (features).
$\overline{r_{zi}}$ average of the correlations between the components and the outside variable.
$\overline{r_{ii}}$ average inter-correlation between components.

8. Display the ranked list of attributes
9. The union operator is applied for filter method ReliefF and wrapper method CFS to create an optimal feature vector

$$\text{OF}_{\text{V}} = \{\text{ReliefF}_v \cup \text{CFS}_v\} \quad (10)$$

10. The selected optimal feature vector is considered for location classification.

The results obtained from the hybrid feature selection method concluded that the features size, standard deviation, information measure of correlation2, dissimilarity, information measure of correlation1, difference entropy are the optimal features for location identification of the aspired foreign body.

### 3.5.2 J48 Decision Tree Algorithm

The main motivation of using J48 decision tree algorithm is its ability of gaining the information of each features presented in the feature set which is used to predict target variables for each type of shape and location class. The basic steps involved in J48 decision tree are discussed in the following.

Basic steps involved the J48 decision tree algorithm

1. Load all the features from optimal feature vector ($OF_{v}$).
2. Calculate entropy for all the features

$$E(T) = \sum_{i=1}^{c} -p_i \log_2 p_i \tag{11}$$

$$E(T,X) = \sum_{c \in X} P(x)E(x) \tag{12}$$

3. The information gain is calculated for every feature, and ranking is applied.

$$\text{Gain}(T,X) = \text{Entropy}(T) - \text{Entropy}(T,X) \tag{13}$$

4. The optimal features are assigned for tree branching.
5. The steps from 2 to 4 continue until features gain highest ROC measures to identify target variables which help to classify a new data.
6. The resultant tree generates a model which helps to predict new instances of data. The prediction accuracy is measured by precision, recall, F-Measure, and ROC. The ROC measure is gaining popularity in the machine learning community rather than other accuracy measures. The following formulas are used for calculating the accuracy measure.
7. Precision is the measure of how many of the returned values are correct.

$$\text{Precision} = \frac{\text{TP}}{(\text{TP} + \text{FP})} \tag{14}$$

8. Recall is the measure of how many of the positives do the model return.

$$\text{Recall} = \frac{\text{TP}}{(\text{TP} + \text{FN})} \tag{15}$$

**Table 6** Benchmark scaling measures of ROC value

| ROC accuracy value | Benchmark measure |
|---|---|
| 1.0 | Perfect prediction |
| 0.9 | Excellent prediction |
| 0.8 | Good prediction |
| 0.7 | Mediocre prediction |
| 0.6 | Poor prediction |
| 0.5 | Random prediction |

9. F-Measure harmonic average of precision and recall.

$$\text{Fmeasure} = \frac{2 * (\text{Precision} \times \text{Recall})}{(\text{Precision} + \text{Recall})} \tag{16}$$

10. Receiver operator characteristic (ROC) area measure

The ROC value is generally calculated based on the sensitivity and specificity of the model. The sensitivity measure is the ability of correct identification that is true positive rate (TPR), whereas specificity is the ability of a test to correctly identify those incorrect that are true negative rate. The model interpretation is based on the calculated ROC value which is shown in Table 6.

### *3.6 Performance Analysis of Target Features to Identify the Location of Aspired Foreign Body*

The experimental test is done with 80 foreign body aspired pediatric radiography image. The anatomic location of the aspired foreign body is described based on its calculated features presented in the FB descriptor knowledge base. The knowledge is trained for predicting the new target shape class by J48 decision tree algorithm. The location is classified as airway, esophagus, left main bronchus and right main bronchus, gastrointestinal region, right diaphragm, left diaphragm, left lung, and right lung. Initially, all the edge-, shape-, and texture-based features are considered for classification of all the location class. Then the features selected by the hybrid feature selection such as standard deviation, size, dissimilarity, difference entropy, information correlation 1 and information correlation 2 are applied for location classification. The accuracy of the classification is determined by the values of ROC measure which is discussed in Table 7.

**Table 7** Classification accuracy by ROC measure

| Type of anatomic location | ROC measure without optimal feature | ROC measure with common optimal feature |
|---|---|---|
| Large intestine | 0.19 | 0.99 |
| Left main bronchus | 0.48 | 0.72 |
| Left lung | 0.38 | 0.99 |
| GI region | 0.42 | 0.51 |
| Airway | 0.27 | 0.68 |
| Esophagus | 0.27 | 0.98 |
| Right diaphragm | 0.08 | 0.97 |
| Right lung | 0.3 | 0.66 |
| Right main bronchus | 0.38 | 0.71 |
| Left diaphragm | 0.16 | 0.71 |

Based on the observation of above Table 7, without introducing the concept of optimal features, classification accuracy is not desirable. To improve the accuracy of classification of anatomic location, the optimal feature selection method is applied. The results obtained from the developed approach reveal that the accuracy rate of classifying some of the location is still challenging. Therefore, the concept of identifying influential features is introduced for each type of location by considering the ROC measures. The ranking is applied from highest to lowest values of the features. The ranked features are considered for further classification. The accuracy of classification of each location is assessed with the help of selected influenced optimal features based on the ranking. The result of the same is discussed in Table 8.

Based on the experimental results, size, cluster shade, sum entropy, and information correlation are identified as the influenced optimal features for classifying the bronchus anatomic location. Entropy, size, sum entropy maximum probability, and energy are identified as the influenced optimal features for classifying the lung anatomic location. Size, cluster shade, sum entropy, difference entropy, and information correlation are identified as the influenced optimal features for classifying the airway and esophagus anatomic location. Standard deviation, elongation, correlation, difference entropy, inverse difference homogenous measure, and size are identified as the influenced optimal features for classifying the gastrointestinal anatomic location. Similarly, size, dispersion, solidity, contrast, cluster shade, dissimilarity, difference variance, and information correlation are the influenced optimal features for diaphragm anatomic location. By applying the influenced optimal feature, the classification achieves better results in terms of ROC measure when considered with applying common optimal features for all the anatomic location of the class which is illustrated in Table 9.

**Table 8** Assessing the optimal features for location identification

| Ranked optimal features/location | Bronchus | Lungs | Airway | Esophagus | GI region | Diaphragm |
|---|---|---|---|---|---|---|
| Size | 1.04 | 0.97 | 0.99 | 0.99 | 0.58 | 0.99 |
| Sum entropy | 0.89 | 0.94 | 0.99 | 0.99 | 0 | 0 |
| Difference entropy | 0.59 | 0 | 0.68 | 0.68 | 0.93 | 0 |
| Inmcorr2 | 0.58 | 0 | 0.68 | 0.68 | 0 | 0 |
| Cluster shade | 1.09 | | 0.99 | 0.99 | 0 | 0 |
| Entropy | 0 | 0.98 | 0 | 0 | 0 | 0 |
| Maximum probability | 0 | 0.92 | 0 | 0 | 0 | 0 |
| Energy1 | 0 | 0.81 | 0 | 0 | 0 | 0 |
| Inmcorr2 | 0 | 0 | 0.68 | 0.68 | 0 | 0 |
| Standard deviation | 0 | 0 | 0 | 0 | 0.95 | 0 |
| Elongation | 0 | 0 | 0 | 0 | 0.94 | 0 |
| Correlation1 | 0 | 0 | 0 | 0 | 0.94 | 0 |
| IDHomom | 0 | 0 | 0 | 0 | 0.81 | 0 |
| Dispersion | 0 | 0 | 0 | 0 | 0 | 0.99 |
| Solidity | 0 | 0 | 0 | 0 | 0 | 0.98 |
| Contrast | 0 | 0 | 0 | 0 | 0 | 0.87 |
| Dissimilarity | 0 | 0 | 0 | 0 | 0 | 0.76 |
| Difference variance | 0 | 0 | 0 | 0 | 0 | 0.74 |

**Table 9** Classification accuracy by ROC measure

| Type of anatomic location | ROC measure with common optimal feature | ROC measure with location-based influenced optimal feature |
|---|---|---|
| Large intestine | 0.99 | 1 |
| Left main bronchus | 0.72 | 0.8 |
| Left lung | 0.99 | 1 |
| GI region | 0.51 | 1 |
| Airway | 0.68 | 1 |
| Esophagus | 0.98 | 1 |
| Right diaphragm | 0.97 | 1 |
| Right lung | 0.66 | 1 |
| Right main bronchus | 0.71 | 0.8 |
| Left diaphragm | 0.71 | 1 |

The ability of the developed anatomic location identification approach is measured with ROC classification accuracy. The performance comparison is made with feature selection such as without optimal features for location identification,

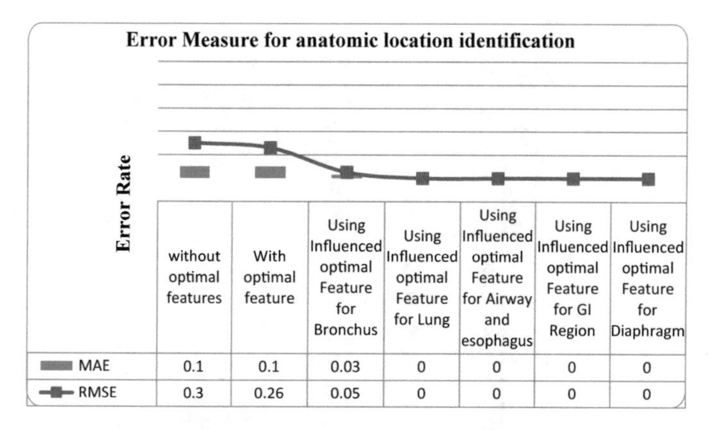


**Fig. 6** Error measure with feature selection and shape determination

with optimal feature selection and with the influenced optimal features obtained from the developed hybrid feature selection method based on the ranking. The root mean square error (RMSE) and mean absolute error (MAE) are defined as the mean of absolute error between identified shape $x_s$ and expected shape $y_s^*$ which are presented in Fig. 6.

Figure 6 shows the error measure for anatomic location identification. As per the ROC accuracy measure and the error measures such as the mean absolute error and root mean squared error obtained from J48 decision tree classification, the location-based influenced optimal feature selection gives better performance than selecting common optimal features for classification. Hence, the interpretation of the anatomic location identification approach gains excellent prediction rate as per the accuracy measure benchmark.

## 4 Experimental Results

The experimentation used 80 radiography images of the pediatric patients. The images were selected randomly but with focused on the pediatric foreign body aspiration. The sample experimental results of edge- and shape-based descriptors are shown in Table 10. Table 11 shows experimental results of texture-based descriptors. Table 12 shows the identified features of sample input of anatomic locations such as airway, esophagus, gastrointestinal region (small and large intestines), diaphragm, lung, and bronchus in the pediatric foreign body identified image.

**Table 10** Experimental results of edge- and shape-based descriptors

| Sample | F1 | F2 | F3 | F4 | F5 | F6 | F7 | F8 | F9 | F10 | F11 | F12 | F13 | F14 | F15 | F16 | F17 |
|---|---|---|---|---|---|---|---|---|---|---|---|---|---|---|---|---|---|
| Sample 1 | 5.9 | 23.36 | 22.92 | 50.03 | 39.00 | 5.00 | Irregular | 1.00 | 1.00 | 3.05 | 0.01 | 10.43 | 0.99 | 0.45 | 10.00 | 0.59 | A |
| Sample 2 | 8.4 | 741 | 10.54 | 12.99 | 0.01 | 77.00 | Polygon | 1.00 | 6.00 | 1.86 | 0.02 | 7.03 | 0.29 | 0.60 | 4.54 | 0.84 | E |
| Sample 3 | 7.79 | 230.00 | 82.28 | 33.47 | 176.00 | 1.00 | Irregular | 35.00 | 0.03 | 0.01 | 0.50 | 0.80 | 0.44 | 1.74 | 82.58 | 0.43 | GI |

*F1*—Size, *F2*—Area, *F3*—Perimeter, *F4*—Standard deviation, *F5*—Minimum intensity, *F6*—Maximum intensity, *F7*—Shape, *F8*—Convex area, *F9*—Compactness, *F10*—Elongation, *F11*—Rectangularity, *F12*—Eccentricity, *F13*—Solidity, *F14*—Dispersion, *F15*—Centroid, *F16*—Thinness ratio, *F17*—Identified location. *A*—Airway, *E*—Esophagus, *GI*—GI region, *RD*—Right diaphragm, *LD*—Left diaphragm, *LMB*—Left main bronchus, *RL*—Right lung, *LL*—Left lung, *RMB*—Right main bronchus

**Table 11** Experimental results of texture-based descriptors

| Sample | T1 | T2 | T3 | T4 | T5 | T6 | T7 | T8 | T9 | T10 | T11 | T12 | T13 | T14 | T15 | T16 | T17 | T18 | T19 | TT20 | TT21 | TT22 |
|---|---|---|---|---|---|---|---|---|---|---|---|---|---|---|---|---|---|---|---|---|---|---|
| Sample 1 | 1.7 | 0.4 | 0.71 | 0.71 | 388.57 | 29.44 | 0.06 | 0.96 | 0.11 | 0.99 | 0.98 | 1.87 | 2.2 | 6.76 | 0.1 | 0.4 | 0.05 | 0.53 | 0.28 | 0.99 | 1.00 | 1.00 |
| Sample 2 | 1.46 | 0.08 | 0.89 | 0.89 | 270.27 | 19.74 | 0.01 | 0.98 | 0.06 | 1 | 0.99 | 1.47 | 2.11 | 5.71 | 0.05 | 0.08 | 0.01 | 0.8 | 0.27 | 1.00 | 1.00 | 1.00 |
| Sample 3 | 1.15 | 0.05 | 0.81 | 0.81 | 85.65 | 6.31 | 0.01 | 0.99 | 0.02 | 1 | 1 | 1.14 | 2.04 | 4.54 | 0.02 | 0.05 | 0.01 | 0.69 | 0.16 | 1.00 | 1.00 | 1.00 |

*TF1*—Autocorrelation, *TF2*—Contrast, *TF3*—Correlation1, *TF4*—Correlation2, *TF5*—Cluster prominence, *TF6*—Cluster shade, *TF7*—Dissimilarity, *TF8*—Energy1, *TF9*—Entropy1, *TF10*—Homogeneity1, *TF11*—Maximum probability, *TF12*—Sum of squares, *TF13*—Sum average, *TF14*—Sum variance, *TF15*—Sum entropy, *TF16*—Difference variance, *TF17*—Difference entropy, *TF18*—Inmcorr1, *TF19*—Inmcorr2, *TF20*—IDHomom, *TF21*—ID normalized, *TF22*—IDMNormalized

**Table 12** Sample experimental results of automatic anatomic location identification approach

| Location | Sample1 airway | Sample1 esophagus | Sample1 GI region | Sample1 diaphragm | Sample1 lung | Sample2 lung | Sample2 bronchus | Sample2 bronchus |
|---|---|---|---|---|---|---|---|---|
| Original image | | | | | | | | |
| Iterative threshold image | | | | | | | | |
| Median filtered image | | | | | | | | |
| Sobel edge detection | | | | | | | | |
| Location identified image | Airway | Esophagus | GI region | Diaphragm | Lung | Lung | Bronchus | Bronchus |

# 5 Conclusions and Future Enhancement

The developed Automatic anatomic location identification approach is efficient for analyzing the pediatric foreign body aspired radiography images for doctors and medical practitioners. Though the proposed approach is developed with aspects such as accuracy and time consumption in mind, there are still a few limitations existing such as the proposed approach is limited to fixed template of the pediatric radiography image for location identification and the knowledge base is designed based on expert suggestion. The proposed hybrid feature selection method gives optimal features for shape and location determination. In experimental results, the ROC performance accuracy for all the location class comparatively gives better results with the existing work of this similar nature on various medical dataset. Hence, the interpretation of the shape determination and AALIA approach gains excellent prediction rate as per the machine learning ROC accuracy measure benchmark. This system can be further extended to large databases where the time consumption for data retrieval and analysis is more. The future work aims to develop automatic anatomic location identification system regardless of fixed anatomic location template of pediatric radiography images.

# References


1. Vasumathy, M., & Thirugnanam, M. (2016). A hybird MIK segmentation technique for diagnosing aspired foreign body on pediatric radiography images. *IIOAB Journal, 7*(1), 210–217.
2. Khan, S. U., Chai, W. Y., See, C. S., & Khan, A. (2016). X-ray image enhancement using a boundary division wiener filter and wavelet-based image fusion approach. *Journal of Information Process System, 12*(1), 35–45.
3. Firoz, R., Ali, M. S., Khan, M. N. U., Hossain, M. K., Islam, M. K., & Shahinuzzaman, M. (2016). Medical image enhancement using morphological transformation. *Journal of Data Analysis and Information Processing, 4*(12), 1–12.
4. Kramer, R. E., & Lerner, D. G. (2015). Management of ingested foreign bodies in children: A clinical report of the NASPGHAN endoscopy committee. *Journal of Pediatric Gastroenterology and Nutrition, 60*(4), 562–574.
5. Kaviani, F., Rashid, R. J., Shahmoradi, Z., & Gholamian, M. (2014). Detection of foreign bodies by spiral computed tomography and cone beam computed tomography in maxillofacial regions. *Journal Dental Research and Dental Clinical Prospects, 8*(3), 166–171.
6. SeikholetKuki, & Gulati, A. (2014). A 'two-in-one' foreign body coin in oesophagus: A case report. *IOSR Journal of Dental and Medical Sciences (IOSR-JDMS), 13*(4), 65–67.
7. Upreti, L., & Gupta, N. (2015). Imaging for diagnosis of foreign body aspiration in children. *Indian Pediatrics, 52*(1), 659–690.
8. Lecron, F., & Benjelloun, M. (2012). Fully automatic vertebra detection in X-ray images based on multi-class SVM. *Medical Imaging, SPIE Proceedings, 8314*(2), 1–8.
9. Juhasz, S., Horvath, A., Nikhazy, L., Horvath, G., & Horvath, A. (2010). Segmentation of anatomical structures on chest radiographs. *International Journal of Medical Science, 4*(3), 1–4.

10. Klinder, T., Ostermann, J., Ehm, M., Franz, A., Kneser, R., & Lorenz, C. (2009). Automated model-based vertebra detection, identification, and segmentation in CT images. *Medical Image Analysis, 13*(3), 471–482.
11. Benjelloun, M., & Mahmoudi, S. (2009). Spine localization in X-ray images using interest point detection. *Journal of Digital Imaging, 22*(3), 309–318.
12. Upadhyaya, E. V., Srivastava, P., Upadhyaya, V. D., Gangopadhyay A. N., Sharma, S. P., Gupta, D. K., & Hassan, Z. (2009). Double coin in esophagus at same location and same alignment—A rare occurrence: A case report. *Cases Journal, 2*(1), 77–58.
13. Pohjalainen, J. (2013). Feature selection methods and their combinations in high-dimensional classification of speaker likability, intelligibility and personality traits. Computer, speech and language. *The Journal of Neuroscience, 3*(9), 93–102.
14. Blum, A. L. (1997) Selection of Relevant Features and Examples in Machine Learning. Artificial Intelligence. *The Journal of Neuroscience, 1*(9), 2393–2402.
15. Choras, R. S. (2007). Image feature extraction techniques and their applications for CBIR and biometrics systems. *International Journal of Biology and Biomedical Engineering, 1*(1), 6–16.
16. Guyon, I. (2003). An introduction to variable and feature selection. *Journal of Machine Learning Research, 3*(1), 1157–1182.
17. ladha, L. (2011). Feature selection methods and Algorithms. *International Journal on Computer Science and Engineering (ijcse), 3*(5), 1787–1797.
18. Janecek, A. G. K. (2007). On the relationship between feature selection and classification accuracy. *Workshop and Conference Proceedings, 4,* 90–105.
19. Vasumathy, M., & Thirugnanam, M. (2016). A Framework for automatic intrude object identification in paediatric foreign body aspired radiography images. *International Journal of Image Mining, 2*(1), 57–67.
20. Gonzalez, R. C., & Woods, R. E. (2008). *Digital image processing* (3rd ed.). Prentice Hall, ISBN: 013168728.
21. Vasumathy, M., & Thirugnanam, M. (2017). Shape determination of aspired foreign body on pediatric radiography images using rule-based approach. *Advanced Image Processing Techniques and Applications, 1*(8), 170–180.

# A Comparative Analysis of Local Pattern Descriptors for Face Recognition

**R. Srinivasa Perumal and P. V. S. S. R. Chandra Mouli**

**Abstract** Face recognition has lot of challenges in biometrics. The challenges are addressed effectively by local pattern descriptors. The idea of local descriptors is to determine the feature vector and then compute the difference between test images with training images by using similarity measure. Based on the observation, local approaches attained better performance rate than other approaches in face recognition. Due to that, researchers made a significant attention on local descriptors for face recognition. For nonlinear subspace, the local descriptors will achieve better result than holistic approach. Local pattern descriptor follows simple procedure to extract the facial features. The steps are face alignment, face representation, and matching. Face alignment is the first step of local descriptor, which is used to divide the image into several blocks. Face representation is used to extract the meaningful information from each region. This local feature extraction method carries discriminant information of the region; it will improve the classification rate and matching rate. Local descriptors extract the discriminant information from the neighbors by setting a threshold value as center pixel value, and it is not capable of extracting the detailed information from microstructure. Finally, matching by classification techniques or distance measure is used to identify or verify the person. The local pattern descriptors are more robust against pose, lighting, and scale variations. This chapter describes the various local pattern descriptors and shows the effectiveness of the descriptor. The results of local pattern descriptors are experimented on standard benchmark databases such as FERET, Extended Yale-B, ORL, CAS-PEAL, LFW, JAFFE, and Cohn–Kannade.

**Keywords** Face recognition · Local pattern descriptor · Feature extraction
Brain computer interfaces · Image processing · Biometrics

---

R. Srinivasa Perumal (✉)
School of Information Technology and Engineering, VIT University, Vellore, India
e-mail: r.srinivasaperumal@vit.ac.in

P. V. S. S. R. Chandra Mouli
School of Computer Science and Engineering, VIT University, Vellore, India
e-mail: chandramouli@vit.ac.in



S. Margret Anouncia and U. K. Wiil (eds.), *Knowledge Computing and its Applications*, https://doi.org/10.1007/978-981-10-8258-0_7

# 1 Introduction

Biometrics is the emerging trends in recent years and most prominent model for identifying and recognizing the individuals. The individuals are identified based on their physiological and behavioral characteristics (such as iris, face, fingerprint, hand geometry, palm print, DNA). Face recognition entails with more advantages than other biometric traits. Many traits needed user cooperation for image acquisition. However, face recognition will acquire the image from distance without any explicit action of the user. Particularly, this is very useful in security and surveillance applications. Face recognition does not carry any health risks, and it is also non-intrusive [1]. Face recognition is used in numerous fields such as access control, criminal justice systems, surveillance, security, human computer interaction, mug shot searching. Some of the face recognition applications [2] are listed in Table 1.

Galton proposed formal method for classifying faces [3]. The method detects the curves and norms from facial profiles and then classifies facial profiles by their deviation from the norm, i.e., the resultant vector of a person that could be matched with vectors of the persons stored in a database. Face recognition can be framed as still (static) or video images in a frame. Verifying or identifying a person or more number of persons in the frame is done by matching with faces stored in a database. Verification and identification are primary tasks in face recognition.

The generic face recognition procedure is shown in Fig. 1. The first step in face recognition is face detection. Face detection identifies the face through exact location and size, facial landmarks, and classification algorithm. Face detection segments the face part from the image by eliminating the background. Face normalization will help to determine the accurate location of face and scale of each detected face parts. The image is normalized with respect to properties of photometrical (color and illumination) and geometrical (pose and size). The representation of face is employed to extract the meaningful information; that is, it categorizes the person with respect to photometrical and geometrical properties. Finally, the face matching computes the similarity score between feature vectors of query image and feature vectors of stored images. Based on the similarity score, the face recognition system made decision whether the person is known or unknown [4].

**Table 1** Face recognition applications

| Areas | Applications |
|---|---|
| Security | ATM machine, airports, border checkpoints, seaports, email authentication |
| Surveillance | Criminal, drug offenders, missing children in the crowd can be monitored by using CCTV |
| Criminal justice | Forensics, mug shot, post-event analysis |
| Multi-media | Human–computer interaction, behavior monitoring at old age and childcare |
| Government | Driving license, passport, national IDs, electoral registration |
| Commercial | Banking, e-commerce, etc. |

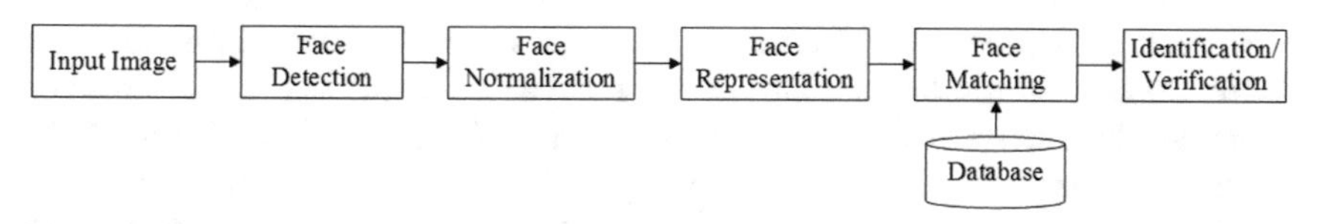


**Fig. 1** Face recognition process

**Table 2** Standard benchmark face database

| Database | No. of subjects | Description |
|---|---|---|
| FERET database [71] | 1199 | Pose, facial expression, illumination, and variation of time |
| Extended Yale-B [72] | 28 | Illumination, pose |
| AT & T (ORL) [74] | 40 | Variation of time, eye glasses, facial expression |
| LFW [73] | 1680 | Pair matching, pose variation, eye glasses, occlusions |
| CAS-PEAL [77] | 1040 | Different types of expressions, accessories, lighting |
| AR face database [78] | 126 | Frontal pose, scarves, occlusions, eye glasses, expressions |
| Cohn–Kanade [76] | 100 | Sequence of facial expressions |
| JAFEE [75] | 10 | Neutral, facial expression, emotion |
| PIE-CMU [79] | 68 | Pose, illumination, facial expression |

Currently, large numbers of databases are available for face recognition. To compare various local descriptor-based algorithm, it is recommended to use a standard benchmark database to prove the efficiency of these algorithms. The database consists of different types of face images and facial expressions. Researchers should choose appropriate databases for their algorithms. Some of the standard benchmark databases are listed in Table 2.

The rest of the paper is organized as follows. Section 2 discusses the challenges and issues of face recognition. Section 3 elucidates the general approaches of face recognition. Section 4 describes the local pattern descriptors. The performance analysis of local pattern descriptors on different databases is done in Sect. 5, and Sect. 6 concludes the work.

## 2 Challenges and Issues in Face Recognition

The human face recognition system has some limitations in identifying the person from a large group of persons in database because a human brain cannot remember everyone accurately. Face recognition system has the capability to handle the large group of persons in database. Human face is not a distinct; there are several factors that cause the appearance of the face to vary. The appearance can be classified into intrinsic and extrinsic factors. Intrinsic factors can be divided subject to interpersonal and intrapersonal characteristics. Interpersonal characteristics focus on different facial appearances of the same person. Intrapersonal characteristics focus on different facial appearances of different persons. Extrinsic factors concentrate on pose, lighting, different orientation of image. Poor quality of image, pose, illumination, and facial expressions are the major challenges for the researchers. This section summarizes the challenges and issues in face recognition system [5, 6].

**Dedicated Process**: Face recognition system is a dedicated process based on the evidence of existing system [7, 8]. Faces are easily remembered by humans, and even face blindness patient can perceive nose, eyes, and mouth.

**Feature or Holistic Analysis**: Both are very crucial in identification of face [9]. Bruce et al. suggested that if any dominant features are present in face, then feature-based method will provide better result than holistic analysis.

**Facial Features**: The current evidence [10] shows that the facial features are very useful in recognizing a person. In addition, the behavioral characteristics of human components will be taken into account, such as outline of the face, eyes, nose, and mouth. The reason behind the selection of these characteristics is that the upper part of the face contains more useful information than the lower part.

**Pose**: For face recognition in unconstrained environment like surveillance system, the camera is mounted at some location to capture the person. In unconstrained environment, capturing a frontal position of a person is very difficult. Even if the person does not look at the camera, various poses can still be captured, such as a frontal part of a face, upside of a face, downside of a face, a partial face image, and various degrees of a face. Pose is the one of the most challenging situations.

**Occlusion**: In a group photograph, the faces may occlude with other persons face or by other objects. Because of that, the face recognition finds difficult to extract the complete facial features from the face image. If the person has a scar, beard, mustache, and wearing the glasses, the system may face the difficulty while extracting the facial features.

**Imaging Conditions**: Based on the sensor characteristics or lens, the quality of image may degrade while forming an image, due to this lighting and intensity variations problem will occur. The orientation of the image depends on the appearance of the face, i.e., rotation, scaling.

**Illumination**: Illumination is one of the most challenging factors in face recognition system. It determines the quality of the image and is related to the lighting problem that exists in the images. In this case, the face image may be very dark or very bright or some facial features may be dark and rest of the facial features are bright. This lighting variation will affect the result.

**Facial Expressions**: Another challenging task is facial expression. The human will express their emotions and feelings in their face such as angry, happy, sad. The appearance of the person with happy and appearance of the person with sad are totally different. Hence, the facial expression is directly affecting the appearance of the face.

# 3 Literature Reviews of Face Recognition Methods

This section discusses the various face recognition techniques for intensity images. Face recognition method is generally classified into two types. They are feature-based and holistic approach. Feature-based method extracts the facial features like outline of the face, nose, eyes, followed by determining the relationship between each facial feature to recognize the person. Holistic method extracts the features from the whole image for recognizing the face [11].

## *3.1 Holistic Approach*

Holistic approach extracts the whole information from face for recognizing the person. The information from face is described by number of features extracted directly from the pixel information of face images. The feature vectors are used to recognize the individuals. This method is divided into statistical and artificial intelligence approaches.

### 3.1.1 Statistical Approaches

The image is represented as a two-dimensional array. The pixel value of each position is directly correlated between the query face and all the faces in the database. This approach [12] is a simple statistical approach method, and it will work under equal scale, equal illumination, and same pose. It is very expensive, and this method has sensitivity in noise, size, lighting conditions, and background. The performance for this direct matching method is comparatively lower, and the dimensionality is also very high [13]. To reduce the dimensionality, several methods are proposed. The methods are obtaining the meaningful descriptor of the face and reduce the dimensionality of the image before recognition.

Sirovich et al. proposed the dimensionality-reduced method, principle component analysis (PCA) [14]. The PCA represents the face patterns by Eigen pictures' coordinate space. Based on his findings, Eigen face represents the feature space by determining the Eigen vectors associated with eigenvalues for recognizing the face [15]. An Eigen face reduces the dimensionality of the original face image drastically. The performance of the eigenface will reduce with scale variations and it can be fairly robust with pose orientation and lighting conditions. In the case of having multiple samples per person, Belhumeur et al. argue that by choosing the projection with maximizes due to total scatter [16]. Fisher proposed linear discriminant analysis, which maximizes between classes and minimizes within class to improve the recognition rate in illumination and pose variation.

PCA normally leads to high-dimensional vector, but independent component analysis (ICA) leads in both higher-order and second-order data to find the feature vector and the vectors are statistically independent. ICA reconstructs the image better than PCA even it is noisy [17]. Due to large size of vectors, it is difficult to determine the accurate feature vector space in PCA. To overcome the large size of vector problem, Yang et al. proposed 2D-PCA to determine the accurate feature vector without converting matrix into vector [18]. Eigenfaces and Fisher face approaches have various extensions. Some PCA-based algorithms are symmetrical PCA [18], adaptively weighted subpattern PCA [19], Kernel PCA [20, 21]. Some LDA-based algorithms are direct weighted LDA [22], singular value decomposition [23], component-based cascade LDA [24], Kernal LDA [25], Gabor LDA [26], Fourier LDA [27]. The drawback of this technique is to work effectively only in linear subspace.

### 3.1.2 Artificial Intelligence-Based Methods

These approaches use tools for face recognition such as machine learning and neural network techniques. PCA and autoassociative memory are used to reduce the dimension [28]. The hierarchical neural network performs better in face recognition automatically even the faces are not trained [29]. Lawrence et al. applied hybrid neural network which combines self-organizing map and local image sampling [30]. Self-organizing map reduces the dimensionality, and it provides rotation, scale, and invariance to translation [31]. Eleyan et al. used PCA to attain feature vector space from face images and classify using feed-forward neural networks [32]. Support vector machine [33] is used for classifying the faces in the face recognition to improve the performance of the system. In hidden Markov model [34], the image is converted into one-dimensional vector to attain the peak value. Many researchers used eigenfaces or Fisher faces for determining the feature vector space, and they used neural network or machine learning algorithm for classification to improve the recognition rate [35].

Holistic approaches do not neglect any information from the image. The pixels are equally important in holistic approach. These techniques require high degree of relationship between the training and test images and are computationally expensive. These approaches fairly perform under huge variations in scale, pose, and illumination conditions [36]. Due to this reason, most of the algorithms under this category are modified or enhanced to improve the recognition rate. Subspace analysis method extracts meaningful discriminative information to achieve higher recognition rate than feature-based approaches [37].

## *3.2 Feature-Based Approaches*

Feature-based approaches determine the facial features of the face and extract the information from the facial features such as eyes, nose, mouth, outline of the face. The relationship between facial features that transforms an image into a vector is computed. Kanade proposed a geometric feature method to extract 16 facial features for matching [38]. Brunelli and Poggio extract a set of geometrical features such as width, length, and position of facial features [11]. Template matching chooses four feature templates automatically, i.e., whole face, eyes, nose, and mouth. Template matching is computationally expensive and complex. Graph matching determines the closest stored graph which employs a dynamic link structure to recognize the invariant object. This process is computationally expensive. Wiskott and Von proposed a rotation invariant and dynamic link matching to recognize the face [39]. Heisele et al. developed a framework for component detection and identification [40]. In component-based method, the facial features are located by landmark extraction, and then, the facial features are aligned to represent the vectors of each component [41].

In feature-based methods, the feature vector is used to recognize the face. These methods are robust to pose variations, size orientation, and lighting conditions. Feature-based methods consume time when comparing the query feature vector with the stored feature vector of the face image. The major drawback of these methods is that the facial feature extraction is difficult. As on date, there is no subsequent method that can compensate the above drawbacks.

## 4 Local Pattern Descriptors for Face Recognition

This section describes the most recent research on local descriptors for face analysis to understand the present issues in the face recognition, facial expression recognition, and face spoofing. The idea of local descriptors is to extract the micropatterns from the image by dividing into blocks. These patterns are further processed to form a feature vector to represent the face image. Dissimilarity measures are used to check the query image with training set for proper identification. Based on the observations from existing systems, local approaches improve the performance rate of face recognition than global approaches [40]. Due to this reason, researchers made a significant attention on local descriptors for face recognition. For nonlinear subspace, the local descriptors achieve better result than holistic approaches. The local descriptors follow three steps and are shown in Fig. 2. The steps are face alignment, face representation, and matching. Face alignment is the first step of the local descriptor, which is used to divide the image into several blocks. Face representation is used to extract the meaningful information from each region. This local feature extraction method carries discriminant information of the region; it will improve the classification rate and matching rate. Finally, matching by classification techniques or distance measures is used to identify or verify the person. The reason for using local pattern descriptors in face recognition is that local approaches are more robust to pose, lighting, and scale variations and gained significant interest in overcoming the limitations of holistic approaches.

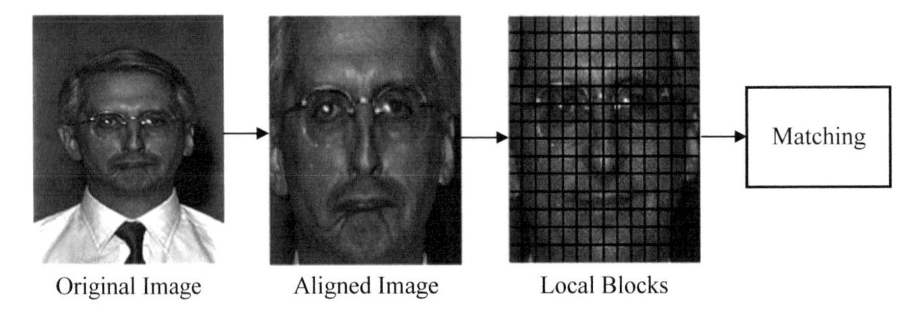


**Fig. 2** Local pattern descriptor steps of face recognition process

## 4.1 Local Binary Pattern

Ojala et al. proposed local binary pattern (LBP) descriptor for texture description, and it is widely used in different applications [42]. LBP operators divide the image into different blocks of size 3 × 3. LBP extracts the discriminant features from each block, assign a label to each pixel value of a block, and set the center pixel value of block as threshold value. The resultant value of LBP is a binary number. The histogram of the block contains discriminant information. Ojala et al. extended LBP operator to recognize the uniform pattern [43]. This method is robust in scale variation. The primary uniform patterns are considered as feature vectors of the LBP such as corners, edges, and spots. The special structure of the image is defined in Eq. (1).

$$\begin{aligned} \mathrm{LBP}_{P,R}(x_c, y_c) &= \sum_{p=0}^{P-1} s(g_p - g_c)2^P \\ s(x) &= \begin{cases} 1, & x \geq 0 \\ 0, & x < 0 \end{cases} \end{aligned} \quad (1)$$

where $P$ represents the number of sampling points on a circle of radius $R$.

LBP was applied in face recognition [44] to extract the feature vector of face image. The face image is divided into several regions and extracts the feature vector from each region. Concatenate the feature vector of each region into one-dimensional vector. Some dissimilarity measure is used to perform the recognition task. An LBP is robust to pose, facial, lighting; it is also computationally efficient and simple. The working mechanism of LBP is shown in Fig. 3.

Variants of LBP have been proposed by many researchers to improve the face recognition rate. These variations are used to improve its discriminative capability, enhance its robustness, and choose its neighborhood and hybrid approaches [45]. To improve the discrimination capability, Jin et al. proposed the improved LBP to extract more structural information by comparing each pixel value with the average intensity value of the region [46]. Complete LBP (CLBP) [47] improves the

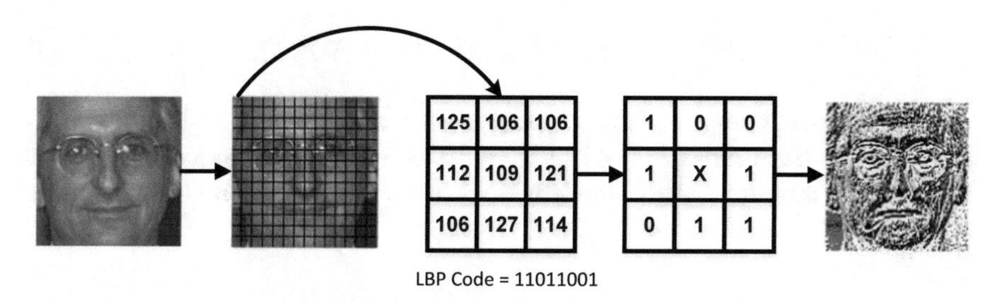


**Fig. 3** Working mechanism of LBP

discriminative power based on the gray value difference of center pixel and both positive and negative signs of its neighbor. Similarly, the Hamming LBP [48] and extended LBP [49, 50] will improve the discriminative capability than LBP operator.

To enhance the robustness, Local ternary pattern (LTP) [51] is replaced for LBP. LTP generates three value codes, and user can specify the threshold value. The code is generated by Eq. (2).

$$s'(u, i_c, t) = \begin{cases} 1 & u \geq i_c + t \\ 0 & |u - i_c| < t \\ -1 & u \leq i_c - t \end{cases} \tag{2}$$

The LTP code is more resistant to noise. The drawback of LTP is defining a threshold value, which is not simple. Multi-block LBP (MB-LBP) [52] captures both micro- and macrostructures by comparing average intensity value of neighboring subregions. The hybrid approaches like local Gabor binary pattern [53] and LBP and SIFT [54] improve the performance, time, and discriminant power.

## 4.2 Local Derivative Pattern

LBP extracts the discriminant information from the neighbors by setting the threshold value as center pixel value is not capable of extracting the detailed information from microstructure. Local derivative pattern (LDeP) generates a binary code from $(n - 1)$th-order derivative directional variation. LBP is first-order non-directional pattern operator because it extracts the information from all the directions. LDeP extracts the higher-order information which contains more meaningful features. The process is used to extract up to $(n - 1)$th-order derivative along 0°, 45°, 90°, and 135° directions denoted as $F'(X)$ where $F(X)$ is an image [55].

Let $X_0$ be a point in $F(X)$, and $X_i$, $i = 1, 2, \ldots, 8$, are the neighbor of $X_0$. The second-order derivative is defined in Eq. (3).

$$\mathrm{LDeP}^2(X) = \left\{\mathrm{LDeP}^2_\alpha(X) | \alpha = 0^\circ, 45^\circ, 90^\circ, 135^\circ\right\} \tag{3}$$

LDeP operator compares the two derivative directions at two neighboring pixels and combines the results of 0°, 45°, 90°, and 135° directions. The resultant bit is a 32-bit binary code for LDeP. The derivative direction comparison is defined using Eq. (4).

$$z(F(X_0), F(X_i)) = \begin{cases} 0, & \text{if } F(X_i) - F(X_0) \leq \mathrm{th} \\ 1, & \text{if } F(X_i) - F(X_0) > \mathrm{th} \end{cases} \tag{4}$$

where $z(;,;)$ is a thresholding function, th represents threshold, and $X_i$, $i$ = 1, 2, …, 8, are the neighbors of $X_0$.

Figure 4 demonstrates the computational steps of LDeP for 0°. The two derivative directions are denoted in light color in the template. In the template, the left side values are monotonically increased, and then, the bit value is 0 otherwise 1. Similarly, for 45°, 90°, and for 135° degrees, LDeP 32-bit binary code is generated. The $n$th-order LDeP is defined in Eq. (5).

$$\mathrm{LDeP}^n(X) = \left\{\mathrm{LDeP}^n_\alpha(X) | \alpha = 0^\circ, 45^\circ, 90^\circ, 135^\circ\right\} \tag{5}$$

LDeP extracts the complete high-order information, and the over detailed information will also cause noise instead of determining the information. In second order and third order, LDeP extracts the discriminant information for face recognition than LBP. The advantages of LDeP than LBP are extracting more

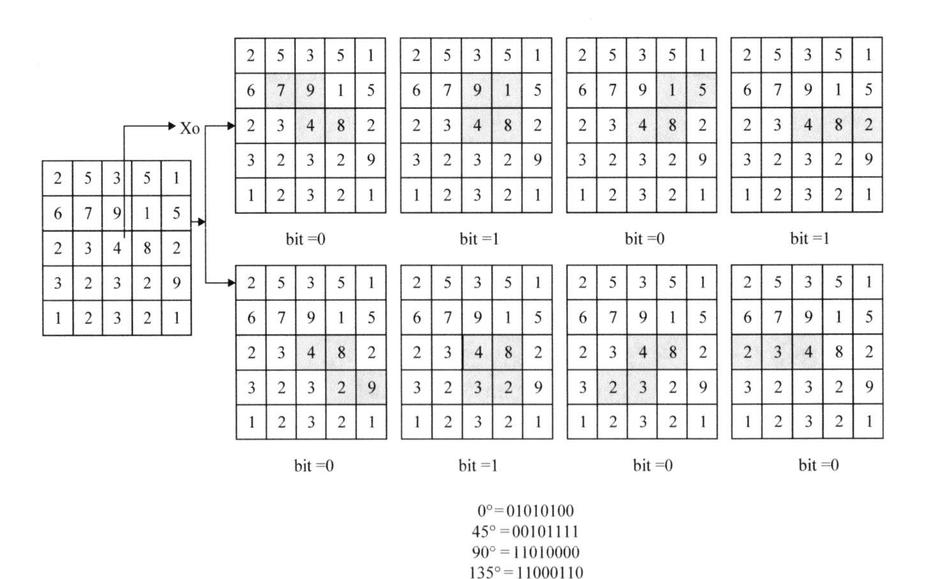


**Fig. 4** Computation steps of local derivative pattern

discriminant information from the image by encoding the nth-order gradient direction, and it encodes the various spatial relationships with its neighbors in the local region.

## 4.3 Local Directional Pattern

Local directional pattern (LDP) [56] determines the edges and spot corners values in eight directions and it extracts the discriminate information from the image. LDP assigns an 8-bit binary code to each pixel in the image. Assign 8-bit binary code of each pixel by comparing the edge response values of each direction. LDP calculates the directional edge response value of a pixel by convoluting with Kirsch mask $(M0, M1\ldots, M7)$. Apply Kirsch mask to obtain edge response value with respective directions. LDP edge response values are not equally important in all directions. The edges or corners occur in high response value. So LDP selects the most prominent direction by selecting top *k* values in a region. The top *k* positions are set as 1, and rest of the positions are 0. It is defined mathematically in Eq. (6).

$$C[f(x,y)] = (c_i = 1) \quad \text{if}\, 0 \le i \le 7 \text{ and } m_i \ge \psi \tag{6}$$

where $\psi = k\text{th}\{M\}$ and $M = \{m_0, m_1, \ldots, m_7\}$.

LDP extracts the features by selecting the most prominent responses from all directions. LDP follows three steps to represent the face. First, original image is encoded by applying LDP operator on the original image. In the second step, histogram is extracted from each local region of the encoded image to represent the face. All the histograms of local regions are concatenated to represent the global representation of face in the final step. Chi-square dissimilarity measure is used to recognize the face from the database. LDP is stable even if image consists of noise and non-monotonic lighting conditions. LDP is more robust to random noise, illumination changes, facial expressions, and pose variations. Figure 5 shows the working mechanism of LDP.

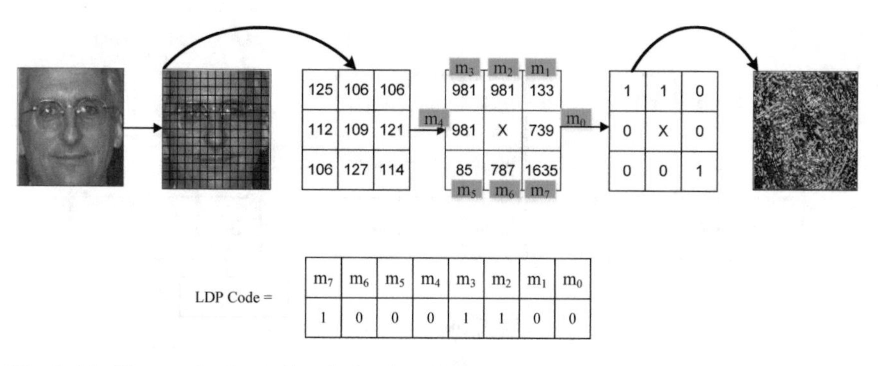


**Fig. 5** Working mechanism of local directional pattern

Kabir et al. proposed a similar descriptor called local directional pattern variance (LDPv), which characterizes the local information as well as spatial structure of LDP [57]. The dimension of the feature vector is reduced using PCA and classifies the faces with SVM. LDPv extracts the useful range of discriminant features of a low-resolution image. Enhanced local directional pattern (EnLDP) [58] follows the same procedure of LDP to extract the edge response values, but EnLDP generates an octal code instead of binary code of LDP. EnLDP selects the top two prominent directions to generate a code. EnLDP is also robust similar to LDP, and it attains more accuracy than LDP.

## 4.4 Local Sign Directional Pattern

LDP misses some directional information because it treats all the directions equally. Due to this, some important information may lose even though LDP is robust to noise and lighting conditions. Local sign directional pattern (LSDP) [59] is introduced to overcome the above problem. LSDP is a 6-bit binary code assigned to each pixel which represents both spatial and intensity transitions. The LSDP computes the edge response values by convoluting with Kirsch masks on original image.

In LSDP, the code can be generated by analyzing the edge responses from convoluting Kirsch masks with face image. The top most positive direction and top most negative directions are used to generate the 6-bit code. Figure 6 illustrates the working procedure of LSDP process.

For each region, LSDP is encoded using the most positive and most negative position values to define the meaningful descriptor. The prominent values are computed by using the sign information; the most significant bit is assigned for positive direction, and least significant bit is used for negative direction. It is defined in Eq. (7).

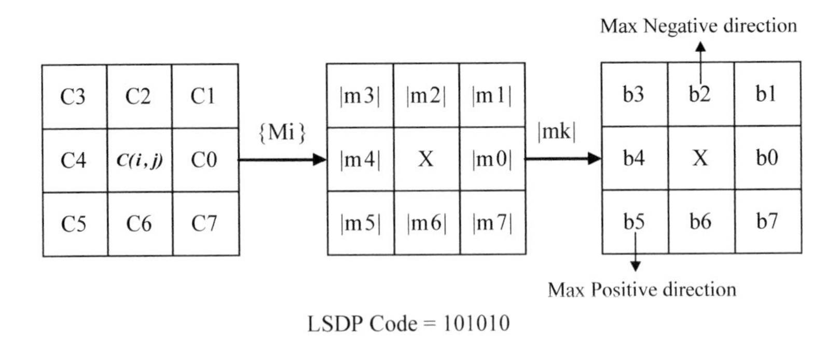


**Fig. 6** Computation steps of LSDP

$$\mathrm{LSDP(x, y) = 8i_{x,y} + j_{x,y}} \tag{7}$$

where $(x, y)$ is the central pixel value of the region, $i_{x,y}$ is the top most positive direction, and $j_{x,y}$ is the top most negative direction defined in Eqs. (8) and (9), respectively.

$$i_{x,y} = \arg_i \max\{\Pi^i(x, y) | 0 \le i \le 7\} \tag{8}$$

$$j_{x,y} = \arg_j \max\{\Pi^j(x, y) | 0 \le j \le 7\} \tag{9}$$

where $\Pi^i$ is the convolution of the image $I$ and the $i$th mask, $M^i$, defined by

$$\Pi^i = I * M^i \tag{10}$$

The encoded image is divided into regions of size 8 × 8 and computes the histogram bins of each region. Finally, concatenate the histogram bins to extract the descriptor of the face image. Compare the query image feature vector with stored image feature vector by chi-square dissimilarity measure. Local Gaussian direction pattern (LGDP) [60] is a novel descriptor to extract the texture and intensity variations (dark to bright and bright to dark). Local Gaussian number pattern (LGNP) generates a 6-bit code similar to LSDP by convoluting with Gaussian mask instead of Kirsch mask to avoid the noise perturbation. LGNP implicitly uses sign direction for generating a code of the each pixel. Local directional number pattern descriptor (LDNP) [61] is a similar descriptor like LDP, which combines the LSDP and LGNP. Local directional number pattern (LDNP) extracts the intensity variations (LGNP) and structural information (LSDP) of the face texture information from eight directions. Figure 7 shows the working mechanism of LDNP.

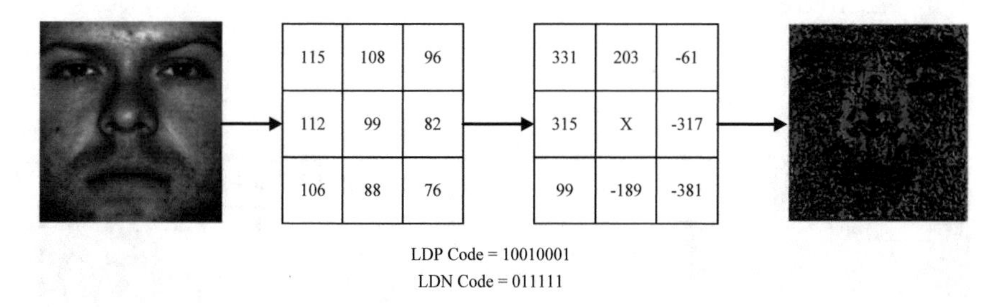


**Fig. 7** Working mechanism of local directional number pattern

### 4.5 Eight Local Directional Patterns

LDP used three prominent directions and LDNP used top most positive direction and negative direction to encode the local information of the face image. Instead of choosing three prominent directions or two dominant directions, eight local directional patterns (ELDP) [62] used all eight directional numbers to encode the local information of the face image. ELDP extracts more edge information by using eight directional numbers. ELDP produces illumination invariant representation by convoluting Kirsch mask to determine the edges and responses. Each edge response has different levels, but they are significant. ELDP generates 8-bit binary code. If the directional numbers of edge magnitude are positive, then the respective position is set as 1 otherwise set as 0. The binary code S(i) is assigned based on Eq. (11).

$$S(i) = \begin{cases} 1, & \Pi^i > 0 \\ 0, & \Pi^i \leq 0 \end{cases} \tag{11}$$

The computational steps of ELDP are demonstrated in Fig. 8. All the state-of-the-art methods use Kirsch mask to extract the directional information to achieve illumination invariant face recognition method. Faraji and Qi used homomorphic eight local directional patterns to represent the insensitivity in illumination [63]. These adaptive homomorphic eight local directional patterns were named as AHELDP. This adaptive filter was used to reduce the influence of illumination, and it was used to enhance the image features for face recognition. Faraji and Qi extended the AHELDP to reduce the computational time and improve the recognition rate, and it is named as CELDP [64]. CELDP followed the procedure of completed LBP, and it constructs the binary code from the absolute value of all eight direction magnitude values for each pixel based on the mean of eight neighbors in the region. The above local directional patterns are preformed effectively under various illumination conditions.

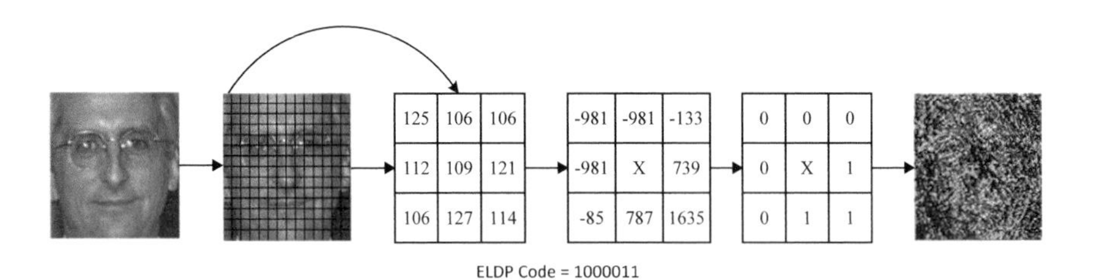


**Fig. 8** Working mechanism of eight local directional pattern

## 4.6 Dimensionality-Reduced Local Directional Pattern

Perumal and Mouli [65] proposed a dimensionality-reduced local directional pattern (DR-LDP). It is an 8-bit code assigned to each subregion of size 3 × 3. This code represents the textural pattern of the block. LDP computes an 8-bit code for each pixel in a 3 × 3 block, whereas the DR-LDP computes a single 8-bit code for the block. DR-LDP computes patterns similar to those of LDP. The difference is that post-processing of LDP patterns obtained for a block is done that reduces to a single 8-bit code. DR-LDP is extended to two levels and is named as two-level dimensionality-reduced local directional pattern (TL-DR-LDP). TL-DR-LDP extracts the local features by dividing the image into regions. The conventional LDP generates a code for each pixel in the input image. This method generates an 8-bit code per region for two levels [66]. DR-LDP works well in all standard benchmark databases. In the case of flat image, DR-LDP and TL-DR-LDP will fail to obtain better result. Figure 9 illustrates working mechanisms of DR-LDP.

## 4.7 Local Gabor Directional Pattern

Ishraque et al. proposed a descriptor for facial expression recognition named the local Gabor directional pattern (LGDP) [67]. The above state-of-the-art methods used Kirsch mask as filter to encode the local structure information of the face

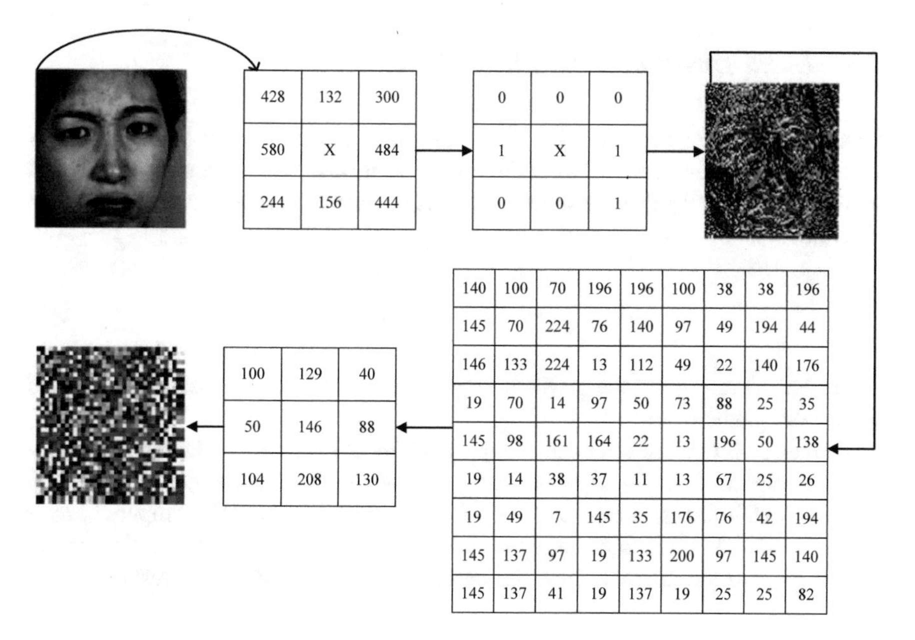


**Fig. 9** Working mechanism of DR-LDP

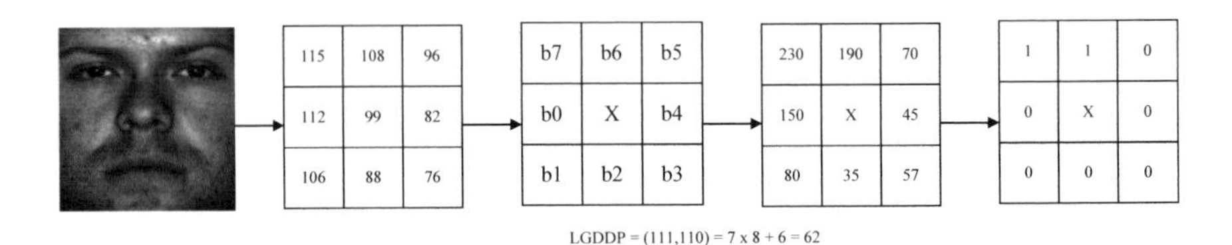


**Fig. 10** Working mechanism of LGDDP

image. LGDP used Gabor wavelet as filter to encode the local structure information of the face image, and it followed the same steps like LDP for generating histogram and face recognition. Lin et al. [68] extended the descriptor for face recognition with dominant direction to improve the recognition rate and reduce the computational complexity. Local Gabor dominant direction pattern (LGDDP) is similar to LGDP except the selection of the prominent direction. LGDDP chooses the prominent direction by comparing with eight neighborhood pixels. The dominant directions are two top most numbers in the eight neighborhoods. Figure 10 demonstrates the procedure of choosing the dominant direction in LGDDP.

Gabor feature outperformed the state-of-the-art methods for face recognition, but it has some disadvantages. The disadvantage of the Gabor feature is that the length of the feature descriptors is too high compared with state-of-the-art methods. Due to high-dimensional feature space, it requires concentrated processing time and lot of memory space to store the features.

## 4.8 *Local Directional Gradient Pattern*

Gradient directional pattern (GDP) enhances the edge information of an image like LBP, LDP. It retains and manages the information constantly than gray level-based methods [69]. GDP generates the direction of the feature vector of each pixel that is determined by Eq. (12).

$$\alpha(x, y) = \tan^{-1}\frac{G_y}{G_x} \tag{12}$$

where $\alpha(x, y)$ represents the gradient direction angle of the pixel $(x, y)$ and $G_x$ and $G_y$ are the gradient vectors.

Gradient vectors are obtained by applying Sobel operator on the source image. The operator is convoluted with original image in both horizontal and vertical kernel to obtain the gradient vectors *Gx* and *Gy*, respectively. The computed image is divided into $3 \times 3$ images and quantizes each neighbor gradient direction angles with respect to the gradient direction angle of the center pixel by threshold.

Chakraborty et al. proposed a descriptor local directional gradient pattern (LDGP) for face recognition [70]. It identifies the relationships between the neighbor's pixels and reference pixel to generate the local information in the high-order derivative space. LDGP reduces the length of the feature vector and achieves even the lower features to improve the performance of the recognition rate in facial expression, constrained pose, and lighting variation. The accuracy in case of unconstrained pose is still challenging in LGDP.

# 5 Comparison of Local Pattern Descriptors

The performance of local pattern descriptors is tested on standard benchmark databases such as FERET [71], Extended Yale-B [72], LFW [73], Databases of Faces (ORL), [74], JAFFEE [75], and CK [76]. In these databases, images are captured under different illumination conditions, with various pose variations and facial expressions.

## *5.1 FERET Database*

The performance of local pattern descriptor method is tested on the FERET database [71] based on the CSU face identification system. FERET database contains 14,501 images of 1010 individuals. The images are captured under various poses, illumination changes, and facial expressions and with aging effects. The database is classified into five sets of images such as Fa, Fb, Fc, dup-I, dup-II. Fa set is used as gallery image set, and rest of the image sets are called as probe images.

- Fa consists of 1010 frontal images.
- Fb consists of 1009 images with different expression variations.
- Fc consists of 194 images with illumination variation.
- dup-I, 722 images are captured after certain period of time.
- dup-II, 234 images are the subset of dup-I.

In FERET database, dup-I and dup-II image sets are very critical since the images are captured at later time from the time of images captured for gallery set. Figure 11 shows the recognition rate of local pattern descriptors in FERET database. From Fig. 11, the TL-DR-LDP has a higher recognition rate compared with the other local pattern descriptors. DR-LDP and TL-DR-LDP have equal recognition rate.

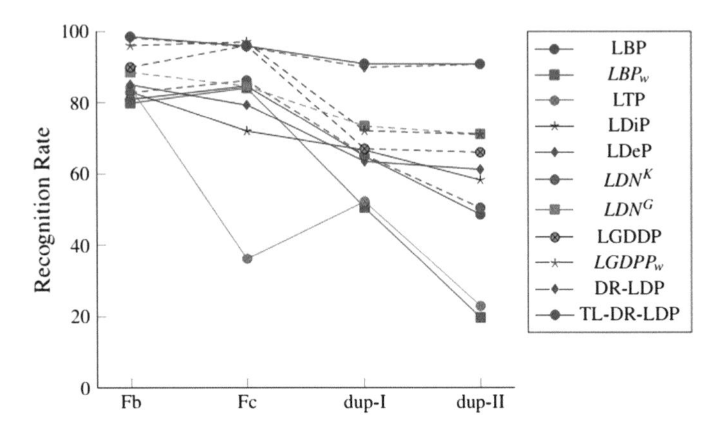


**Fig. 11** Recognition rate of local pattern descriptors on FERET database

## 5.2 *Extended YALE-B*

Extended Yale-B database [72] consists of images with illumination variation.

The database contains 16,128 images of 28 subjects: each subject with nine poses in 64 different illumination conditions. The images are categorized into five sets. The sets are labeled as Sub1, Sub2, Sub3, Sub4, and Sub5. Sub1 is used as gallery image set, and rest of the image sets are used for probe images. The subsets are categorized based on the angle of lighting. The angle less than or equal to 12° is labeled as Sub1, between 13° and 25° labeled as Sub2, between 26° and 50° labeled as Sub3, between 51° and 77° labeled as Sub4, and greater than 77° is labeled as Sub5. Figure 12 shows the recognition rate of local pattern descriptors on Extended Yale-B database. As can be seen in Fig. 12, the CELDP achieved higher recognition rate compared with the other local pattern descriptors in all categories.

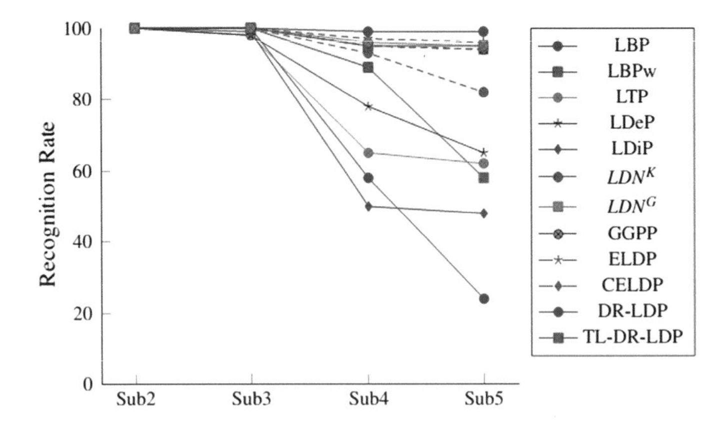


**Fig. 12** Recognition rate of local pattern descriptors on extended Yale-B database

### 5.3 ORL Database

ORL database [74] consists of 400 images of 40 subjects captured under different lighting conditions, different periods of time, facial expressions, and facial details. Each subject contains ten images out of which two are randomly selected for training and the rest for testing set. In ORL database, 80 images are used as gallery images and remaining 320 images are used as probe images. Table 3 shows the recognition rate of local pattern descriptors in ORL database. As can be seen in Table 3, the TL-DR-LDP has a higher recognition rate compared with the other local pattern descriptors.

### 5.4 LFW Database

LFW database [73] contains 13,233 images of 5,749 subjects. On this database, the experiments were conducted on 1,680 subjects only. These subjects are chosen based on the number of images available for each subject. At least two or more images should be present for each image to be considered for training. The remaining subjects are not used in the experiment. 1,680 subjects are classified into four sets based on the number of images that occur in each subject. The subimage sets are Sub1, Sub2, Sub3, and Sub4. The numbers of images with more than 20 in a subject are labeled as Sub1. Number of images with more than 30 in a subject is labeled as Sub2. Sub3 has the number of images more than 40 per subject, and Sub4 has more than 50 images per subject. Each subset is grouped into five subsets. In that group, four groups are used as gallery set and remaining group is used for probe set. Figure 13 shows the recognition rate of local pattern descriptors in LFW database. As can be seen in Fig. 13, the local directional number pattern with Gaussian mask has a higher recognition rate compared with the other local pattern descriptors in all categories.

**Table 3** Recognition rate of local pattern descriptor on ORL database

| Method | Recognition rate |
|---|---|
| LBP | 87.80 |
| LDiP | 88.50 |
| LDeP | 89.25 |
| LDN(K) | 90.30 |
| LDN(G) | 92.39 |
| LDGP | 97.75 |
| DR-LDP | 97.62 |
| TL-DR-LDP | 98.45 |

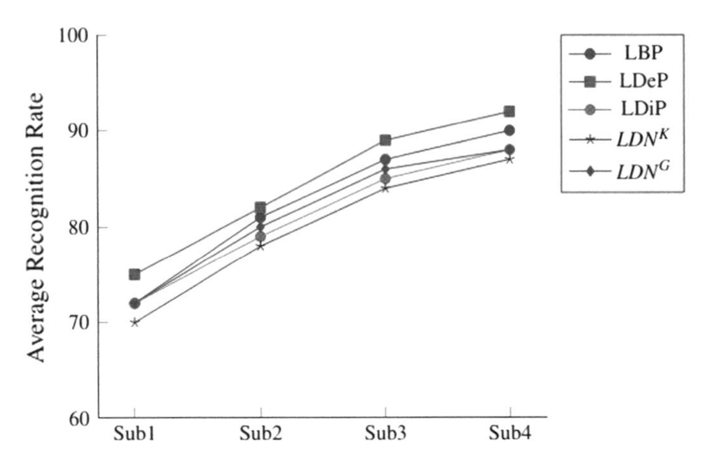


**Fig. 13** Recognition rate of local pattern descriptors on LFW database

## 5.5 *CAS-PEAL Database*

CAS-PEAL [77] is a facial expression database. It consists of 30,863 images of 1040 subjects (595 males and 445 females) with different expressions, pose, accessory, lighting, time, background, and distance. Normal images in 1040 subjects are labeled as gallery image set, and rest are in probe image set. The local pattern descriptor provides good results in different environments. Table 4 shows the recognition rate of local pattern descriptors in CAS-PEAL database. As can be seen in Table 4, the LDNP with Gaussian mask has a higher recognition rate compared with the other local pattern descriptors [61].

## 5.6 *Cohn–Kanade Expression Database*

The Cohn–Kannade (CK) databases [76] contain 100 university students. The students are all in the age of 18–30. Fifteen of the students were African American, 65% were female, and 3% were Asian or Latino. Each subject consists of six

**Table 4** Recognition rate of local pattern descriptors on CAS-PEAL database

| Methods | Accessory | Age | Background | Distance | Expression | Lighting |
|---|---|---|---|---|---|---|
| LBP | 75.06 | 89.39 | 98.73 | 97.45 | 87.45 | 14.62 |
| LDP | 78.21 | 90.91 | 99.64 | 96.73 | 87.58 | 17.83 |
| LDeP | 79.15 | 87.81 | 97.84 | 95.43 | 86.38 | 32.83 |
| LDN(K) | 80.00 | 95.45 | 97.83 | 98.18 | 84.20 | 27.99 |
| LDN(G) | 82.14 | 93.94 | 99.46 | 97.09 | 87.26 | 39.81 |

**Table 5** Recognition rate of local pattern descriptors on CK database

| Methods | Six-class | | | Seven-class | | |
|---|---|---|---|---|---|---|
| | Linear | Polynomial | RBF | Linear | Polynomial | RBF |
| Gabor [80] | 89.4 ± 3.0 | 89.4 ± 3.0 | 89.8 ± 3.1 | 86.6 ± 4.1 | 86.6 ± 4.1 | 89.8 ± 3.6 |
| LBP [81] | 91.5 ± 3.1 | 91.5 ± 3.1 | 92.6 ± 2.9 | 88.1 ± 3.8 | 88.1 ± 3.8 | 88.9 ± 3.5 |
| LDP [82] | 94.9 ± 1.2 | 94.9 ± 1.2 | 98.5 ± 1.4 | 92.8 ± 1.7 | 92.8 ± 1.7 | 94.3 ± 3.9 |
| LDN [61] | 98.4 ± 1.4 | 99.1 ± 0.7 | 99.2 ± 0.8 | 92.3 ± 3.0 | 95.1 ± 4.1 | 94.8 ± 3.1 |
| DR-LDP | 96.4 ± 1.3 | 97.2 ± 1.3 | 98.7 ± 0.9 | 91.5 ± 1.7 | 91.9 ± 1.7 | 93.4 ± 2.9 |

prototypic emotions of the person with 23 facial expressions. Image sequences from neutral to target display were digitized into different scales like 640 × 480 or 640 × 490. In this paper, we selected 408 image sequences from 96 subjects, each of which was labeled as one of the six basic emotions. For six-class expression, the three most expressive image frames were taken from each image sequence that resulted in 1224 expression images. To build seven-class expression databases, the first frame (neutral expression) from all 408 image sequences was selected. The resulted set consists of 1632 images (Table 5).

## 5.7 *JAFFE Database*

The Japanese Female Facial Expression (JAFFE) database [75] consists of 213 images of females using 10 subjects. Each image has a size of 256 × 256 with almost the same number of images for each categories of expression. Usually, the database consists of frontal pose image and their hair was tied back to expose all the expressive areas of the face (Table 6).

From the results, we noticed that the recognition rate of JAFFE database is relatively lesser than that of CK database because in JAFFE database the expression is mapped incorrectly or expressed inaccurately. Those samples are used for training or testing, and the recognition rate may affect.

**Table 6** Recognition rate of local pattern descriptors on JAFFE database

| Methods | Six-class | | | Seven-class | | |
|---|---|---|---|---|---|---|
| | Linear | Polynomial | RBF | Linear | Polynomial | RBF |
| Gabor [80] | 85.1 ± 5.0 | 85.1 ± 5.0 | 85.8 ± 4.1 | 79.7 ± 4.2 | 79.7 ± 4.2 | 80.8 ± 3.7 |
| LBP [81] | 86.7 ± 4.1 | 86.7 ± 4.1 | 87.5 ± 5.1 | 80.7 ± 5.5 | 80.7 ± 5.5 | 81.9 ± 5.2 |
| LDP [82] | 89.9 ± 5.2 | 89.9 ± 5.2 | 90.1 ± 4.9 | 84.9 ± 4.7 | 94.9 ± 4.7 | 85.4 ± 4.0 |
| LDN [61] | 92.9 ± 1.7 | 93.4 ± 2.2 | 92.3 ± 1.7 | 90.1 ± 3.0 | 91.1 ± 4.1 | 89.2 ± 3.1 |
| DR-LDP | 91.5 ± 2.1 | 92.4 ± 1.5 | 93.3 ± 1.2 | 91.5 ± 3.7 | 91.9 ± 3.5 | 92.1 ± 3.9 |

# 6 Conclusion

Nowadays, the face recognition realized substantial attention in biometrics, and it is challenging in constrained and unconstrained environment. Researchers believed that the local pattern matching is more suitable for face recognition. The local pattern extracts the significant facial features with different rotations and scaling than holistic approaches. In this chapter, the local pattern descriptors are experimented on several standard benchmark databases to determine the best method. The performance rate of each method is estimated on standard benchmark databases such as FERET, Extended Yale-B, ORL, CAS-PEAL, LFW, JAFFE, and Cohn–Kanade. The local pattern descriptors are more robust to random noise, pose, illumination, aging, and facial expressions. The experimental results show that the local pattern descriptors attained promised performance in face recognition.

# References

1. Jain, A. K., Flynn, P., & Ross, A. A. (2007). *Handbook of biometrics*. Berlin: Springer.
2. Jafri, R., & Arabnia, H. R. (2009). A survey of face recognition techniques. *Journal of Information Processing Systems, 5*(2), 41–68.
3. Galton, F. (1889). Personal identification and description. *Journal of Anthropological Institute of Great Britain and Ireland,* 177–191.
4. Jain, A. K., & Li, S. Z. (2005). *Handbook of face recognition*, vol. 1. Berlin: Springer.
5. Hatem, H., Beiji, Z., & Majeed, R. (2015). A survey of feature base methods for human face detection. *International Journal of Control and Automation, 8*(5), 61–78.
6. Zhao, W., Chellappa, R., Phillips, P. J., & Rosenfeld, A. (2003). Face recognition: A literature survey. *ACM computing surveys (CSUR), 35*(4), 399–458.
7. Biederman, I., & Kalocsai, P. (1998). Neural and psychophysical analysis of object and face recognition. *NATO ASI Series F Computer and Systems Sciences, 163,* 3–25.
8. Ellis, H. (1986). Introduction to aspects of face processing: Ten questions in need of answers. In *Aspects of face processing*, pp. 3–13. Berlin: Springer.
9. Bruce, V., Hancock, P. J., & Burton, A. M. (1998). Human face perception and identification. In: *Face recognition*, pp. 51–72. Springer.
10. Sagiv, N., & Bentin, S. (2001). Structural encoding of human and schematic faces: Holistic and part based processes. *Journal of Cognitive Neuroscience, 13*(7), 937–951.
11. Brunelli, R., & Poggio, T. (1993). Face recognition: Features versus templates. *IEEE Transactions on Pattern Analysis & Machine Intelligence,* 1042–1052.
12. Baron, R. J. (1981). Mechanisms of human facial recognition. *International Journal Man-Machine Studies, 15*(2), 137–178.
13. Huang, R. J. J. (1998). *Detection strategies for face recognition using learning and evolution*. Ph.D. thesis, George Mason University (1998).
14. Sirovich, L., & Kirby, M. (1987). Low-dimensional procedure for the characterization of human faces. *JOSA A, 4*(3), 519–524.
15. Turk, M., & Pentland, A. (1991). Eigenfaces for recognition. *Journal of Cognitive Neuroscience, 3*(1), 71–86.
16. Belhumeur, P. N., Hespanha, J. P., & Kriegman, D. J. (1996). Eigenfaces vs. fisherfaces: Recognition using class specific linear projection. In *Computer Vision (ECCV'96)*, pp. 43–58. Berlin: Springer.

17. Bartlett, M. S., Movellan, J. R., & Sejnowski, T. J. (2002). Face recognition by independent component analysis. *IEEE Transactions on Neural Networks, 13*(6), 1450–1464.
18. Yang, J., Zhang, D., Frangi, A. F., & Yang, J. (2004). Two-dimensional PCA: A new approach to appearance based face representation and recognition. *IEEE Transactions on Pattern Analysis and Machine Intelligence, 26*(1), 131–137.
19. Tan, K., & Chen, S. (2005). Adaptively weighted sub-pattern PCA for face recognition. *Neuro computing, 64,* 505–511.
20. Nhat, V. D. M., & Lee, S. (2005) An improvement on LDA algorithm for face recognition. In: *Advances in neural networks*–ISNN 2005, pp. 1016–1021. Berlin: Springer.
21. Sun, N., Wang, H., Ji, Z., Zou, C., & Zhao, L. (2008). An efficient algorithm for kernel two dimensional principal component analysis. *Neural Computing and Applications, 17*(1), 59–64.
22. Zhou, D., & Yang, X. (2004). Face recognition using direct-weighted LDA. In*: PRICAI 2004: Trends in Artificial Intelligence*, pp. 760–768. Berlin: Springer.
23. Howland, P., & Park, H. (2004). Generalizing discriminant analysis using the generalized singular value decomposition. *IEEE Transactions on Pattern Analysis and Machine Intelligence, 26*(8), 995–1006.
24. Zhang, W., Shan, S., Gao, W., Chang, Y., & Cao, B. (2005). Component-based cascade linear discriminant analysis for face recognition. In *Advances in biometric person authentication*, pp. 288–295. Berlin: Springer.
25. Liu, Q., Lu, H., & Ma, S. (2004). Improving kernel fisher discriminant analysis for face recognition. *IEEE Transactions on Circuits and Systems for Video Technology, 14*(1), 42–49.
26. Pang, Y., Zhang, L., Li, M., Liu, Z., & Ma, W. (2005). A novel Gabor-LDA based face recognition method. In *Advances in multimedia information processing-PCM 2004*, pp. 352–358. Berlin: Springer.
27. Jing, X. Y., Tang, Y. Y., & Zhang, D. (2005). A fourier–LDA approach for image recognition. *Pattern Recognition, 38*(3), 453–457.
28. DeMers, D., Cottrell, G., & et al. (1993). Non-linear dimensionality reduction. In *Advances in neural information processing systems*, pp. 580–580.
29. Weng, J. J., Ahuja, N., & Huang, T. S. (1993). Learning recognition and segmentation of 3-D objects from 2-D images. In *Fourth International Conference on Computer Vision, 1993. Proceedings*, pp. 121–128. IEEE.
30. Lawrence, S., Giles, C. L., Tsoi, A. C., & Back, A. D. (1997). Face recognition: A convolutional neural network approach. *IEEE Transactions on Neural Networks, 8*(1), 98–113.
31. Kohonen, T. (1998). The self-organizing map. *Neurocomputing, 21*(1), 1–6.
32. Eleyan, A., & Demirel, H. (2005). Face recognition system based on PCA and feedforward neural networks. In: *Computational intelligence and bioinspired systems*, pp. 935–942. Berlin: Springer.
33. Li, Z., & Tang, S. (2002). Face recognition using improved pairwise coupling support vector machines. In *Proceedings of the 9th International Conference on Neural Information Processing*, ICONIP'02, vol. 2, pp. 876–880. IEEE.
34. Samaria, F. S. (1994). *Face recognition using hidden Markov models*. Ph.D. thesis, University of Cambridge.
35. Li, B., & Yin, H. (2005). Face recognition using rbf neural networks and wavelet transform. In *Advances in neural networks,* ISNN 2005, pp. 105–111. Berlin: Springer.
36. Beumier, C., & Acheroy, M. (2000). Automatic 3D face authentication. *Image and Vision Computing, 18*(4), 315–321.
37. Yang, Q., & Tang, X. (2005). Recent advances in subspace analysis for face recognition. In *Advances in biometric person authentication*, pp. 275–287. Berlin: Springer.
38. Kanade, T. (1973). *Picture processing system by computer complex and recognition of human faces.* Ph.D. thesis, Kyoto University.
39. Wiskott, L., & Von Der Malsburg, C. (1996). Recognizing faces by dynamic link matching. *Neuro Image, 4*(3), S14–S18.

40. Heisele, B., Serre, T., & Poggio, T. (2007). A component-based framework for face detection and identification. *International Journal of Computer Vision, 74*(2), 167–181.
41. Bonnen, K., Klare, B. F., & Jain, A. K. (2013). Component-based representation in automated face recognition. *IEEE Transactions on Information Forensics and Security, 8*(1), 239–253.
42. Ojala, T., Pietikainen, M., & Harwood, D. (1996). A comparative study of texture measures with classification based on featured distributions. *Pattern Recognition, 29*(1), 51–59.
43. Ojala, T., Pietikainen, M., & Maenpaa, T. (2002). Multiresolution gray-scale and rotation invariant texture classification with local binary patterns. *IEEE Transactions on Pattern Analysis and Machine Intelligence, 24*(7), 971–987.
44. Ahonen, T., Hadid, A., & Pietikainen, M. (2006). Face description with local binary patterns: Application to face recognition. *IEEE Transactions on Pattern Analysis and Machine Intelligence, 28*(12), 2037–2041.
45. Huang, D., Shan, C., Ardabilian, M., Wang, Y., & Chen, L. (2011). Local binary patterns and its application to facial image analysis: A survey. *IEEE Transactions on Systems, Man, and Cybernetics, Part C: Applications and Reviews, 41*(6), 765–781.
46. Jin, H., Liu, Q., Lu, H., & Tong, X. (2004) Face detection using improved LBP under bayesian framework. In *2004 IEEE First Symposium on Multi-agent security and survivability*, pp. 306–309.
47. Guo, Z., Zhang, L., & Zhang, D. (2010). A completed modeling of local binary pattern operator for texture classification. *IEEE Transactions on Image Processing, 19*(6), 1657–1663.
48. Yang, H., & Wang, Y. (2007). A LBP-based face recognition method with hamming distance constraint. In *Fourth International Conference on Image and Graphics, ICIG 2007*, pp. 645–649. IEEE.
49. Huang, D., Wang, Y., & Wang, Y. (2007). A robust method for near infrared face recognition based on extended local binary pattern. In *Advances in visual computing*, pp. 437–446. Berlin: Springer.
50. Huang, Y., Wang, Y., & Tan, T. (2006). Combining statistics of geometrical and correlative features for 3D face recognition. In *BMVC*, pp. 879–888. Citeseer.
51. Tan, X., & Triggs, B. (2010). Enhanced local texture feature sets for face recognition under difficult lighting conditions. *IEEE Transactions on Image Processing, 19*(6), 1635–1650.
52. Liao, S., Zhu, X., Lei, Z., Zhang, L., & Li, S. Z. (2007). Learning multi-scale block local binary patterns for face recognition. In *Advances in biometrics*, pp. 828–837. Berlin: Springer.
53. Zhang, W., Shan, S., Gao, W., Chen, X., & Zhang, H. (2005). Local Gabor binary pattern histogram sequence (LGBPHS): A novel non-statistical model for face representation and recognition. In *Tenth IEEE International Conference on Computer Vision*, pp. 786–791. IEEE.
54. Heikkila, M., Pietikäinen, M., & Schmid, C. (2009). Description of interest regions with local binary patterns. *Pattern Recognition, 42*(3), 425–436.
55. Zhang, B., Gao, Y., Zhao, S., & Liu, J. (2010). Local derivative pattern versus local binary pattern: Face recognition with high-order local pattern descriptor. *IEEE Transactions on Image Processing, 19*(2), 533–544.
56. Jabid, T., Kabir, M. H., & Chae, O. (2010). Local directional pattern (LDP) for face recognition. In *2010 Digest of technical papers international conference on consumer electronics (ICCE)*.
57. Kabir, M. H., Jabid, T., & Chae, O. (2012). Local directional pattern variance (LDPv): A robust feature descriptor for facial expression recognition. *The International Arab Journal of Information Technology, 9*(4), 382–391.
58. Zhong, F., & Zhang, J. (2013). Face recognition with enhanced local directional patterns. *Neuro computing, 119,* 375–384.
59. Castillo, J. A. R., Rivera, A. R., & Chae, O. (2012). Facial expression recognition based on local sign directional pattern. In *19th IEEE International Conference on Image Processing (ICIP)*, pp 2613–2616.

60. Ramirez Rivera, A., Rojas, J., & Chae, O. (2012). Local Gaussian directional pattern for face recognition. In: *21st International Conference on Pattern Recognition (ICPR)*, pp. 1000–1003. IEEE.
61. Ramirez Rivera, A., Rojas Castillo, J., & Chae, O. (2013). Local directional number pattern for face analysis: Face and expression recognition. *IEEE Transactions on Image Processing, 22*(5), 1740–1752.
62. Faraji, M. R., & Qi, X. (2015). Face recognition under illumination variations based on eight local directional patterns. *IET Biometrics, 4*(1), 10–17.
63. Faraji, M. R., & Qi, X. (2014). Face recognition under varying illumination based on adaptive homomorphic eight local directional patterns. *IET Computer Vision, 9*(3), 390–399.
64. Faraji, M. R., & Qi, X. (2016). Face recognition under varying illuminations using logarithmic fractal dimension-based complete eight local directional patterns. *Neurocomputing, 199,* 16–30.
65. Srinivasa Perumal, R., & Chandra Mouli, P. (2016). Dimensionality reduced local directional pattern (DR-LDP) for face recognition. *Expert Systems with Applications, 63,* 66–73.
66. Srinivasa Perumal, R., & Chandra Mouli, P. (2016). Two-level dimensionality reduced local directional pattern for face recognition. *International Journal of Biometrics, 8*(1), 52–64.
67. Ishraque, S. Z., Banna, A. H., & Chae, O. (2012). Local Gabor directional pattern for facial expression recognition. In *15th International Conference on Computer and Information Technology (ICCIT)*, pp. 164–167. IEEE.
68. Liu, J., Jing, X., Sun, S., & Lian, Z. (2015). Local Gabor dominant direction pattern for face recognition. *Chinese Journal of Electronics,* 245–250.
69. Ahmed, F. (2012). Gradient directional pattern: A robust feature descriptor for facial expression recognition. *Electronics Letters, 48*(19), 1203–1204.
70. Chakraborty, S., Singh, S. K., & Chakraborty, P. (2015). Local directional gradient pattern: A local descriptor for face recognition. *Multimedia Tools and Applications,* 1–16.
71. Phillips, P. J., Moon, H., Rizvi, S., & Rauss, P. J. (2000). The feret evaluation methodology for face recognition algorithms. *IEEE Transactions on Pattern Analysis and Machine Intelligence, 22*(10), 1090–1104.
72. Lee, K. C., Ho, J., & Kriegman, D. J. (2005). Acquiring linear subspaces for face recognition under variable lighting. *IEEE Transactions on Pattern Analysis and Machine Intelligence, 27* (5), 684–698.
73. Huang, G. B., Ramesh, M., Berg, T., & Learned-Miller, E. (2007). Labeled faces in the wild: A database for studying face recognition in unconstrained environments. Technical Report 07-49, University of Massachusetts, Amherst.
74. AT&T Laboratories (2002). Database of faces. http://www.cl.cam.ac.uk/research/dtg/attarchive/facedatabase.html.
75. Lyons, M., Akamatsu, S., Kamachi, M., & Gyoba, J. (1998). Coding facial expressions with Gabor wavelets. In *Third IEEE International Conference on Automatic Face and Gesture Recognition, 1998. Proceedings*, pp. 200–205. IEEE.
76. Kanade, T., Cohn, J.F., & Tian, Y. (2000). Comprehensive database for facial expression analysis. In: *Proceedings of Fourth IEEE International Conference on Automatic Face and Gesture Recognition*, pp. 46–53. IEEE.
77. Gao, W., Cao, B., Shan, S., Chen, X., Zhou, D., Zhang, X., et al. (2008). The CAS-PEAL largescale Chinese face database and baseline evaluations. *IEEE Transactions on Systems, Man and Cybernetics, Part A: Systems and Humans, 38*(1), 149–161.
78. Martinez, A. M. (1998). The AR face database. CVC Technical Report 24.
79. Sim, T., Baker, S., & Bsat, M. (2002). The CMU pose, illumination, and expression (PIE) database. In *Fifth IEEE International Conference on Automatic Face and Gesture Recognition, 2002. Proceedings*, pp. 46–51. IEEE.
80. Bartlett, M. S., Littlewort, G., Frank, M., Lainscsek, C., Fasel, I., & Movellan, J. (2005). Recognizing facial expression: machine learning and application to spontaneous behavior. In *IEEE Computer Society Conference on Computer Vision and Pattern Recognition (CVPR'05)*, IEEE, Issue 2, pp. 568–573.

81. Shan, C., Gong, S., & McOwan, P. W. (2009). Facial expression recognition based on local binary patterns: A comprehensive study. *Image and Vision Computing, 27*(6), 803–816.
82. Jabid, T., Kabir, M. H., & Chae, O. (2010). Robust facial expression recognition based on local directional pattern. *ETRI Journal, 32*(5), 784–794.

# Part II
# Knowledge Engineering Process in Information Management

# Agent-Based Architecture for Developing Recommender System in Libraries

**Snehalata B. Shirude and Satish R. Kolhe**

**Abstract** The volume of data in terms of library resources is big and continuously increasing, therefore manipulation, searching, and suggesting these resources to user have become very difficult. The automation in libraries along with intelligence can solve the problem by easy and correct recommendation system. The intelligence can be incorporated into the system by implementing agent-based recommender system for libraries. An efficient intelligent agent-based library recommender system is developed to generate suggestions to end user by understanding his/her interest automatically by learning his/her profile. The manual recommendation process includes understanding the need of the user, solutions available; decision has to be made about which solution(s) can satisfy the user's need and finally provide the solution(s) by filtering. The needs of user and available solutions are adaptive. It is clear that recommendation process involves the task of decision making and performing action based upon perception of recommender. The term software agent can be referred as a self-contained program capable of controlling its own decision making and acting, based on its perception of its environment, in pursuit of one or more objectives (Jennings and Wooldridge in IEE Rev 42(1):17–20, [1]). Therefore the framework of recommender system is designed using intelligent agent. The framework consists of agent, viz. profile agent, content-based agent, and collaborative agent. Profile agent automatically extracts the input from profile of user. Users may also have tried to search some resources. The search query and the results of the search are used by profile agent while collecting more information about active user. The task of identification of subject area(s) of active user is performed by profile agent. Then profile agent updates the user profile. Content-based agent and collaborative agent perform the main tasks of filtering and providing recommendations. In general, the perfect simulation of the natural recommendation process that human being performs to recommend to someone can provide better results. In this scenario, the main problem to the researchers is to think about the framework which can provide such perfect simulation. The main objective of the proposed architecture is to develop an

S. B. Shirude (✉) · S. R. Kolhe
School of Computer Sciences, North Maharashtra University, Jalgaon, India
e-mail: snehalata.shirude@gmail.com



S. Margret Anouncia and U. K. Wiil (eds.), *Knowledge Computing and its Applications*, https://doi.org/10.1007/978-981-10-8258-0_8

agent-based recommender system for providing effective and intelligent use of library resources such as finding right book(s), relevant research journal papers and articles. The various issues with the dataset and library recommender system are identified. There are many recommendation domain and applications where content and metadata play a key role can be seen in the literature. In domains such as movies, the relationship between content and usage data has seen thorough investigation done already but for many other domains such as books, news, scientific articles, and Web pages, it is not still known if and how these data sources should be combined to provide the best recommendation performance. This has motivated to develop recommender system using library domain specifically. Some of the datasets such as Book-Crossing dataset, Techlens + dataset, ACL anthology reference corpus (Bird in The ACL anthology reference corpus: a reference dataset for bibliographic research in computational linguistics [2], Torres et al. in Proceedings of the 4th ACM/IEEE-CS joint conference on digital libraries, pp. 228–236 [3], Ziegler in current trends in database technology-EDBT 2004 workshops [4]) are available. They are not very useful since only the use of title of the book is not sufficient to improve the performance (Ziegler in current trends in database technology-EDBT 2004 workshops, pp. 78–89 [4]), not made publically available (Torres et al. in Proceedings of the 4th ACM/IEEE-CS joint conference on digital libraries, pp. 228–236 [3]), limited domain (NLP only) (Bird in The ACL anthology reference corpus: a reference dataset for bibliographic research in computational linguistics [2]). Since no benchmarking dataset is suitable, the proposed research contributes by creating dataset for library recommender systems. Some of the library resources are added from library of congress using Z39.50 protocol. Others are from catalogues of various publishers', viz. PHI learning, Laxmi, Pearson education, Cambridge, and McGraw Hill publications. User profiles, Library Resources, and Knowledge base (ACM Computing Classification System 2012 used as Ontology) are important components of dataset. Profile agent is a goal-based agent which auto extracts more information about user and updates profile by removing noise. The results for the task of identification of user's interested subject area(s) are improved using n-Gram and Jaccard's measure (Kern et al. recommending scientific literature: Comparing use-cases and algorithms [5]). The comparison made with the similar work in the literature gives that 'implicit interest extraction of the user,' 'implicit feedback mechanism,' and 'weighted profiles updation using semantic concepts in ACM CCS 2012' are novel factors of profile agent (Kim and Fox in Digital libraries: international collaboration and cross-fertilization. Springer, Berlin, pp. 533–542 [6]). Library recommender agent is utility-based agent implemented using hybrid approach (combining results of content based and collaborative agent). Agent performs tasks such as classifying library resources, filtering and providing recommendations. Library resources are classified among the category in ACM CCS 2012. The task of classification is well performed using PU learning (Naïve Bayes classifier) since no negative records exist in any real-world library (Wang and Blei in Proceedings of the 17th ACM SIGKDD international conference on knowledge discovery and data mining, pp. 448–456 [7]). Filtering library resources require to measure similarity between user profile and library resources. It is found that cosine distance outperformed with other measures such as Euclidean, Manhattan, and Chebyshev. Singular value decomposition technique helped to reduce the final computations needs to perform while providing recommendations. The ability to provide semantically related library resources by retrieval of related concepts in ACM CCS 2012 and weight assignment have encouraged the performance of library recommender agent (Morales-del-Castillo et al. in Int J Technol Enhanced Learn 2(3):227–240 [8]). The results of the library recommender system are evaluated using precision, recall, and F1. This proposed agent-based library recommender system improved performance up to the precision 81.66, recall 83.19, and F1 82.28 percentages (Morales-del-Castillo et al. in Int J Technol Enhanced Learn 2(3):227–240 [8], Porcel et al. in international conference on education and new learning technologies [9], Tejeda-Lorente et al. in Procedia Comput Sci 31:1036–1043 [10]). The research concludes that the discovery of the user interest and retrieval of semantically related recommendations are the important factors on which overall performance of recommender system depends. Integration of intelligent agent-based recommender system in existing library system is focused as the future work.

**Keywords** Library recommender system · Intelligent agent · Cosine similarity
PU learning · Naïve Bayes · Jaccard's measure · N-gram · Singular value decomposition · Weighed profiles · Ranked concepts · Semantic similarity
Implicit feedback · Profile generation · Profile exploitation · Content-based filtering · Collaborative filtering

## 1 Introduction

Digital libraries are designed to fulfill need of information of specific type of users. Accordingly, one can observe kinds of libraries such as historical digital libraries, educational digital libraries, research digital libraries, cultural digital libraries. Though there are specific libraries available, they create individual information overload due to their broadness. Digital library personalization can give solution to this problem [11]. This solution is in the form of some tools and techniques which can provide suggestions for relevant library resources to digital library user. These tools and techniques are referred as recommender system for digital library [12].

Large volume of information is one of the major problems for academic library users to obtain the relevant library the relevant library resources in libraries. Development of recommender system is the solution to solve this problem. Most of the work related to generating recommendations is found in e-commerce, music, tourism, social networks, television, and news area. Ready datasets are available for these areas [13]. After the literature study, it is found that no benchmark datasets are available for research in library field. The study gives the idea of the approaches and techniques used for the development of recommender system in the field of e-commerce, music, tourism, social networks, television, and news. There is lack of efficient recommender system for libraries [14, 15]. Therefore, there is necessity of developing a recommender system for digital library. There is possibility of improvement in recommendations for library users. This is the main motivation behind proposing the recommender system using intelligent agent for the efficient use of libraries. The agent-based architecture is used to design the framework of recommender system. The taxonomy of general recommender system includes two main phases such as profile generation and maintenance and profile exploitation. Profile generation and maintenance include steps such as user profile representation, initial profile generation, profile learning technique, relevance feedback. Profile exploitation includes steps such as information filtering method, user profile–item matching technique, user profile matching technique, and profile adaptation technique [13]. The manual recommendation process includes understanding the need of the user, solutions available, decision has to be made about which solution(s) can satisfy the user's need and finally provide the solution(s) by filtering. The needs of user and available solutions are adaptive. It is clear that recommendation process involves the task of decision making and performing action based upon perception of recommender. The software agent is a program which is capable of decision making and perceives the environment [1]. In this work, the prototype of a library recommender system is proposed with the objective to provide effective and intelligent use of library resources such as finding right book(s), relevant research journal papers and articles using approaches like collaborative filtering, content-based filtering, opinion-based filtering. The proposed framework has the agent-based architecture. The agent-based architecture is most suitable because the recommendation process involves the task of decision making and performing action based upon the perception of recommender. The architecture satisfying the objective involves the implementation of profile agent and library recommender agent (content based and collaborative agent). The role of the agents is clearly defined via Performance Measure Environment Actuators Sensors (PEAS) measures and the task environment.

## 2 Related Work

Related work explores the present state of the work in the field of developing recommender system for library resources. The literature is viewed from different angles such as approach used, technique applied for different tasks, viz. similarity computation, classification, etc., method for relevance feedback and performance of the system. BipTip, ExLibris bX, Techlens, Foxtrot, Fab, LIBRA are some of the existing solutions to the field [16]. The goal of CBRecSys2014, which is workshop on new trends in content-based recommender systems, was organized with the goal to provide the platform for the papers dedicated to all aspects and new trends in the development of recommender system. There are many recommendation domains and applications where content and metadata play a key role. The relationship between content and usage data has seen through investigation in domains such as movies but for many other domains, such as books, news, scientific articles, and Web pages, it is not still known if and how these data sources should be combined to provide the best recommendation performance. This is motivating to the researchers in the field of recommender system using library domain specifically [17].

Software agents are used in diverse fields [1, 18–20]. J. M. Morales-del-Castillo, E. Peis, and E. Herrera-Viedma have presented a multi-agent and recommender system prototype for digital libraries designed to be used by the e-scholars community [8]. The conceptual framework of a multi-agent-based recommender system is proposed by U. Pakdeetrakulwong, and P. Wongthongtham, This framework provides active support to access and utilize knowledge and project information in the software engineering ontology [21]. The framework of argumentation enabled interest-based personalized recommender system using information management agent, user agent, and recommender agent is proposed. Fuzzy c-means algorithm is used to determine other similar books [22].

The study for datasets in the domain of library field is performed. Book-Crossing dataset consists of three tables for users, books, and ratings. Book-Title, Book-Author, Year-of-Publication, and Publisher are the fields of table for books. Ratings are expressed on a scale from 1 to 10 in the table for ratings. The title of the book provided in table for books is not sufficient to generate better recommendations [4]. In Techlens+, a dataset with papers extracted from CiteSeer is created for performing the experiments. The dataset created is not made publically available [3]. Some other systems found in the literature do not provide the dataset likewise. The ACL anthology reference corpus is available as a reference dataset for bibliographic research in computational linguistics. It is the digital archive of conference and journal papers in natural language processing and computational linguistics. The corpus provides very limited domain of NLP and computational linguistic only [2]. The structure of the ACL ARC dataset is referred while adding the conference and journal papers in dataset constructed. The literature survey for datasets suggests no such benchmarking dataset is available. This puts need of creation of such dataset for developing library recommender system. The review of the recommender system in library domain is summarized and given in Table 1.

Since the recommendation process involves the task of decision making and performing action based upon perception of the recommender, the agent-based approach is the most suitable in the development. Though there are some previous works identified in field of recommender system development for library, performance improvement is possible. The most identified difficulty is lack of existing datasets with rich data about library resources with table of contents, index, abstract, keywords, etc. The improvement in performance of filtering is possible by considering semantically related keywords and weight assignment, use of ontology, and richer dataset of library resources.

**Table 1** Summary of review performed

| Authors | Semantic similarity | Use of ontology | Feedback from users | Findings |
|---|---|---|---|---|
| Bedi and Vashisth [22] | Semantically related terms are not identified | No use of ontology | Feedback is achieved via comments the user is given | The 84% users have given positive response to the use of system. The value of *N* in top-N is set as 5 |
| Tejeda-Lorente et al. [10] | Semantics are defined using fuzzy subsets | No use of ontology | Rating criteria are defined in feedback phase | System reaches to precision value 69.13 |
| Cao et al. [30] | Semantically related terms are not identified | No use of ontology | Feedback is explicitly taken as input | GMM-SVM proved better for classification of feedback from users |
| Kern et al. [5] | Semantically related terms are not identified | No use of ontology | Feedback mechanism is not described | System achieves the performance up to the precision 70%, Citation network, usage clicks can improve results |
| Pakdeetrakulwong and Wongthongtham [21] | Semantic annotation phase is implemented | Ontology is used | Feedback is suggested in future work | System is developed only for active software engineering ontology |
| Wang and Blei [7] | LDA is used to obtain semantic concepts | No use of ontology | Feedback mechanism is not described | The limits of the themes to which articles are arranged are not specified |
| Sugiyama and Kan [25] | Semantically related terms are not identified | No use of ontology | Feedback mechanism is not described | System is implemented specifically for research papers. Citation to papers is considered to obtain the interest of user |
| Pudota et al. [32] | Semantic annotation phase is implemented | Ontology is used | Relevance feedback is taken from user to evaluate the system performance | Key phrases are extracted by ontology mining. The ontology is specific for software engineering only |
| Porcel et al. [33] | Semantics are computed using fuzzy modeling | No use of ontology | Opinion about the recommended items is asked user to get feedback | System achieves average precision 63.52, recall 67.94, and *f*1 65.05 |

(continued)

**Table 1** (continued)

| Authors | Semantic similarity | Use of ontology | Feedback from users | Findings |
|---|---|---|---|---|
| Kodakateri Pudhiyaveetil et al. [34] | Semantic concepts are retrieved | Use of ACM CCS | Feedback mechanism is not described | System gives idea that first three concepts are useful. It is developed for collection of research papers |
| Morales-del-Castillo et al. [8] | Semantic relations are computed | No use of ontology | Feedback is based on users assessment | System reaches to precision value 50.00 |
| Middleton et al. [31] | Collaborative filtering finds similar users using OntoCoPI | Use of AKT ontology | Feedback mechanism is not described | System reaches to precision value 84.00; the precision is computed for five users |

# 3 Framework of the Recommender System

This section describes the proposed system, architecture of the recommender system implemented, and the dataset used during implementation of the system.

## 3.1 Proposed System

In proposed system architecture, the library user has to register in the system to become the member. System generates initial profile into XML format for each user. Profile agent obtains the interests of logged in user and provides it to the system for updation of the profiles. There are library resources such as journal articles, books, theses into the dataset. The main task of the system is to provide the recommendations to the active user as per the interests of that user. Therefore, by exploiting the active user profile, filtering is performed to generate the recommendations. Content based, collaborative, and hybrid are three approaches for information filtering. In content-based filtering, the interest of user is matched with all library resources in dataset. Collaborative filtering groups the similar user and accordingly the recommendations are provided. The proposed system is using hybrid approach for filtering with content-based and collaborative filtering techniques. Filtering uses similarity measures to find the closeness between the library resources and active user profile. The ranked list of library resources as the recommendations is provided to the active user via user interface.

### *3.2 Architecture of the Recommender System*

The architectural design of the framework presented in Fig. 1 is based on the proposed system given in the previous section. The framework consists of profile agent, content-based agent, and collaborative agent. User, data storage, library resources, and knowledge base are the important components within framework. Profile agent performs main task of identification of interest of active user. Library recommender agent performs the main tasks of filtering and providing recommendations. The recommendation performance of the system is evaluated using precision and recall which in turn gives accuracy of the recommender system.

### *3.3 Dataset*

The dataset includes library resources, user profiles, and use of knowledge base ACM CCS 2012.

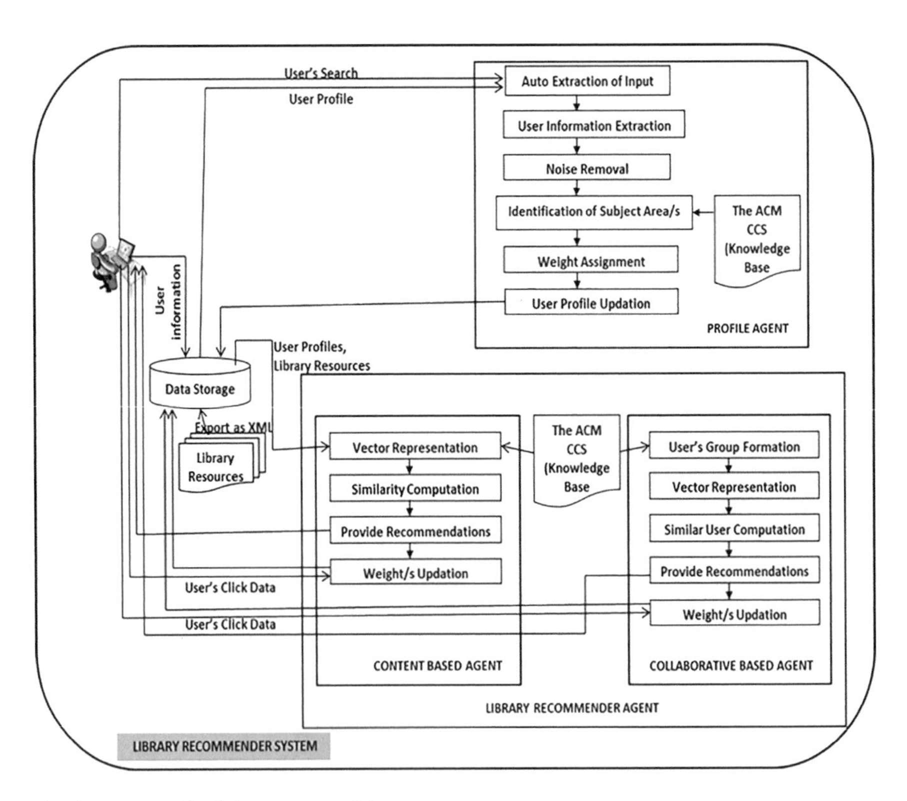


**Fig. 1** Framework of the system architecture

### 3.3.1 Library Resources

Some of the library resources are added into the dataset from Library of Congress using Z39.50 protocol. Library of congress has the collection of recent science and technology books. They relate to the area such as cluster analysis, coding theory, computer networks evaluation, distribution, knowledge management, and business computer programs are downloaded (Library of Congress). These library resources are in MARC format. These are exported to XML format using Koha admin module (Koha). Additions to this, the resources are added to the dataset from catalogues of PHI learning, Pearson education, Cambridge, McGraw Hill publications, etc. One of the important fields of records of these library resources is table of contents. Collection of conference and journal papers from computer science field is added into the dataset in XML form. The records of journal articles contain abstract which is helpful while training the system. There are 705 library resources added into the dataset in last experiments performed.

### 3.3.2 User Profiles

User profiles stored in XML format. The general syntax of user profile in XML format is given below.

*<User ID="“”><FirstName></FirstName><LastName></LastName><MemberType> </MemberType><Department></Department><Course></Course><Subject> </Subject><EmailAddress></EmailAddress><ResearchTopic></ResearchTopic> <PublishedWork></PublishedWork></User>*

### 3.3.3 ACM Computing Classification System 2012

ACM Computing Classification System 2012 is used as knowledge base [23] as no standard ontology available for computer science subject domain. The Simple Knowledge Organization System (SKOS) format is used. This classification scheme has 14 top concepts [23]. Each top concept has broader and narrower concepts. Due to this, the semantically similar concepts extraction is made possible.

## 4 Profile Agent

Figure 1 in Sect. 3.2 gives the architecture of profile agent. The architecture gives the steps of a program which understand the need of the user and decides the subject area(s) in which user has interests. This program makes decision of selection of the relevant keywords and performs action of user profile updation. This is based upon the perception of the program about the interests of user. The program resembles to the notion of intelligent agent; therefore, the program is referred as profile agent. The profile agent automatically extracts the input and updates the user profiles. User gets enrolled in the library recommender system and becomes the user of the system. This process creates a user profile in XML format. Details of interests of the user is collected, noise is removed and updated into user profile.

Library recommender system provides a way to search for a topic by keyword-based retrieval of library resources. The keywords depict the interest of the user. The keywords are added into the user profile after noise removal. Profile agent makes identification of subject area(s). These are the areas in which user have interests. The task is performed using n-Gram and Jaccard's similarity. Use of 2-gram and 3-gram allows grouping two or three consecutive words while computing similarity. The ACM's Computing Classification System is used as knowledge base. Knowledge base has 14 categories containing subcategories. These categories are treated as subject areas. The results are evaluated using precision, recall, and F1. The information related to the user reflects some strong and some weak interests in various area(s). The weights are assigned to the keywords extracted from subject area(s), and profiles are updated accordingly. The agent program which combines some sort of goal information with the information about the results of possible actions in order to choose actions that achieve the goal is referred as the goal-based agent. Goal information describes situations that are desirable. This allows an agent a way to choose among multiple possibilities, selecting the one which reaches a goal state. Resembling to the definition, profile agent performs the action of updating the information into user profile which is done by adding the keywords reflecting about the interests of active user and weight assignment. To reach toward the goal state, agent senses the environment having Web pages containing information about the user and result of search performed by the user. The action of the agent refers to the ACM CCS categories which are the multiple possibilities and profile agent selects similar ones. Therefore, the profile agent is goal-based agent.

## *4.1 PEAS Description*

The task environment specification along with PEAS description for profile agent is given in this subsection. Table 2 gives the summary of PEAS description of the task environment for the profile agent.

The goal of profile agent is to identify the subject area(s) of user in which active user has interests and assigns the weights to the identified subject areas. Table 3 provides the task environment and its characteristics.

The documents giving information about the active user consists of noise (e.g., stop words, html tags). The sensor cannot tell whether the particular keyword is useful to identify the interest of active user. Therefore, the environment is partially

**Table 2** PEAS description of the task environment for profile agent

| Agent | Performance measure | Environment | Actuators | Sensors |
|---|---|---|---|---|
| Profile agent | Correctness in identification of subject area(s) of user and weight assignment | Web pages containing information about user, search result performed by the user, knowledge base | Ability to follow links, fill in forms, display-related information | Ability to parse the Web pages, documents |

**Table 3** Task environment and their characteristics for profile agent

| Task environment | Observable | Deterministic | Episodic | Static | Discrete | Agents |
|---|---|---|---|---|---|---|
| Profile agent | Partially observable | Partly deterministic | Sequential | Semi-dynamic | Discrete | Single |

observable. It is partly deterministic because the system enters into the next state which has identified interest of active user determined by the current state which is knowledge learned about active user via sensing environment and action of updation of profile. The environment is sequential as the further task of generating recommendations is based on the action performed by profile agent. The environment cannot change itself, but it may get changed due to the action performed by the agent. Hence, it is semi-dynamic rather than static or dynamic. The state of the environment is discrete and does not change continuously. The agent is working in a single-agent environment.

## 4.2 *Profile Generation*

The user interface of the system provides a screen to users for entering basic information. The basic information includes name of member, date of birth, member type (teacher/research student/student), name of the department/school, subject, designation, e-mail address. The initial profile of user is generated by adding basic information into XML file. The user enrolls to the system to get the recommendations by filling the library registration form. More information is required about the user. The information is extracted by profile agent using Google APIs.

## 4.3 *Identification of Subject Area/s*

The domain of subject area(s) includes the concepts from ACM Computing Classification System 2012 [23]. The SKOS format of ACM CCS 2012 has 14 top concepts, and each top concept has broader and narrower concepts. The semantically similar concepts are traced with the use of broader and narrower concepts in the hierarchy. Identification of subject area(s) of user is performed using n-gram and Jaccard's similarity. This study and experiment are used to implement collaborative filtering while providing recommendations. Since the user may have interests in multiple subjects or concepts, the classification of user is done into multiple classes. The use of n-gram allows having two and three consecutive words as a single item of vector. Jaccard's measure compares vectors of profile and concepts. The profile agent provides encouraging results which correctly identify the subject areas in which user have interests. The performance of the content-based recommender agent is improved with this work since the semantically related concepts along with the weights are added into the profile. This work also improves the performance of collaborative recommender agent by grouping the users having similar interests [24]. The performance of profile agent depends on the retrieval of more correct keywords of user's interest as well as more richness of keywords into concept hierarchies.

# 5 Library Recommender Agent

Library recommender agent performs the main tasks of filtering and providing recommendations. It uses hybrid approach that combines results generated by two agents: content-based agent and collaborative agent. Recommender agent makes use of user profiles to filter the library resources. Content-based agent reads user profile and generates recommendations by filtering library resources according to the user's interest. The similarity is computed between vectors representing user's interest and library resources. Collaborative agent performs the task of user's group formation according to the interest. Users having similar interests are in one group. It generates recommendations by matching them with similar users. Library recommender agent is utility-based agent. The utility function of the agent is measuring how much the user is satisfied with the provided recommendations.

## *5.1 Content-Based Agent*

Figure 1 given in Sect. 3.2 shows the architecture of content-based agent. Content-based agent reads user profile and library resources. User profiles and library resources are represented as vectors containing term frequencies for every significant keyword [25]. In library resources, most of the book records have table of contents; journal articles include abstract and keywords. This provides rich set of keywords available while computing the similarity. More correct recommendations are provided by the addition of semantically equivalent keywords into the vectors. Ontology is used to add these semantically related keywords. ACM Computing Classification 2012 is used as static ontology with 14 categories [23].

Machine learning is used to identify the category of the library resource. The results of the machine learning experiment performed using different techniques such as rough set rule extraction (Genetic 10 fold, LEM 10 fold, Covering 10 fold), decision tree (11 leaf nodes 10 fold, 500 leaf nodes 10 fold, 1163 leaf nodes 10 fold), k-nearest neighbor (Distance-based 10 fold, Accuracy-based 10 fold, LOOCV leave one out cross validation approach) [26], Naïve Bayes Positive Unlabeled (PU) learning approach are given in Table 5. The identified category provided a way to obtain semantically related keywords for the library resources. The addition of these semantically related keywords into the vectors of respective library resources improved the results of content-based agent. Similarity is computed between user profiles and library resources. Cosine distance, Euclidean distance, Manhattan distance, and Chebyshev distances are used for similarity computing [24]. The size of the generated vectors for similarity computation is very large. Therefore, singular value decomposition (SVD) is used to delete insignificant dimensions. Library resources which are near to user interests are recommended by filtering. User may like or dislike a particular resource. User selects some library resources from the provided ones. This knowledge is used for data weights calculation and updation performed by the content-based agent. The agent is rational utility-based agent which chooses the action so that to maximize the expected utility which measures how much the user is satisfied with the provided recommendations.

### 5.1.1 PEAS Description

Summary of PEAS description [27] of the task environment for the content-based agent is given in Table 4.

The goal of the content-based agent is to generate recommendations by matching interest of user with the contents of library resources. The architectural design of content-based agent is shown in Fig. 1 in Sect. 3.2 which describes the steps performed by the agent to reach toward the goal. To reach the goal, the content-based agent performs subtasks such as identification of category of the library resources, filtering library resources records by measuring similarity between user profiles and library resources, assigning weight values to the library resources. Table 5 provides the task environment and its characteristics.

**Table 4** PEAS description of the task environment for content-based agent

| Agent | Performance measure | Environment | Actuators | Sensors |
|---|---|---|---|---|
| Content-based agent | Correctness of provided recommendations which depends on correct identification of category of library resources, similarity computation, weight assignment | User profile, knowledge base, library resources | Ability to follow the links in user interface and display the information | Ability to parse documents and Web pages |

**Table 5** Task environment and their characteristics for content-based agent

| Task environment | Observable | Deterministic | Episodic | Static | Discrete | Agents |
|---|---|---|---|---|---|---|
| Content based-agent | Partially observable | Deterministic | Sequential | Semi-dynamic | Discrete | Single |

The task environment is partially observable since the library resources, user profiles documents consist of noise in the form of stop words. The sensor cannot tell whether the particular keyword is useful or not in identification of category of library resources. Initially, the current state of the system is such that profile agent has created the profile of the active user and user expects the recommendations. The next state of the system is the agent which has provided recommendations to the user. At any instance of time, the current state of the system can provide the next state into which the system enters and hence environment is deterministic. It is sequential performing actions one after another. The environment is semi-dynamic since it cannot change itself but get changed by the action of agent during weight assignment task. The state of the environment is discrete and does not change continuously. The agent is working in a single-agent environment.

### 5.1.2 Classification of Library Resources

The variety of library resources suggests the need of classifying and grouping them which resembles to the idea of arranging similar library records into the common shelf. The library resources are classified into fourteen categories specified in ACM Computing Classification System 2012. Table 6 provides the summary of the results obtained using various machine learning techniques. It is observed that Naïve Bayes classifier using PU learning approach provides better results.

## *5.2 Collaborative Agent*

Figure 1 given in Sect. 3.2 shows the architecture of collaborative agent. Collaborative agent recommends those library resources to the active user which is provided to the similar user having similar interests. This agent forms the users' group by identifying similar users. The goal of the collaborative agent is to generate recommendations by matching interest of user with the interests of other similar past users. The architectural design of collaborative agent is shown in Fig. 1 which describes the steps performed by the agent to reach toward the goal. To reach the goal, the collaborative agent performs subtasks such as identifying similar users, filtering library resources records by measuring similarity between profiles of similar past users and library resources, assigning weight values to the library resources. Performance measure of this agent depends on correct identification of

**Table 6** Summary of results for classification of library resources

| Machine learning technique | Precision (%) | Recall (%) | *F*1 (%) | Coverage (%) |
|---|---|---|---|---|
| Rough Set Rule Extraction (Genetic 10 Fold) | 71.96 | 59.09 | 64.72 | 100 |
| Rough Set Rule Extraction (LEM 10 Fold) | 82.76 | 71.80 | 76.90 | 57.80 |
| Rough Set Rule Extraction (Covering 10 Fold) | 71.88 | 58.30 | 64.38 | 16.10 |
| Decision Tree (11 Leaf Nodes 10 Fold) | 88.05 | 76.90 | 82.10 | 46.10 |
| Decision Tree (100 Leaf Nodes 10 Fold) | 75.43 | 63.80 | 69.13 | 61.70 |
| Decision Tree (500 Leaf Nodes 10 Fold) | 77.89 | 64.50 | 70.57 | 53.90 |
| Decision Tree (1163 Leaf Nodes 10 Fold) | 76.58 | 66.60 | 71.25 | 55.20 |
| K-Nearest Neighbor (Distance-based 10 Fold) | 74.40 | 60.32 | 66.47 | 100 |
| K-Nearest Neighbor (Accuracy-based 10 Fold) | 73.78 | 60.91 | 66.60 | 100 |
| K-Nearest Neighbor (LOOCV Approach) | 65.52 | 65.52 | 65.52 | 100 |
| Naïve Bayes (PU Learning Approach) | 90.05 | 90.05 | 90.05 | 100 |

similar users, similarity computation for filtering library resources, and weight assignment. The correctness is measured using precision, recall, *f*1 measure. Similarity is measured using cosine distance. The agent program assigns weight to the library resources which are recommended by the expert user. The agent works in the environment consisting database of user profiles in XML format, knowledge base [23], library resources in XML format. The agent acts in environment using the interface of the system. While achieving the goal, the agent refers to the user profiles, library resources, past history, click-through data, and feedback gained from provided recommendations. Therefore, the actuators are made able to follow the links in user interface, displaying of information such as provided recommendations. Sensor of the agent has ability to parse the documents and Web pages. These documents are user profiles, library resources, click-through data, and knowledge base. The Web pages showing provided recommendations are parsed by sensor of the agent to update the database. The properties of the task environment and its characteristics are similar to the content-based agent. The task environment is partially observable since the library resources, user profiles documents consist of noise in the form of stop words. The sensor cannot tell whether the particular keyword is useful or not in identification of category of library resources. Initially, the current state of the system is such that profile agent has created the profile of the active user and user expects the recommendations. The next state of the system is the agent has provided recommendations to the user. At any instance of time, the current state of the system can provide the next state into which the system enters and hence environment is deterministic. It is sequential performing actions one after another. The environment is semi-dynamic since it cannot change itself but get changed by the action of agent during weight assignment task. The state of the environment is discrete and does not change continuously. The agent is working in a single-agent environment. Formation of users group is performed with the use of subject area(s) in which user(s) has interest which is identified by the profile agent.

# 6 Library Recommender System

Library recommender system combines the results generated by content-based agent and collaborative agent. This overall approach describing the final recommendations generation to the user is given in this section. The accuracy achieved by the system is discussed in the results section.

## 6.1 Agent-Based Library Recommender System

The architectural design given in Fig. 1 and given in Sect. 3.2 is implemented to develop the library recommender system. The summary of the tasks performed by the profile agent is below.

1. Represent users as the vectors $U_1$, $U_2$, $U_3$, …, $U_N$, where $N$ denotes the total number of users who have completed registration process.
2. Create $N$ number of .xml files to represent the user profile.
3. Acquire the knowledge about each active user.
4. Add keywords from knowledge to user profile after removal of noise and stop words.
5. Represent vector Ui as string of $K$ keywords Ui = (ui1, ui2,…, ui$K$).
6. Represent all top concepts in ACM CCS 2012 using vector Cj = (cj1, cj2, cj3, …, cj$N$) where $N$ denotes the number of keywords exists within the concept $j$.
7. Identify subject areas in which users have interests such as JS ($U1$, $C1$), JS ($U1$, $C2$), JS ($U1$, $C3$),…, JS ($U1$, $C11$) where JS computes similarity between $U1$ and all concepts $C1$, $C2$, $C3$,…, $C11$. Vectors of user and concepts are represented using 2-gram and 3-gram. This is performed for all $U1$, $U2$, $U3$, …, $U_N$.
8. The distance values returned by JS are represented as the matrix JDistance.

$$\text{JDistance} = \begin{bmatrix} \text{JS }(U1, C1) & \cdots & \text{JS }(U1, C11) \\ \vdots & \cdots & \vdots \\ \text{JS }(U_N, C1) & \cdots & \text{JS }(U_N, C11) \end{bmatrix}$$

Each row of the matrix represents the distance values with all top concepts in the hierarchy of ACM CCS 2012 for the particular user.

9. User profile is updated with the addition of semantic concepts by assigning the weights.

The summary of the tasks performed by the content-based library recommender agent is given below.

1. All the library resources are classified into categories defined in ACM CCS 2012 using machine learning techniques.
2. Vector space is generated which consists of unique keywords from all available library resources. The semantically related keywords are added into the vector space with use of library resources classification along with weight assignment referring to the rules used by profile agent.
3. The user for which content-based agent is supposed to generate recommendation is the active user of the system. The vector of active user Ui is represented Ui = (ui1, ui2,…, ui*K*), where ui1, ui2,…, ui*K* are the term frequencies counted referring to all keywords from vector space. Each library resource is also represented as the vector Li = (li1, li2,…, li*K*) same way.
4. Similarity is measured between the active user profile and all the library resources using suitable distance measures. While performing the step from the matrix of the library resources, the insignificant dimensions are deleted using singular value dimensions. Cosine distance is proved giving more correct results. Library resources are arranged categorywise in the decreasing value of similarity.
5. Active user gets recommendations of by retrieving top-*N* library resources from each class. Here the value of *N* varies in a range, maximum up to existing library resources in particular class. The *N* is the number of best library resources recommended by the system. The recommended library resources are listed in the form of links which the user can explore to see more details.
6. Content-based agent assigns weight with value 2 to the library resource if the user explored the link of that resource, otherwise no weight is assigned. The weight values for each library resources are initialized to 1.
7. While regenerating the recommendations, the library resources are arranged categorywise in the decreasing value of similarity as well as weight values. Due to that the library resources having same similarity values are get arranged by the order of weight values.

The recommendations provided by the content-based agent to the particular user are added to the recommendations provided by collaborative agent. Collaborative agent provides recommendations by finding similar users to the active user. The tasks performed by the collaborative agent are listed stepwise below.

1. Collaborative agent refers to the matrix

$$\text{JDistance} = \begin{bmatrix} \text{JS}(U1, C1) & \cdots & \text{JS}(U1, C11) \\ \vdots & \cdots & \vdots \\ \text{JS}\ (U_N, C1) & \cdots & \text{JS}\ (U_N, C11) \end{bmatrix}$$

   generated by the profile agent. In this matrix, each row represents the distance values with all top concepts in the hierarchy of ACM CCS 2012 for the particular user. The distance values are arranged by decreasing values of similarity and maximum first top four values are obtained which give the subject areas in user have more interest. For every user, an array is generated which stores the subject areas of interest of that particular user.
2. System generates array of eleven array lists for each concept in the ACM CCS 2012 such as $\text{GC1}, \text{GC2}, \text{GC3}, \ldots, \text{GC11}$. Each array list is get added by the user who has interest in that subject area.
3. Collaborative agent asks each user to rate the library resources which are provided as the recommendations to the user by library recommender agent. User rates each of the resource as high, middle, or low (represented by 3, 2, or 1). Default value is 1. A function Rating $\text{Value}(U, L)$ is written which returns the rating assigned by the user $U$ to the particular library $L$.
4. Collaborative agent decides ratings for any new user Rating $\text{Value}(\text{NewU}, L)$ by computing the weighted sum of the product of similarity between candidate users (users belonging to the same group) and Rating $\text{Value}(\text{Candidate User}, L)$ which is divided by the total number of candidates users from the particular user group says (GC).

$$\text{Rating Value}(\text{NewU}, L) = \frac{\sum_{\text{CUser} \in \text{GC}} \text{Similarity}(\text{NewU}, \text{CUser}) \times \text{Rating Value}(\text{CUser}, L)}{\text{Total number of CUsers in that group}}$$

   where CUser is CandidateUser. Similarity(NewU, CUser) is calculated using cosine similarity. The similarity value never comes to value zero because NewU and CUser belong to same group GC (GC1, GC2, GC3, GC4….GC11). Rating $\text{Value}(\text{NewU}, L)$ is converted into integer. Similarly, the agent performs for any old user while generating re-recommendations.
5. The library resources are arranged according to the rating values and provided them as the recommendations to the respective user.

Thus, the content-based filter is based on the contents of user profile while collaborative filter performs using ratings of the similar users. The library recommender agent combines the recommendations provided by content-based agent and collaborative agent. It displays the library resource only once by removing duplicates.

## 6.2 Experiment

The framework of proposed agent-based library recommender system is implemented using C# and XML. The user interface of the system provides a screen to users for entering basic information. The basic information includes name of member, date of birth, member type (teacher/research student/student), name of the department/school, subject, designation, e-mail address. The initial profile of user is generated by adding basic information into XML file. The user enrolls to the system to get the recommendations by filling the library registration form. More information is required about the user. The information is extracted by profile agent using Google APIs. Packages Google::Search and REST: Google::Search are used to search more information about the user in World Wide Web over the Internet. The values of the field: name of the member, name of the department/school, subject, and designation are given as the query value to the method Web of Google::Search. The list of Uniform Resource Identifiers (URIs) is obtained as the result. Every file identified by each URI from the list is parsed to get the contents only. The package MyParser from the HTML::Parser module is used to design the parser. For category of user 'postgraduate student,' syllabus is extracted to add more information into the profile. The syllabi are found in PDF format on university website. The curriculum vitae of teachers are also found in PDF format on university Web site. The content within the PDF files is converted into text using package CAM::PDF. The text is giving the more information about the active user. The noise such as stop words, punctuation symbols is removed from the text. The keywords within the text are added to the profile of active user.

Section 6.1 gives algorithm depicting the generation of recommendations by agent-based library recommender system using hybrid approach.

## 6.3 Evaluation Metric

The overall performance of the library recommender system is evaluated for 25 users. The precision, recall, and F1 are widely used measures to evaluate whether a recommender system properly recommends library resources that the user will consider relevant [28]. The confusion matrix is built by computing the values specified in Table 7.

**Table 7** Values to compute in confusion matrix

| | Recommended | Not recommended |
|---|---|---|
| Relevant | Nrr | Nrn |
| Irrelevant | Nir | Nin |

where

Nrr Number of relevant recommended library resources
Nrn Number of relevant not recommended library resources
Nir Number of irrelevant recommended library resources
Nin Number of irrelevant not recommended library resources

$$\text{Precision} = \frac{\text{Nrr}}{\text{Nrr} + \text{Nir}} \quad \text{Recall} = \frac{\text{Nrr}}{\text{Nrr} + \text{Nrn}}$$
$$F1 = \frac{2 \times \text{Precision} \times \text{Recall}}{\text{Recall} + \text{Precision}}$$

## 6.4 Results and Discussion

The performance of the system is evaluated for the three approaches. First the performance of the recommender agent is evaluated using only the cosine similarity measure without referring to any ontology. It is observed that the system reaches to precision of 67.54. Richer user profiles are updated by adding the semantically related keywords. This is performed with the use of ontology. A small ontology is designed for experimentation using the protégé editor which consists of few concepts related to computer science subject [29]. It increases the precision to 73.53. This proved that the system performs well by the addition of semantically related keywords. It is found that ACM Computing Classification System 2012 can be used as the better ontology for computer science field. With the use of ACM CCS 2012, the recommender agent has classified library resources. The profile agent makes use of ACM CCS 2012 for the identification of subject areas in which user has interests. The experiment using ACM CCS 2012 gives the better precision 81.66 among the three approaches. Table 8 gives the summary of the results for generation of recommendations for the sample 25 users using above three approaches.

## 6.5 Comparison of the Results

The performance and features of the library recommender system are compared with the other similar works performed in this area.

The hybrid recommendation strategy is used to provide recommendations in the development of university library recommender system. The system uses the 2-tuple fuzzy linguistic approach. To evaluate the system 200 research resources and 30 users are considered. System obtained 69.13, 66.63, and 67.65 as average values for precision, recall, and $F1$, respectively [10]. Gaussian mixture model (GMM) using probability density function and Gaussian distribution is implemented. The feedback of the users is classified into positive, medium.

**Table 8** Evaluation of 25 users of the library recommender system

| Users | Only cosine similarity approach (%) | | | Ontology (self-created using protégé) (%) | | | Ontology (ACM CCS) (%) | | |
|---|---|---|---|---|---|---|---|---|---|
| | Precision | Recall | *F*1 | Precision | Recall | *F*1 | Precision | Recall | *F*1 |
| User1 | 92.31 | 82.76 | 87.27 | 91.30 | 72.41 | 80.77 | 93.10 | 87.10 | 90.00 |
| User2 | 50.00 | 33.33 | 40.00 | 87.50 | 87.50 | 87.50 | 90.00 | 90.00 | 90.00 |
| User3 | 44.44 | 85.71 | 58.54 | 73.08 | 90.48 | 80.85 | 86.36 | 90.48 | 88.37 |
| User4 | 41.94 | 81.25 | 55.32 | 70.00 | 91.30 | 79.25 | 80.77 | 91.30 | 85.71 |
| User5 | 90.00 | 90.00 | 90.00 | 81.82 | 81.82 | 81.82 | 81.82 | 81.82 | 81.82 |
| User6 | 41.67 | 71.43 | 52.63 | 64.29 | 64.29 | 64.29 | 76.92 | 66.67 | 71.43 |
| User7 | 55.00 | 78.57 | 64.71 | 66.67 | 80.00 | 72.73 | 75.00 | 80.00 | 77.42 |
| User8 | 45.45 | 50.00 | 47.62 | 68.75 | 78.57 | 73.33 | 81.25 | 81.25 | 81.25 |
| User9 | 62.50 | 83.33 | 71.43 | 62.79 | 79.41 | 70.13 | 81.58 | 81.58 | 81.58 |
| User10 | 62.50 | 62.50 | 62.50 | 69.23 | 69.23 | 69.23 | 78.57 | 84.62 | 81.48 |
| User11 | 84.62 | 73.33 | 78.57 | 84.62 | 78.57 | 81.48 | 84.62 | 84.62 | 84.62 |
| User12 | 78.38 | 72.50 | 75.32 | 78.38 | 72.50 | 75.32 | 78.38 | 78.38 | 78.38 |
| User13 | 77.27 | 65.38 | 70.83 | 77.27 | 73.91 | 75.56 | 77.27 | 85.00 | 80.95 |
| User14 | 69.70 | 76.67 | 73.02 | 69.70 | 76.67 | 73.02 | 84.62 | 75.86 | 80.00 |
| User15 | 82.76 | 77.42 | 80.00 | 82.76 | 85.71 | 84.21 | 92.31 | 85.71 | 88.89 |
| User16 | 67.86 | 90.48 | 77.55 | 67.86 | 90.48 | 77.55 | 80.77 | 91.30 | 85.71 |
| User17 | 70.83 | 85.00 | 77.27 | 70.83 | 85.00 | 77.27 | 89.29 | 86.21 | 87.72 |
| User18 | 69.57 | 64.00 | 66.67 | 69.57 | 64.00 | 66.67 | 81.82 | 75.00 | 78.26 |
| User19 | 77.78 | 63.64 | 70.00 | 77.78 | 63.64 | 70.00 | 87.10 | 77.14 | 81.82 |
| User20 | 72.73 | 72.73 | 72.73 | 72.73 | 85.71 | 78.69 | 72.73 | 85.71 | 78.69 |
| User21 | 70.00 | 60.00 | 64.62 | 70.00 | 75.00 | 72.41 | 70.00 | 75.00 | 72.41 |
| User22 | 68.18 | 78.95 | 73.17 | 68.18 | 78.95 | 73.17 | 76.19 | 84.21 | 80.00 |
| User23 | 71.43 | 88.24 | 78.95 | 71.43 | 88.24 | 78.95 | 83.33 | 86.96 | 85.11 |
| User24 | 66.67 | 80.00 | 72.73 | 66.67 | 80.00 | 72.73 | 71.43 | 83.33 | 76.92 |
| User25 | 75.00 | 75.00 | 75.00 | 75.00 | 85.71 | 80.00 | 86.36 | 90.48 | 88.37 |
| Average | 67.54 | 73.69 | 69.46 | 73.53 | 79.16 | 75.88 | 81.66 | 83.19 | 82.28 |

Library records from Hubei University consisting 30 user's marks of 100 books are used. It is observed that GMM-SVM performs better for the classification of feedback of user's data [30]. Content, collaborative, and hybrid approaches are used to generate the recommendations. Stemming, stop-word removal, n-grams, word bi-grams are employed. Research articles from Mendeley's Web site are used in the dataset. The system achieves the performance up to the precision 70%. Citation network, usage clicks can be integrated into the system to improve the performance [5]. Collaborative filtering, probabilistic topic modeling, latent dirichlet allocation (LDA), collaborative topic regression techniques are used. The recommendations are based on content and other user's ratings. Articles are arranged in the themes such as 'biology,' 'physics,' 'statistics'. Research articles (title and abstract) and users' data from CiteUlike and Mendeley are used. It is observed that the limits of

**Table 9** Comparison of the proposed system

| Approach | Precision (%) | Recall (%) | *F*1 (%) |
|---|---|---|---|
| Tejeda-Lorente et al. [10] | 69.13 | 66.63 | 67.65 |
| Morales-del-Castillo et al. [8] | 50.00 | 70.66 | 58.19 |
| Porcel et al. [9] | 63.52 | 67.94 | 65.05 |
| Proposed recommender system | 81.66 | 83.19 | 82.28 |

the themes to which articles are arranged are not specified. The system is evaluated for varying number of recommended articles [7]. Different fuzzy linguistic modeling approaches are applied. The architecture is defined using three software agent (interface, task, and information agents). E-LIS open access repository is extracted to collect the RSS feeds. Sample 12 researches from University of Granada are considered as the users. Archival science consisting of 96 concepts is used as the domain to obtain semantic relations. The prototype system is evaluated which gives average precision, recall, and *f*1 values as 50.00, 70.66, and 58.19, respectively [8]. The Quickstep recommender system is proposed which recommends the online research papers. K-NN is used to classify the research papers. The tool OntoCoPI which uses AKT ontology is used to find the group of people having common interests. System effectively tries to solve the cold start problem using the new user algorithm which given the precision to 84.00 with error rate 55.00. The system is evaluated with five users [31]. Our proposed agent-based library recommender system gives performance with the precision 81.66, recall 83.19, and *F*1 82.28. The comparison of the proposed system in terms of precision, recall, and *F*1 is given in Table 9.

It shows that proposed recommender system improved the accuracy of the recommendations. Middleton et al. [31] achieved the precision to 84.00 by performing collaborative filtering. But the evaluation is performed using five users only. The average accuracy of the recommendations may vary with more number of users.

## 7 Conclusion

The research work provides framework for the library recommender system with the use of intelligent agent. The role of the agents is defined via PEAS measures. Profile agent performed the task of identification of subject area(s) in which user has interests. The results of the task are improved with the use of 2–3 g and Jaccard's measure. Use of 2–3 g allowed combining two or three consecutive words as a single item of vector. Weighed profiles increase the performance of library recommender agent. It is found that cosine similarity measure is better way to compute similarity. The vectors of keywords play important role, hence use of ontology is necessary to add semantically related keywords and it improves the performance.

The variety of library resources suggests need of classifying and grouping them which resembles to the idea of arranging similar library records into the common shelf. The classification of library resources is performed by the classifiers such as PU learning (Naïve Bayes classifier), *k*-nearest neighbor, decision tree, rough set classifier. In all these experiments, PU learning approach performed better. Decision tree and rough set classifier have the low coverage. Rating concept used to obtain relevant feedback improved the results. Hybrid approach combining results of content-based agent and collaborative agent satisfy the need of users by providing right library resources to users. The effectiveness of the recommender system is directly proportional to the ability to identify interest of user and retrieval of recommendations with respect to identified interests.

## References

1. Jennings, N., & Wooldridge, M. (1996). Software agents. *IEE Review, 42*(1), 17–20.
2. Bird, S. (2008). The acl anthology reference corpus: A reference dataset for bibliographic research in computational linguistics.
3. Torres, R., McNee, S. M., Abel, M., Konstan, J. A., & Riedl, J. (2004). Enhancing digital libraries with TechLens+. In *Proceedings of the 4th ACM/IEEE-CS Joint Conference on Digital Libraries* (pp. 228–236). ACM.
4. Ziegler, C. N. (2005). Semantic web recommender systems. In *Current trends in database technology-EDBT 2004 workshops* (pp. 78–89). Berlin: Springer.
5. Kern, R., Jack, K., & Granitzer, M. (2014). Recommending scientific literature: Comparing use-cases and algorithms. arXiv:1409.1357.
6. Kim, S., & Fox, E. A. (2005). Interest-based user grouping model for collaborative filtering in digital libraries. In *Digital libraries: international collaboration and cross-fertilization* (pp. 533–542). Berlin: Springer.
7. Wang, C., & Blei, D. M. (2011). Collaborative topic modeling for recommending scientific articles. In *Proceedings of the 17th ACM SIGKDD International Conference on Knowledge Discovery and Data Mining* (pp. 448–456). ACM.
8. Morales-del-Castillo, J. M., Peis, E., & Herrera-Viedma, E. (2010). A filtering and recommender system for e-scholars. *International Journal of Technology Enhanced Learning, 2*(3), 227–240.
9. Porcel, C., Lizarte, M. J., & Herrera-Viedma, E. (2009). A linguistic recommender system for university digital libraries to help students in their learning processes. In *International Conference on Education and New Learning Technologies* (EDULEARN 09).
10. Tejeda-Lorente, A., Bernabé-Moreno, J., Porcel, C., & Herrera-Viedma, E. (2014). Integrating quality criteria in a fuzzy linguistic recommender system for digital libraries. *Procedia Computer Science, 31,* 1036–1043.
11. Neuhold, E. et al. (2003). Personalization in digital libraries: An extended view. In *Digital Libraries: Technology and Management of Indigenous Knowledge for Global Access: 6th International Conference on Asian Digital Libraries*, ICADL 2003, pp. 1–16. Kuala Lumpur, Malaysia, Dec. 2003. Berlin: Springer. http://dx.doi.org/10.1007/b94517.
12. Ricci, F., Rokach, L., Shapira, B., Kantor, P. B. (Eds.) (2011). *Recommender systems hand-book.* New York: Springer.
13. Montaner, M., López, B., & De La Rosa, J. L. (2003). A taxonomy of recommender agents on the internet. *Artificial Intelligence Review, 19*(4), 285–330.

14. Bobadilla, J., Ortega, F., Hernando, A., & Gutiérrez, A. (2013). Recommender systems survey. *Knowledge-Based Systems, 46,* 109–132.
15. Castro-Schez, J. J., Miguel, R., Vallejo, D., & López-López, L. M. (2011). A highly adaptive recommender system based on fuzzy logic for B2C e-commerce portals. *Expert Systems with Applications, 38*(3), 2441–2454.
16. Gottwald, S., & Koch, T. (2011). Recommender systems for libraries. ACM Recommender Systems 2011 Chicago.
17. Bogers, T., Koolen, M., & Cantador, I. (2014). Workshop on new trends in content-based recommender systems: (CBRecSys 2014). In *Proceedings of the 8th ACM Conference on Recommender Systems* (pp. 379–380). ACM.
18. Mönnich, M., & Spiering, M. (2008). Adding value to the library catalog by implementing a recommendation system. *D-Lib Magazine, 14*(5), 4.
19. Prakash, N. (2004). Intelligent search in digital libraries. 2nd Convention PLANNER, Manipur University, Imphal Copyright INFLIBNET Centre, Ahmedabad.
20. Prakasam, S. (2010). An agent-based intelligent system to enhance e-learning through mining techniques. *International Journal on Computer Science and Engineering.*
21. Pakdeetrakulwong, U., & Wongthongtham, P. (2013). State of the art of a multi-agent based recommender system for active software engineering ontology. *International Journal of Digital Information and Wireless Communications, 3*(4), 29–42.
22. Bedi, P., & Vashisth, P. (2015). Argumentation-enabled interest-based personalised recommender system. *Journal of Experimental & Theoretical Artificial Intelligence, 27*(2), 199–226.
23. https://www.acm.org/about/class/2012/.
24. Shirude, S. B., & Kolhe, S. R. (2014). Measuring similarity between user profile and library book. In *International Conference on Information Systems and Computer Networks* (ISCON), IEEE.
25. Sugiyama, K., & Kan, M. Y. (2010). Scholarly paper recommendation via user's recent research interests. In *Proceedings of the 10th Annual Joint Conference on Digital Libraries* (pp. 29–38). ACM.
26. Shirude S. B. & Kolhe S. R. (2016). Machine learning using k-nearest neighbour for library resources classification in agent based library recommender system. In *Advances in computing and management.* Berlin: Springer.
27. Russell, S. J., & Norvig, P. (1995). *Artificial intelligence: A modern approach.* Upper Saddle River: Prentice-Hall Inc.
28. Gunawardana, A., & Shani, G. (2009). A survey of accuracy evaluation metrics of recommendation tasks. *The Journal of Machine Learning Research, 10,* 2935–2962.
29. Shirude, S. B., & Kolhe, S. R. (2012). A library recommender system using cosine similarity measure and ontology based measure. *Advances in Computational Research, 4*(1), 91–94.
30. Cao, J., Guo, Y., Dong, C., & Liu, P. (2014). Personalized recommendation for digital library using Gaussian mixture model. *Journal of Networks, 9*(10), 2775–2781.
31. Middleton, S. E., Alani, H., & De Roure, D. C. (2002). Exploiting synergy between ontologies and recommender systems. cs/0204012.
32. Pudota, N., Dattolo, A., Baruzzo, A., Ferrara, F., & Tasso, C. (2010). Automatic key phrase extraction and ontology mining for content based tag recommendation. *International Journal of Intelligent Systems, 25*(12), 1158–1186.
33. Porcel, C., Moreno, J. M., & Herrera-Viedma, E. (2009). A multi-disciplinar recommender system to advice research resources in university digital libraries. *Expert Systems with Applications, 36*(10), 12520–12528.
34. Kodakateri Pudhiyaveetil, A., Gauch, S., Luong, H., & Eno, J. (2009). Conceptual recommender system for CiteSeerX. In *Proceedings of the Third ACM Conference on Recommender Systems* (pp. 241–244). ACM.
35. http://koha-community.org.
36. http://www.loc.gov/rr/business/toc/tocsci.html.

37. Neumann, A. W. (2007). Motivating and supporting user interaction with recommender systems (pp. 428–439). Berlin: Springer.
38. Shirude, S. B., & Kolhe, S. R. (2014). Identifying subject area/s of user using n-Gram and Jaccard's similarity in profile agent of library recommender system. In *Proceedings of the 2014 International Conference on Information and Communication Technology for Competitive Strategies* (ICTCS '14). New York: ACM. Article 23, 6 p. http://dx.doi.org/10.1145/2677855.2677878.

# Knowledge Management Using Recommender Systems

**S. S. Sandhu and B. K. Tripathy**

**Abstract** Knowledge is defined as the practical or theoretical comprehension of a subject. It refers to the skills, information, and facts acquired over time through education and/or experience. Knowledge management plays a vital role in the industry today. Knowledge that cannot be shared or communicated with others is mostly redundant and becomes actionable and useful only when shared. Knowledge management refers to a set of processes developed specifically for the purpose of creating, storing, disseminating, and applying knowledge. The idea here is to give an organization the capability to learn from its environment and to incorporate the acquired knowledge into its business processes so as to streamline them and increase their efficiency. With the amount of data/information increasing exponentially, discerning what information is relevant becomes tougher by the day and as a result, knowledge management systems are gaining importance. Recommender systems are a subcategory of information filtering systems. These seek to predict the probability of a user preferring a particular item out of a given set of items. To aid in the knowledge retrieval and dissemination processes of knowledge management systems, the use of intelligent techniques is on the rise. Recommender systems form one such category of intelligent techniques. This chapter presents an overview of the different works done to incorporate recommender systems into the domain of knowledge management. Applications in the scientific, engineering, and industrial knowledge management contexts have been discussed.

S. S. Sandhu (✉) · B. K. Tripathy
School of Computing Science and Engineering, VIT University,
Vellore 632014, Tamil Nadu, India
e-mail: sabhijiit@gmail.com

B. K. Tripathy
e-mail: tripathybk@vit.ac.in





S. Margret Anouncia and U. K. Wiil (eds.), *Knowledge Computing and its Applications*, https://doi.org/10.1007/978-981-10-8258-0_9

# 1 Introduction

The following passages introduce certain keywords concerning the topic of discussion.

## 1.1 Data, Information, and Knowledge

To understand the term "knowledge," it is first important to also understand the terms "data" and "information," and how they all relate to and differ from each other. Laudon and Laudon [1] define data as "a flow of events or transactions captured by an organizations system that by itself is useful for transacting but little else." In other words, data is a collection of facts or figures which can be used for the purpose of making inferences. It is the raw material from which information is obtained. According to Robert M. Losee's "Discipline Independent Definition of Information" [2], information can be envisaged as the "Value attached or instantiated to a characteristic or variable returned by a function or produced by a process." It is the data that, within a framework, has been recorded, classified organized into categories of understanding and related, or interpreted within a framework so that meaning emerges. Lastly, knowledge is interpreting the information available at hand to find patterns, rules, and contexts (Fig. 1).

Knowledge can be classified using many different labels. It can be a cognitive or even a physiological event, that might have an intangible form, taking place inside peoples' heads, or is also available in many tangible forms such as those that are either shared during lectures or stored in records and libraries. Firms also store it in the form of employee know-how and business processes. Knowledge is of two types: One is the tacit knowledge which is undocumented and resides in the minds of employees. The other is explicit knowledge which refers to the documented part. Knowledge is generally believed to be located somewhere in specific business

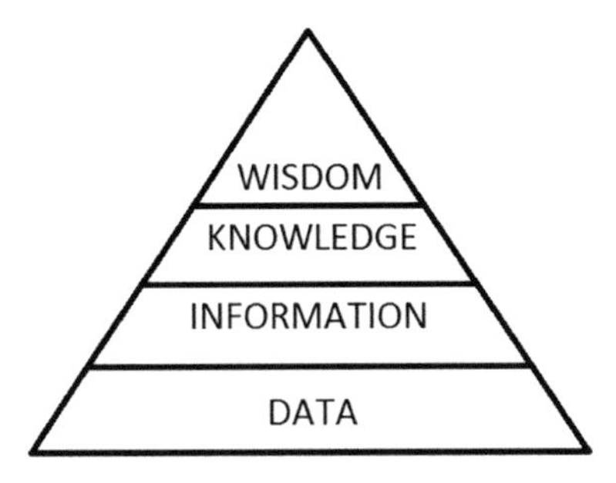


**Fig. 1** DIKW pyramid (also known as the "knowledge hierarchy," is used to represent the functional and structural relationships between data, information, knowledge, and wisdom. It defines information in terms of data and knowledge in terms of information [54])

processes or in human brains. It can also reside in unstructured or structured documents, in voice mail, or e-mail, as well as in graphics. Knowledge is "sticky" in that it is rather slow and very difficult to transfer knowledge from one person to the other. This is because most knowledge is not universally applicable, having been obtained from information and data for a particular process. Finally, knowledge has both situational and contextual characteristics [1, 3].

## 1.2 Concept of Knowledge Management

Alavi and Leidner [4] present the various implications and the different perspectives of knowledge that have on knowledge management (Table 1).

Knowledge management refers to the use of certain tools, methods, and procedures to efficiently manage the resources and information within institutions such as commercial organizations. Quintas et al. [5] state that "knowledge management is a process of continually managing knowledge of all kinds and requires a company-wide strategy which comprises policy, implementation, monitoring, and evaluation." They further add that "Such a policy should ensure that knowledge is available when and where needed and can be acquired from external as well as internal sources." The knowledge management process has five steps [1].

1. Information systems activities
2. Knowledge management systems
3. Organizational and management tasks
4. Assessment and evaluation
5. Acquiring data and information.

**Table 1** Different perspectives of knowledge and their implications for knowledge management

| Perspective | Implication on KM |
|---|---|
| Knowledge from data and information | KM is focused on granting individuals exposure to information that might be potentially useful as well as smoothening the information assimilation process |
| Knowledge in terms of "state of mind" | KM attempts to enhance the learning and comprehension of an individual by providing information |
| Knowledge as an "object" | KM's primary concern here is to build and manage knowledge stocks |
| Knowledge as a "process" | KM focuses on the flow of knowledge and the processes of creation, distribution, and sharing of knowledge |
| Knowledge in terms of "access to information" | The key KM issue here is to organize the processes of accessing and retrieving content |
| Knowledge in terms of "capability" | KM is concerned with understanding strategic know-how and building core competencies |

This table has been referred from [4]

Knowledge management is focused on creating/extracting value from organizations intangible assets. The driving force behind knowledge management is the aim to create a process that can value a firm/organization's intangible property to allow for its effective usage. The idea is to provide an environment that is conducive to sharing knowledge as opposed to simply keeping it to oneself [6].

## 1.3 Knowledge Management Systems

Knowledge management systems are a collection of business processes devised to allow an organization to create, store, share, and use knowledge. These collections are IT-based systems that are used to improve the capability of organizations to learn/gain information from their environment and incorporate this acquired information into their business processes.

Knowledge management systems are of three major types:

- Enterprise-wide knowledge management systems
- Knowledge work systems (KWS)
- Intelligent techniques.

Davenport et al. in "Successful knowledge management projects" [7] talk about the objectives of knowledge management projects along with the conditions to be met and the factors that come into play for a knowledge management project to be counted as a success.

Gold et al. in "Knowledge management: An organizational capabilities perspective" [8] present their findings from a research conducted that examines from a perspective of an organizations capabilities, the issue of effective knowledge management.

Numerous knowledge management frameworks have been devised over the years. Some are given below along with their descriptions.

- By Van Heijst et al. [9]: Develop, Consolidate, Distribute, Combine.
- By The National Technical University of Athens, Greece [10]: Get context, Organize knowledge according to knowledge management goals, Strategize development and distribution, Culture.
- By American Management Systems [11]: Find knowledge, Organize it and Share.
- By Arthur Andersen Consulting [12]: Evaluate the target/problem, Define what the role of knowledge is and Create a knowledge strategy for the specified target, Identify technologies and processes needed for implementation, and lastly, Implement feedback mechanisms.
- By Dataware technologies [13]: Identify problem, Prepare for the change, Create a KM team, Perform the audit and analysis of knowledge, Define the main features of proposed solution, Implement the building blocks, and finally Link the knowledge to people.

- By Ernst and Young [14]: Generate knowledge, Represent it, Codify it and finally, Application of knowledge.
- By Holsapple and Joshi [15]: Acquire knowledge, Select required knowledge, Asses, Target and Deposit knowledge, Use it, Generate knowledge, Externalize knowledge.
- By Liebowitz [6]: Transform the information into knowledge, then Identify and Verify knowledge, Capture and Secure it, Organize it, Retrieve and Apply it, Combine it, Learn knowledge, Create knowledge (Loop from step 3), Sell it.
- By Marquardt [16]: Knowledge Acquisition, Creation, Transfer and Utilization and Storage.
- By O'Dell [17]: Identify, Collect, Adapt, Organize, Apply, Share, and Create knowledge.
- By Ruggles [18]: Create, Acquire, Synthesize, Fuse and Adapt knowledge, Capture and Represent knowledge, Transfer it.
- By The Mutual Group [19]: Gather information, Learn/Gain knowledge, Transfer it and Act accordingly.
- By American Productivity and Quality Center [20]: Find knowledge, Filter and Format it, Forward it to relevant people and Get feedback.
- By Van der Spek and Spijkervet [21]: Develop new knowledge, Secure it along with existing knowledge, Distribute these, and Combine all available knowledge.
- By Wielinga et al. [22]: Inventory, Represent and Classify knowledge, Create models for knowledge development and knowledge results and resources, Combine, Consolidate, Integrate, Develop and Distribute knowledge.
- By Wiig [23]: Create and Source knowledge, Compile and Transform it, Disseminate knowledge and Conduct Value Realization.
- By Knowledge Associates [24]: Acquire, Develop, Retain, and Share knowledge.
- By The Knowledge Research Institute [25]: Leverage the already existing knowledge followed by Creating new Knowledge and Capturing and Storing it, then Organize and Transform it and finally Deploy the knowledge.

## *1.4 Recommender Systems*

To understand what "Recommender Systems" are, we first take a look at the term "Information Filtering System." Information filtering systems are systems that aim to expose users to only that information that is useful/relevant to them. They remove extraneous data from incoming streams of data using computerized or (semi) automated methods.

Recommender systems are a class of information filtering systems that seek to predict the probability of a user choosing an item by assigning ratings. They identify recommendations autonomously for individuals based on their previous search (and purchase) history as well as based on other people's behavior. These recommendations are made to ease the decision-making process such as what song to listen to, what video to watch, what product to buy, or what news to read. Recommender systems serve to increase the interaction between the user and UI so as to provide a richer experience [26, 27].

Recommender systems are mainly aimed at people that lack enough experience or proficiency to navigate through the potentially massive amount of results a Web site, for example, might return for a given search string. Recommendations are both of personal, which are usually unique to each individual, and generalized nature, which are simpler to generate and are typically along the likes of top ten movies/ songs of all time, etc [28].

Ricci et al., in the Recommender Systems Handbook [28], put forth several reasons why service providers might choose to employ the technology:

- *To increase the number of items sold.* As the recommended item has a good chance of fitting the user's needs, this goal is likely to be achieved.
- *To sell more diverse items.* Less popular material from a user's preferred genre/ choice can also be advertised to the same using recommender systems.
- *To increase customer satisfaction.* With the presence of a properly designed UI, customer experience can be improved and this along with interesting and relevant recommendations serves to increase customer satisfaction.
- *To increase customer fidelity.* Increased customer satisfaction directly translates to increased customer fidelity. The more time a user spends on the site, the more refined their preference information becomes. This allows the recommendations to become highly customized to match preferences of the user.
- *To better understand what the user wants.* This important function of recommender systems can be leveraged to many other applications as well. Knowledge of user preferences can be reused for other purposes rather than just recommending.

Herlocker et al. [29] put forth eleven "End-User goals and tasks" that recommender systems help to achieve.

- *Annotation in context*: Certain specific existing content is emphasized upon based on the user's preference history.
- *Find all good items*: This recommends each, every item that can fulfill the requirement of user. Here, it is not enough to only recommend some good items.
- *Find some good items*: Here, only some items are recommended according to their rankings which show how likely it is for the user to use them.
- *Just browsing*: This is when the user is going through the product catalog without any intention of making a purchase. The recommender system should suggest products that would most likely fall within the user's area of interest.
- *Improve user profile*: This is a necessary task to achieve personalized recommendations. The user should be capable of providing information about his or her likes and dislikes to the recommender system.
- *Expressing self*: Users should be allowed to express their satisfaction or lack of it on the Web site. Though not a part of the recommendation process directly, it counts in terms of customer satisfaction.

- *Helping others*: Like the previous point, this is also to do with ratings and feedback as some users are happy to contribute information for the benefit of others.
- *Influence others*: In online recommender systems, there might be some malicious users that would like to influence others into purchasing some products.
- *Find credible recommender*: Some functionality can be offered to allow the user to test the quality of recommendation of the system.
- *Recommend a sequence*: Here, recommending a series of items, like a movie or TV series, a discography, or a collection of books is the primary goal.
- *Recommend a bundle*: Here, a bundle of goods that fit well as a group are recommended, for example, a tourism voucher may comprise a number of travel destinations located close by.

There are three basic approaches that are adopted to design most recommender systems [30]. Two of the more common one's, Collaborative filtering and Content-based filtering, have been discussed here.

**Collaborative filtering** models recommendations based on a user's prior behavior. These recommendations can be solely based on an individual's behavior or can also incorporate from the behavior of other users with similar traits and to much better effect. For example, an online music player uses a collaborative recommender system to suggest new songs. How it does this is, it uses information from the myriad subscribers who regularly listen to songs and groups the users together based on their listening preferences. For example, subscribers listening to mostly similar genres are grouped together. From their information, a list of most popular songs/artists/genres is then prepared and to any particular user in that group, the music player recommends the most popular songs/artists/genres that he or she neither listens to nor follows. Refer to Fig. 2 for a pictorial representation of the collaborative filtering logic.

**Content-based filtering** methods compare the item's description with a profile of the user's preferences to decide whether to recommend that particular item or not. Here, items are described using keywords and the profile of an user is built

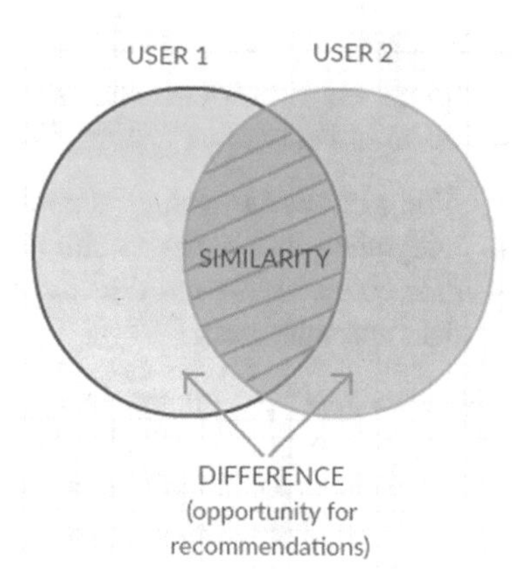


**Fig. 2** A sparse representation of the collaborative filtering approach

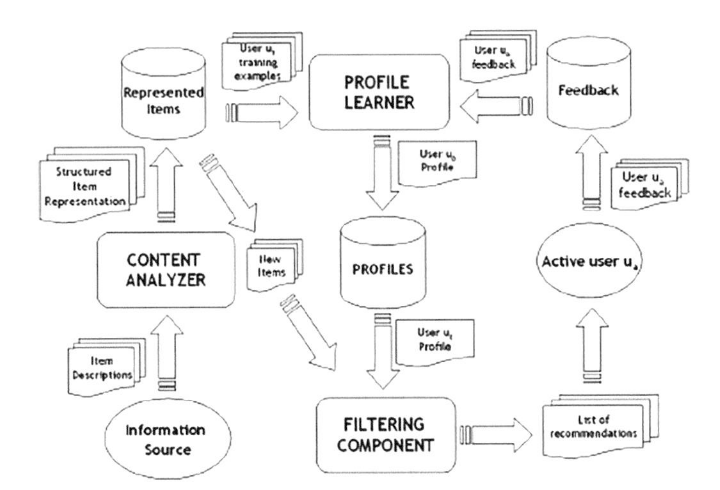


**Fig. 3** Architecture for a content-based filtering recommender system (Figure was originally proposed by Felfernig et al. [30])

using selected keywords that are indicative of the user's preference history. In other words, recommendations here are of items that either belong to preference history of the user or are similar to his current search interest. Content-based filtering involves comparing various candidate items with the user's previously rated items, and the best matching items are returned. Refer to Fig. 3.

As stated before, recommender systems assign ratings to items to predict their chances of being selected by the user. There are several different data mining techniques that can be used to assign these ratings. The data mining process has, broadly speaking, three main steps. *Data preprocessing*, followed by *data analysis,* and lastly *result interpretation.*

**Data Preprocessing** involves processes to "prepare" the data in order to make it suitable for application of the machine learning techniques in the data analysis step. The major tasks in the data preprocessing stage are data cleaning, data integration, data reduction, data transformation, and data discretization. Data preprocessing is important as it improves the accuracy of the input data, makes sure it's complete and consistent, and enhances its believability and interpretability. The major tasks here are as follows [31]:

- filling in missing values, smoothening the noisy data, and removing outliers;
- integrating data across multiple files and databases;
- data compression and dimensionality reduction;
- data normalization.

**Data Analysis** in recommender systems is concerned with classifying the items into different labels (classes). This involves mapping each element based on its characteristic features to an appropriate label. For example, a book recommender system might classify a book as popular or unpopular based on the number of reads it has.

While there are various classifiers out there, we will limit our discussion to supervised and unsupervised classification. Supervised classification is when both the input and the output variables are known, and the algorithm is concerned with learning the mapping from the input to the output. The goal is to get an approximate mapping function such that for any new input data, the output can be correctly predicted. Since the outputs here are already known, the process is known as supervised learning. Unsupervised classification occurs when only the input data is given, and corresponding outputs are unknown. The aim of unsupervised learning is to gain intuition about the underlying structure or distribution of the data in order to learn more about it. Unsupervised learning algorithms are so called because they do not have any correct answers, and there is no supervision of the results obtained. The algorithms are expected to model or represent the data unassisted and on their own.

Some supervised and unsupervised classification algorithms are as follows:

- *Nearest neighbors algorithms* are supervised learning algorithms that find the most closely related data points for the point to be classified and assign its class label on the basis of those "nearest neighbors." The idea is to assign a point to whatever class label is predominant in the neighborhood of the area where it lies.
- *Decision trees* are supervised learning algorithms whose goal is to correctly predict the value of the target variable given several input variables. These are of two types, classification trees where the predicted outcome is the class label or regression trees, where the predicted outcome is a real number.
- *Rule-based classifiers* are supervised learning algorithms that work using the "if…then…else" construct. Here, the condition or rule antecedent is an expression made by attribute conjunctions. The rule consequent is an outcome or classification that can either be positive or negative.
- *Bayesian classifiers* are supervised learning algorithms that use a probabilistic framework for the classification process. They make use of the Bayes theorem and conditional probability. Here, the uncertainty in relationships is modeled using probability. Each attribute and class label is considered as a random variable for the Bayesian classifiers problem. They seek to predict the correct label by finding the class that gives maximum posterior probability for a given data.
- *Artificial neural networks* are supervised learning algorithms that seek to model the structure of the biological brain. They are made up of an assembly of interconnected nodes called neurons with weighted links. The neurons are analogous to the brain's axons. Layers and layers of these neurons form a network that after being trained with an adequate amount of data can learn the classification problems. These are designed to perform nonlinear classification task as well as stay robust in the face of partial system failure.
- *Support vector machines* are supervised learning algorithms that are discriminative classifiers that are defined by a separating hyperplane. That is, the input data is separated in such a way that the optimal hyperplane maximizes the

margin between itself and the data points. The rationale behind support vector machines is that if the margin is maximized, future misclassification of unknown data becomes less likely.

- *K-Means* is an unsupervised learning algorithm that seeks to partition a data set into *k* contiguous clusters in such a way that the within-cluster sum of squares is minimized. It works by randomly initializing the cluster centroids, assigning points to the closest centroid and then repeatedly revising the centroid allocation till the sum-of-squares metric is not minimized. This ensures high intra-cluster but low inter-cluster similarity.
- *Association rule mining* is another unsupervised learning algorithm that finds rules for predicting occurence of an item in a transaction based on other items of the same transaction. Association rules are also "if…then" construct based that generate relationships between seemingly unrelated items/data. To create the rules, the data is frequently analyzed for if…then patterns, and the most important relationships are identified using the confidence and support as the two criteria.

**Result Interpretation** is where the recommendations are generated based on the results from the data analysis step. The classification methods assign the "ratings" to the data, and these ratings then influence what products are recommended.

## 2 Chapter Structure

In the subsections above, key concepts related to the topic of discussions like knowledge management systems and recommender systems along with the difference between the terms data, information and knowledge have been introduced. The rest of the chapter is organized as follows: in the section "Recommender Systems in Knowledge Management," a few related works have been discussed in detail and an analysis of each has been presented. In the section "Other Works," a list other experiments/projects/publications on the same matter have been presented. Lastly, in the section "Scope of Future Work," some scenarios/projects that can be worked on in the future have also been discussed. Following that, a list of all the sources referred to in the chapter, including research papers, books, Web sites, is given.

## 3 Recommender Systems in Knowledge Management

The development of knowledge management systems suffers from the fact that both the organizational and personal environments are continually changing. Recommender systems can be of help in this context as they can help to better cope with the complexity and size of knowledge structures [32–34]. They can be of use in the following contexts:

- *Understanding of Knowledge Base*: A proper understanding of the structure and basic elements of a knowledge base is imperative for its efficient development and maintenance [35].
- *Testing and Debugging of Knowledge Base*: Recommender systems can be used to recommend the minimal sets of changes required to restore consistency of knowledge bases thereby improving the efficiency of the testing and debugging process [33, 35].
- *Refactoring of Knowledge Base:* Recommender systems can be used to recommend relevant refactoring (structural changes while semantics are preserved) of the knowledge base from time to time to ensure the understandability and maintainability of the knowledge base are maintained.
- *Recommender Systems in Databases*: Recommender systems can also be used in the process of information search by improving the accessibility of databases by recommending queries [32].

Above given are the different contexts in which recommender systems can be used to enable better, easier use, and maintenance of knowledge management systems. We now take a look at the different scenarios where recommender systems have been incorporated in projects to aid with knowledge management.

Various fields have seen the implementation of recommender systems for knowledge management. We discuss examples from the engineering and academic environment, tourism, and hospitality, expertise systems in the industrial and the scientific settings as well. Following are some knowledge management systems that have been implemented using recommender systems. Presented is a chronologically ordered list of some of the work done so far.

## 3.1 By Natalie Glance et al.

The authors, in 1998, proposed an information technology system called the "Knowledge Pump" [36] for the purpose of supporting and connecting online communities and repositories. Its aim is to create an environment that is conducive to the creation, flow, and use of knowledge and to ensure that the right information reaches the correct people and on time. Another objective is to map repository content and community networks.

Knowledge Pump has three different design perspectives:

- Designing for the user: the user can make general recommendations or review and review any repository item. The reviews and recommendations are then collected and distributed to those who are judged to find these useful.
- Designing for the community: it provides a standard channel to the users for communication and resource sharing and also supports communities.

– Designing for the organization: using Knowledge Pump, the various overlapping communities in an organization are mapped to bring together the various disjoint parts of an organization.

Knowledge Pump implements the client–server architecture shown in Fig. 4. The client is Java-based and runs on Web browsers. The client talks to the Java-based server that performs a number of important functions such as running the collaborative filtering algorithm periodically, providing an interface to system administration and building "What's recommended?" pages for each user. These pages are saved for later and delivered to the user via the HTTP server. All the data is stored in a database which is accessed by yet another server. Any communication with the database takes place through the database server.

The Knowledge Pump uses recommender systems for item classification. It makes use of "community-centered collaborative filtering" to estimate what level of interest a user would show for the unread items in their multiple domains of interest. Here, a partial view of the social network constructed from user-input lists of trusted users bootstraps the collaborative filter. In such a bootstrapping, by giving a higher weightage to the opinions of a user's closest contacts, the system performs good from the beginning.

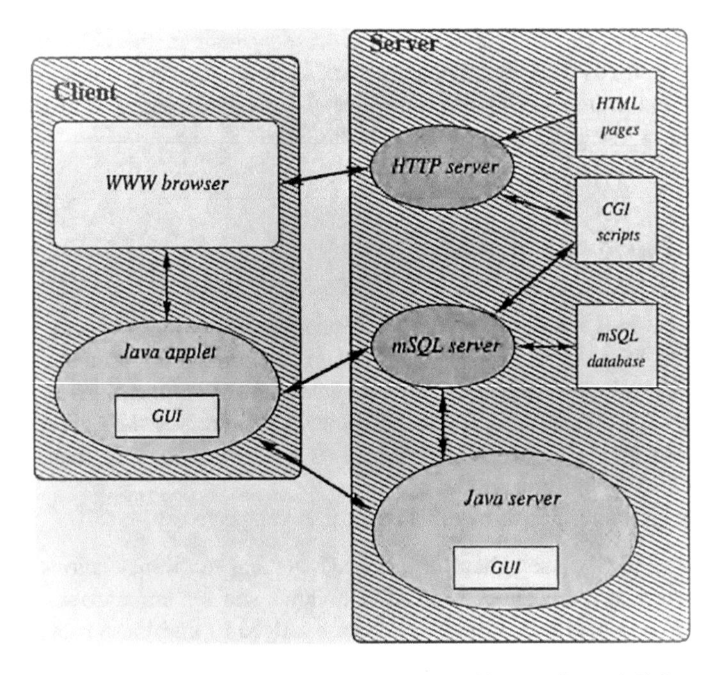


**Fig. 4** Knowledge Pump architectural overview as presented by Natalie et al. [36]

The system has the following characteristics:

- Portability: means a platform-independent code. This meant using Java and suggested the application should be built to ride on top of the Web, have no browser interactions like plug-ins or cookies.
- Ease-of-use: Very low cost of installation is also achieved through Java.
- Immediate value: This implies providing a set of functionalities that are valuable above and beyond recommendations, which was achieved by building basic document management tools like search and retrieval over bookmarked items.

## *3.2 By Dawit Yimam et al.*

The authors, in 2000, proposed the Dynamic Expertise Modeling from Organizational Information Resources (DEMOIR) framework for developing and testing expertise modeling algorithms [37–39].

Expert finders or expertise recommenders are specialized knowledge management systems that help in finding out and continually recording the "expertise" of the "experts" in an organization. They then make this expertise available to other users using recommender systems, enabling them to solve problems that exceed their personal capabilities.

The proposed framework builds upon two types of expertise models: one is the aggregated expertise (AE) model while the other is the personal agent (PA)-based model. These two approaches both have trade-offs and thus are tough to evaluate/compare. While distributed "PA" models offer easy privacy maintenance, they suffer from the limitation of only relying on personal sources of information. Here, due to the expertise data being distributed, limited accessibility and sub-optimal utilization are issues. Scalability is also an issue as having an individual expertise modeling agent for every person overloads the network. "AE" models overcome the above-mentioned shortcomings of the personal agent-based models by allowing for an open, multi-purpose exploitation of the expertise information. Manipulation of said information and monitoring of a wide range of sources for latest data are also provided. Lastly, aggregated models also facilitate the use of both statistical expertise and knowledge-based modeling. The drawbacks though are the lack of localization, privacy of expertise data as well as a compromise on the privacy of individual experts.

The proposed DEMOIR architecture (refer Fig. 5) integrates both the aggregated and personal agent-based expertise models. It is a centralized expertise model-based modular architecture that also incorporates distributed clients as well as decentralized expertise indicator source gathering and extraction. It does so by separating functions like extraction of expertise indicators, source gathering, and expertise modeling and assigns them to specialized components for separate implementation.

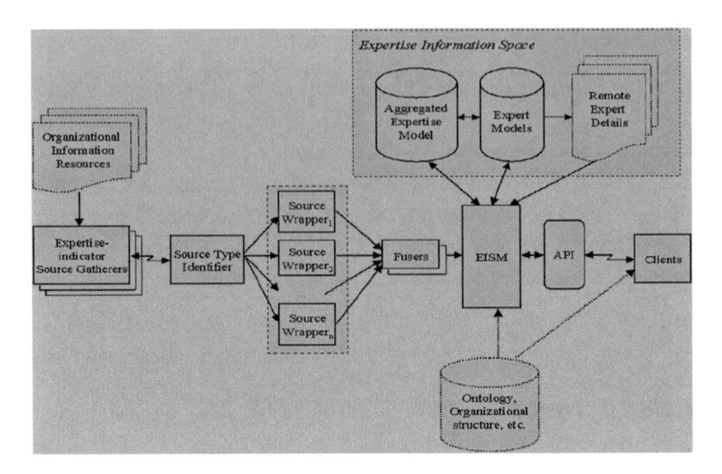


**Fig. 5** DEMOIR architecture, as proposed by Yimam et al. [37, 38]

Its components can be grouped into three general functions:

- **Expertise indicator source gathering**. Performed by robots and personal agents that have data source recognition logic built into them.
- **Expertise modeling**. DEMOIR has four components to meet the user's needs. These are a source type identifier, source wrappers, fusers, and the expertise information space manager (EISM).
- **Expertise model exploitation**. Achieved using API's that provide a customized usage experience.

The characteristics of the architecture are:

- All modules are configurable and extensible, thus allowing reusability and interoperability.
- It has a combination of both centralized and distributed monitoring of expertise data, thus tackling the heterogeneity gap and privacy problems.
- It captures how the sources relate to experts and where the expertise evidence came from, and factors this information into the expert modeling process.
- It provides structure for integration of domain knowledge with statistical and heuristic methods at all steps.

## 3.3 By David McDonald et al.

The authors, in 2000, presented an architecture for the Expertise Recommender system (ER) [40] and showed an implementation as well. It implements a client–server architecture. It can support both simple clients (e.g., Web-based interface) and clients tailored to support specific features of the ER server.

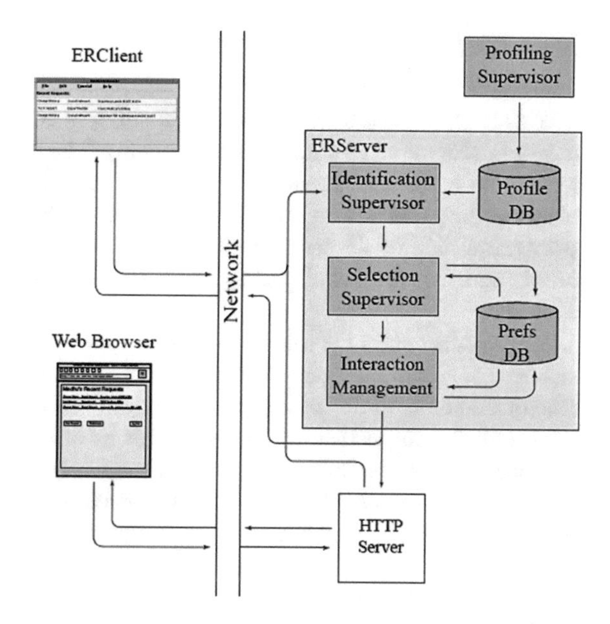


**Fig. 6** Architecture of the expertise recommender system as given by McDonald et al. [40]

The ER model has a pipe and filter architecture (refer Fig. 6). It is a collection of easily extensible heuristic models, their data stores and high-level supervisors. The supervisors provide connections and general services like identification, profiling, selection, and interaction management to facilitate a specific implementation. The underlying heuristic modules are coordinated by the supervisors to provide the required services. The databases are used for providing storage for user profiles and their various preferences. All the different portions are glued together using the ER server which also handles details of servicing requests and managing connections. Client–server interactions are handled via a protocol implemented by the server.

The advantages of this system are:

- It is a flexible and open architecture that can address the various types of organizational environments.
- It has more robust organization-specific implementation by separating the technical aspects of making good recommendations from social and collaborative aspects of matching individuals.
- It uses a different approach to the ratings for creating and maintaining user profiles. Here, organizationally relevant data sources are used for profile creation and these profiles are more suited for automated expertise location.

## 3.4 By Joaquin Delgado et al.

The authors, in 2002, implemented the "TripMatcher" system [41] which is a recommender system for travel and hospitality. The chief motive behind this was to reduce the amount of needless expenditure on sales effort to assist customers in planning their travel. In a traditional business model, travel providers hire experts with knowledge of destinations they offer to assist customers and hope that customers can find the travel package of their choice out of the thousands of offers available.

TripMatcher plays the role of an experienced online salesperson. It interacts with customers to learn their preferences and then displays highly personalized, targeted recommendations and relevant and customized content. This allows travel providers to effectively address their customers' needs by offering destination, itineraries, and products tailor-made for their customers' preferences.

The recommender achieves the above mentioned by building a knowledge base using a decision tree which is built two complimentary sources. First, it uses content and ratings provided by experts and second, automatically generated ratings through text-mining of product descriptions.

The system performs the following steps/components for generating predictions:

- Content filters
- Calculating matching function
- Retrieval of result items information from the database
- Event-based calculation
- Offline attribute-based collaborative filtering.

A customized version of TripMatcher was implemented for ski-europe.com. The results of the implementation were positive with an increased conversion rate of interested people to customers. From the case study, it was evidenced that people using this system were more likely to request assistance in purchasing their travel arrangements. The increased conversion rate was seen constantly for a period of four months pointing to the consistency of the system.

## 3.5 By Eduardo Barbosa et al.

The authors introduced the "MISIR" recommendation module [42] into the GCC [43] in 2007 to facilitate the evaluation and dissemination of any kind of explicit scientific knowledge. The GCC was created for the purpose of aiding knowledge management in educational institutions. As numerous explicit sources of scientific knowledge, such as formulae, experiments, models, theses, publications and technical reports exist, and the number of these is also growing, to ease the process of sifting through the vast reservoir of knowledge, MISIR was incorporated into it to the GCC.

There are two stages in the MISIR approach:

1. The prediction stage: Here, the algorithm searches for documents that have not been evaluated by the user. Based on each rating given for that document by other users and on profile similarities, the algorithm calculates a grade for each non-evaluated document found.
2. The recommendation stage: Here, all the predicted grades are verified and documents with grades higher than a given threshold/reference value are selected.

As discussed earlier, the process of knowledge management has the following steps: (1) Identification, (2) Capture, (3) Selection and Validation, (4) Organization and Storage, (5) Dissemination, (6) Application, and (7) Creation.

Evaluating MISIR on this model, the following observations were made:

- It helps in the process of *identification* as the approach is directly related to the identification of explicit knowledge at an institution.
- It also helps identify individual competencies.
- As it emphasizes explicit knowledge in document form which is more useful to the user, it helps maintain essential competencies in an organization via the *capture* process.
- As MISIR tries to show only the most relevant results to the user, its approach is directly related to the process of *Selection and Validation.*
- As recommendation stage is concerned with showing/spreading relevant knowledge, it relates to the *Dissemination process.*
- In relation to the *Application* stage, the MISIR approach helps in the application of the new knowledge in various research scenarios as successful experiments, design projects, or practice, etc.
- As the proposed solution always attempts to make new, useful, and relevant knowledge available, it also helps the process of *creation* of new knowledge.

In Fig. 7 which is given below (the figure has been referred to from [42]), we get a visual representation of the knowledge transformation process performed by MISIR.

### 3.6 By Worasit Choochaiwattana

In 2015, the author performed a comparative analysis of item-based and tag-based recommendations to determine which approach is better suited for the task of automated knowledge dissemination [44].

The item-based recommendation mechanism, as described in Fig. 8, surveys the usage pattern of the other users of the system and makes recommendations on the basis of matches between the interests of the user and others. There are four main components that are used to implement this: *set of users, similarity measurement,*

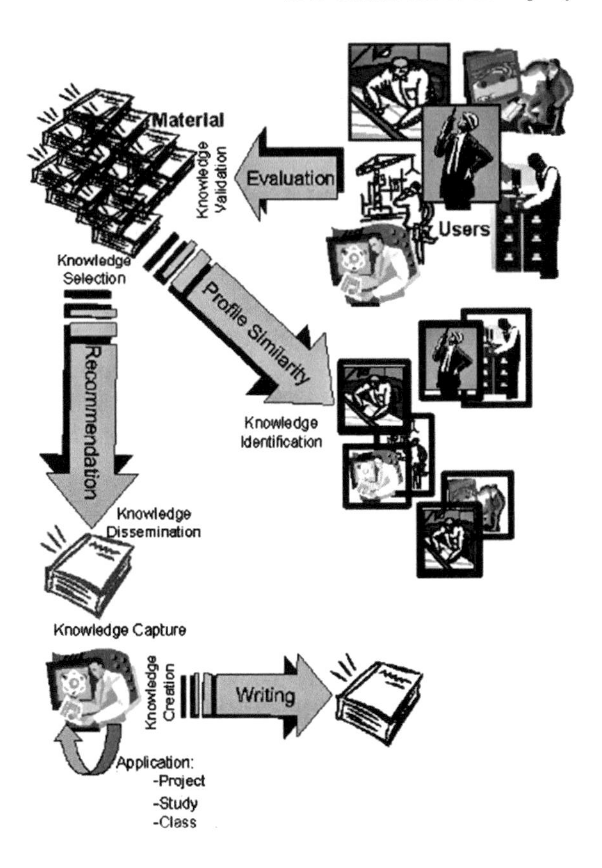


**Fig. 7** MISIR knowledge transformation process

*set of interaction with knowledge items, and knowledge corpus.* Cosine similarity scores are calculated between each pair of users and the users that have similarity scores greater than or equal to a given threshold, will be placed in a group together, to identify alike users, who exhibit interest in similar knowledge items.

The tag-based recommendation mechanism, as described in Fig. 9, on the other hand, recommends knowledge items by making use of knowledge tags. The main components of this mechanism are: *set of users, set of new/unvisited knowledge items, set of interaction with knowledge items, set of users' knowledge tag, and set of tags from new/unvisited knowledge items.* A cosine similarity is calculated here also to recommend knowledge items.

The experiment was conducted to compare the efficacy of the two recommendation mechanisms: one being a collaborative filtering technique (item-based), while the other is a content-based filtering technique (tag-based) in augmenting the automatic knowledge dissemination services in a knowledge management system. The performance of each system was assessed by calculating its percentage of accuracy (Eq. 1), with the mechanism with higher accuracy being more effective.

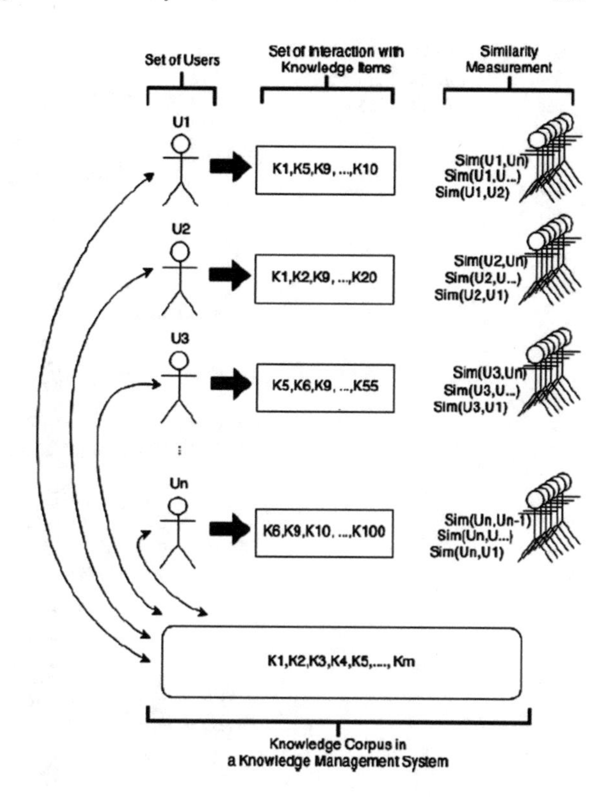


**Fig. 8** Item-based recommendation mechanism by Choochaiwattana [44]

The test data was collected over a period of almost around half a year and was loaded into a KMS that had been embedded with the two proposed recommendation mechanisms. A selective group of people was chosen to use the KMS. A confusion matrix was created to measure the accuracy results for both the mechanisms.

$$\text{Accuracy} = [(\text{True\_Positives} + \text{True\_negatives})/\text{Total\_results}] \times 100 \quad (1)$$

From the experimental results, it was observed that the tag-based recommendation mechanism (i.e., content-based filtering) provides a higher accuracy and thus performs comparatively better in the knowledge item recommendation task. The reason for this is that user's interests are better represented by a set of knowledge tags. If we only use a set of interactions to represent the user's interests, it would at best be a rough idea only. There is no guarantee the user will always have similar knowledge interests.

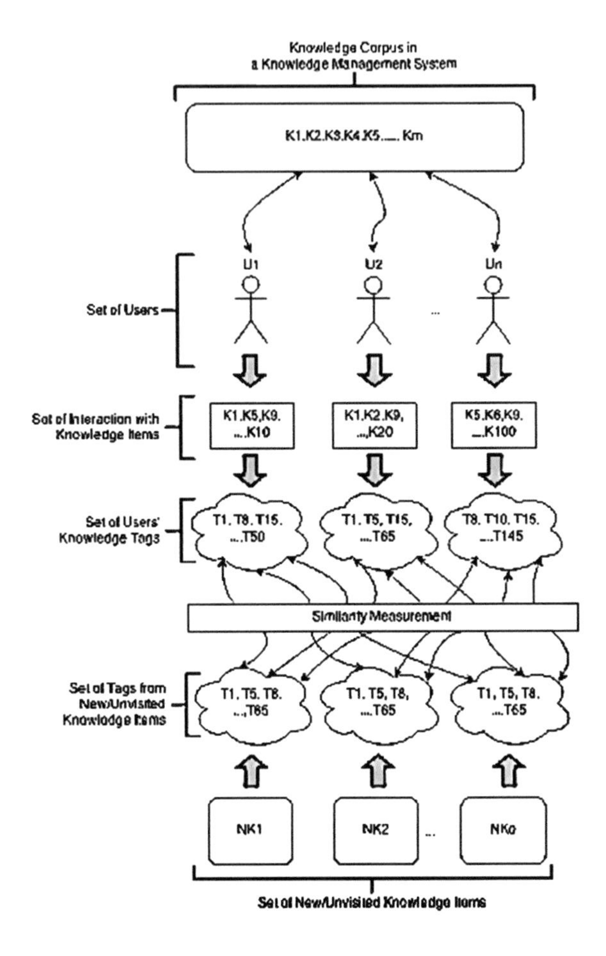


**Fig. 9** Tag-based recommendation mechanism given by Choochaiwattana [44]

## 4 Other Works

- Linton et al. [45] describe how recommender systems can be used to enable the continuous acquisition of knowledge and individualized tutoring of application software across an organization. They propose a recommender system as a means of facilitating organization-wide learning.
- Skrzypczyk et al. [46] talk about how recommender systems can be used in the academic area to support researchers and students in their studies. It discusses the possibility of including recommender systems in the domain of personal knowledge management through different methods and techniques.
- Zhen et al. [47] tackle the problem of personal knowledge management by proposing a recommender system for the purpose of knowledge sharing in a collaborative environment.

- Verbert et al. [48] talk about how recommender systems offer a promising approach when it comes to facilitating learning and teaching tasks. In this paper, they present a framework for identifying the relevant context dimensions in "Technology Enhanced Applications" (TEL) and present an analysis of the existing TEL-based recommender systems along the identified dimensions.
- Mehrpoor et al. [49] present "Intelligent services" which is a recommender system for knowledge representation in the industry. The goal was to improve the accessibility of information and knowledge to make it easier for the stake-holders' to collaborate in the decision-making process.
- Ahlers et al. [50] examine how semantic Context-aware Recommender Systems can be applied in a design-centric engineering domain to improve interaction and navigation.
- Stenmark [51] conducted an empirical study on the use of recommender systems in organizations in order to make use of the tacit knowledge available. Using Polanyi's theories, the author shows how Intranet documents can be used to make tacit knowledge tangible without needing to make it explicit.
- Mehrpoor et al. [52] proposed a recommender system tailor-made for an engineering setting. The aim was to make use of both collaborative and content-based filtering along with semantic technologies to improve knowledge accessibility and provide relevant and accurate recommendations.

## 5 Scope of Future Work

A number of issues/challenges regarding recommender systems and their application to the field of knowledge management still remain open for research and exploration. The following problems still need to be tackled [34]:

- *More focus on the user perspective.* As of now, most recommender systems are focused on increasing the business revenues of companies. Approaches that pay more attention to the customer and user support are needed.
- *Sharing of recommended knowledge.* A dearth of the required recommendation knowledge is one of the major reasons why recommender systems, as of now, are not customer oriented. More customer-oriented recommender systems will act as personal assistants and require the presence of global object information for this to succeed.
- *Presence of context-aware recommenders.* New technologies should make full use of the infrastructure of mobile services to become more aware contextually. Data such as the geographical location, movement, calendar, and social network information can be exploited to provide intelligent recommendations.
- *Unobtrusive preference identification.* User preference knowledge is the most important information for generating relevant predictions. Eliciting user preferences in an unobtrusive way is a major paradigm that is faced when developing new systems.

Recommender systems need to break away from the "Filter Bubble" to able to aid their users in identifying, developing, understanding, and exploring their unique preferences. Recommender systems need to focus on how to integrate self-actualization while recommending so as to become more supportive of the consumer/human decision-making process rather than simply replacing it. A balance needs to be achieved such that recommendations are neither too personalized that they become intrusive and at the same time are not very generic that they do not account for the user's distinct taste [53].

Also, other topics for further investigation include:

- How within organizations, we can identify and exploit expertise indicator sources [39]?
- How to structure and represent expertise and expert models [39]?
- How to apply inference algorithms and rules on expert and expertise relationships [39]?
- How to support users to search, analyze, and exploit the expertise information available [39]?
- What are the integration and contextual issues for the proper deployment and positioning of expert systems in organizations [39]?
- How much a recommendation can be trusted and to ensure that recommendations are absolutely relevant and accurate as usage of information in the wrong context can have big repercussions to research [42]?
- How to improve techniques for user knowledge interest representation [44]?
- Development of context sensors to automate the acquisition of context dimensions for learning [22].
- How to tackle privacy protection challenges when capturing and using contextual data for recommendation [22]?

# 6 Conclusion

In this chapter, the use of recommender systems in the domain of knowledge management has been discussed. Some related works that use recommender systems to facilitate knowledge management in different industries/fields like tourism, academia, scientific research, engineering have been discussed in detail. Despite the fact that attempts at using RS's for KM have been going on for the better part of two decades now, a lot of work/research still needs to be done, as indicated by the above passage. As Etienne Wenger has very wisely quoted, "Knowledge Management will never work until corporations realize it's not about how you capture knowledge but how you create and leverage it." and the use of recommender systems to this effect seems to be apt.

## References


1. Laudon, K. C., & Laudon, J. P. (2004). *Management information systems: Managing the digital firm* (p. 8). New Jersey.
2. Losee, R. M. (1997). A discipline independent definition of information. *Journal of the American Society for Information Science (1986–1998), 48*(3), 254.
3. http://www.stevedenning.com/Knowledge-Management/what-is-knowledge.aspx.
4. Leidner, D., & Alavi, M. (2001). Review: Knowledge management and knowledge management systems: Conceptual foundations and research. *INSEAD. MIS Quarterly, 25* (1), 107–136.
5. Quintas, P., Lefere, P., & Jones, G. (1997). Knowledge management: A strategic agenda. *Long Range Planning, 30*(3), 322385–322391.
6. Liebowitz, J. (1999). *Building organizational intelligence: A knowledge management primer* (vol. 1). Boca Raton: CRC press.
7. Davenport, T. H., De Long, D. W., & Beers, M. C. (1998). Successful knowledge management projects. *Sloan Management Review, 39*(2), 43.
8. Gold, A. H., & Arvind Malhotra, A. H. S. (2001). Knowledge management: An organizational capabilities perspective. *Journal of Management Information Systems, 18*(1), 185–214.
9. Van Heijst, G., Van Der Spek, R., & Kruizinga, E. (1997). Corporate memories as a tool for knowledge management. *Expert Systems With Applications, 13*(1), 41–54.
10. Apostolou, D., & Mentzas, G. (1999). Managing corporate knowledge: A comparative analysis of experiences in consulting firms. Part 1. *Knowledge and Process Management, 6*(3), 129.
11. Rubenstein-Montano, B., Liebowitz, J., Buchwalter, J., McCaw, D., Newman, B., Rebeck, K., et al. (2001). A systems thinking framework for knowledge management. *Decision Support Systems, 31*(1), 5–16.
12. Ahlers, D., & MehrAndersen, A. (1997). Business consulting: Knowledge strategies. *DB/OL*, http://www.arthurandersen.com/aabc.
13. Dataware Technologies. (1998). Seven steps to implementing knowledge management in your organization. *Corporate Executive Briefing*. http://www.dataware.com.
14. Ernst & Young. (1999). http://www.ey.com/consulting/kbb/k2work.asp.
15. Holsapple, C. W., & Joshi, K. D. (2002). Knowledge management: A threefold framework. *The Information Society, 18*(1), 47–64.
16. Marquardt, M. J. (1996). *Building the learning organization: A systems approach to quantum improvement and global success*. NY: McGraw-Hill Companies.
17. O'Dell, C. (1996, December). A current review of knowledge management best practice. In *Conference on knowledge management and the transfer of best practices, Business Intelligence, London*.
18. Ruggles, R. (2009). *Knowledge management tools*. Routledge.
19. Saint-Onge, H. (1998). Knowledge management. In: *Proceedings of the 1998 New York Business Information Technology Conference,* November, TFPL, New York.
20. Steier, D. M., Huffman, S. B., & Kalish, D. I. (1997). *AAAI spring symposium on AI in knowledge management*. New York, NY: PriceWaterhouse-Coopers.
21. Van der Spek, R., & Spijkervet, A. (1997). Knowledge management: Dealing intelligently with knowledge. *Knowledge Management and Its Integrative Elements*, 31–59.
22. Wielinga, B., Sandberg, J., & Schreiber, G. (1997). Methods and techniques for knowledge management: What has knowledge engineering to offer? *Expert Systems with Applications, 13*(1), 73–84.
23. Wiig, K. (1993). *Knowledge management foundations: Thinking about-how people and organizations create, represent, and use knowledge*. Arlington, Texas: Schema.
24. Young, R. (1999). Knowledge management overview: from information to knowledge. *Knowledge Associates.* Available at www.knowledgeassociates.com.

25. Wiig, K. (1998). *The role of knowledge based system in knowledge management*. Workshop on knowledge management and at US Dept. of Labor.
26. https://en.wikipedia.org/wiki/Recommender_system.
27. Resnick, P., & Varian, H. R. (1997). Recommender systems. *Communications of the ACM, 40*(3), 56–58.
28. Ricci, F., Rokach, L., & Shapira, B. (2011). Introduction to recommender systems handbook. In *Recommender systems handbook* (pp. 1–35). Springer US.
29. Herlocker, J. L., Konstan, J. A., Terveen, L. G., & Riedl, J. T. (2004). Evaluating collaborative filtering recommender systems. *ACM Transactions on Information Systems (TOIS), 22*(1), 5–53.
30. Felfernig, A., Jeran, M., Ninaus, G., Reinfrank, F., Reiterer, S., & Stettinger, M. (2014). Basic approaches in recommendation systems. In *Recommendation Systems in Software Engineering* (pp. 15–37). Springer Berlin Heidelberg.
31. Han, J., Pei, J., & Kamber, M. (2011). *Data mining: Concepts and techniques*. Amsterdam: Elsevier.
32. Chesbrough, H. W. (2006). *Open innovation: The new imperative for creating and profiting from technology*. Harvard Business Press.
33. Felfernig, A., Friedrich, G., Schubert, M., Mandl, M., Mairitsch, M., & Teppan, E. (2009, July). Plausible repairs for inconsistent requirements. In *IJCAI* (vol. 9, pp. 791–796).
34. Felfernig, A., Jeran, M., Ninaus, G., Reinfrank, F., & Reiterer, S. (2013). Toward the next generation of recommender systems: applications and research challenges. In *Multimedia services in intelligent environments* (pp. 81–98). Springer International Publishing.
35. Felfernig, A., Reinfrank, F., & Ninaus, G. (2012, December). Resolving anomalies in configuration knowledge bases. In *International Symposium on Methodologies for Intelligent Systems* (pp. 311–320). Berlin, Heidelberg: Springer.
36. Glance, N., Arregui, D., & Dardenne, M. (1998). Knowledge pump: Supporting the flow and use of knowledge. *Information Technology for Knowledge Management, 3*.
37. Yimam, D., & Kobsa, A. (2000). Centralization vs. decentralization issues in internet-based knowledge management systems: Experiences from expert recommender systems. In *TWIST2000,* Irvine, CA.
38. Yimam, D., & Kobsa, A. (2000). Demoir: A hybrid architecture for expertise modeling and recommender systems. In *IEEE 9th International Workshops on Enabling Technologies: Infrastructure for collaborative enterprises, 2000. WET ICE 2000. Proceedings* (pp. 67–74). IEEE.
39. Yimam, D., & Kobsa, A. (2003). Expert-finding systems for organizations: Problem and domain analysis and the DEMOIR approach. *Journal of Organizational Computing and Electronic Commerce, 13*(1), 1–24.
40. McDonald, D. W., & Ackerman, M. S. (2000, December). Expertise recommender: a flexible recommendation system and architecture. In *Proceedings of the 2000 ACM conference on Computer supported cooperative work* (pp. 231–240). New York: ACM.
41. Delgado, J. A., & Davidson, R. (2002). *Knowledge bases and user profiling in travel and hospitality recommender systems*.
42. Barbosa, E., Oliveira, J., Maia, L., & De Souza, J. M. (2007, April). Using recommendation systems for explicit knowledge dissemination and profiling identification for scientific and engineering contexts. In *11th International Conference on Computer Supported Cooperative Work in Design. CSCWD 2007* (pp. 715–721). IEEE.
43. Oliveira, J., Souza, J. D., Miranda, R., & Rodrigues, S. (2005). GCC: An environment for knowledge management in scientific research and higher education centres. In *Proceedings of I-KNOW'05* (pp. 633–640).
44. Choochaiwattana, W. (2015). A comparison between item-based and tag-based recommendation on a knowledge management system: A preliminary investigation. *International Journal of Information and Education Technology, 5*(10), 754.
45. Linton, F., Joy, D., Schaefer, H. P., & Charron, A. (2000). OWL: A recommender system for organization-wide learning. *Educational Technology & Society, 3*(1), 62–76.

46. Skrzypczyk, W., Bleimann, U., Wentzel, C., & Clarke, N. (2009). How recommender systems applied in personal knowledge management environments can improve learning processes.
47. Zhen, L., Song, H. T., & He, J. T. (2012). Recommender systems for personal knowledge management in collaborative environments. *Expert Systems with Applications, 39*(16), 12536–12542.
48. Verbert, K., Manouselis, N., Ochoa, X., Wolpers, M., Drachsler, H., Bosnic, I., et al. (2012). Context-aware recommender systems for learning: A survey and future challenges. *IEEE Transactions on Learning Technologies, 5*(4), 318–335.
49. Mehrpoor, M., Gjarde, A., & Sivertsen, O. I. (2014, June). Intelligent services: A semantic recommender system for knowledge representation in industry. In *2014 International ICE Conference on Engineering, Technology and Innovation (ICE)* (pp. 1–6). IEEE.
50. Ahlers, D., & Mehrpoor, M. (2014). Semantic social recommendations in knowledge-based engineering. In *HT (Doctoral Consortium/Late-breaking Results/Workshops).*
51. Stenmark, D. (2000). Leveraging tacit organizational knowledge. *Journal of Management Information Systems, 17*(3), 9–24.
52. Mehrpoor, M., Gulla, J. A., Ahlers, D., Kristensen, K., Ghodrat, S., & Sivertsen, O. I. (2015, September). Using process ontologies to contextualize recommender systems in engineering projects for knowledge access improvement. In *European Conference on Knowledge Management* (p. 524). Academic Conferences International Limited.
53. https://www.quora.com/What-is-the-future-of-recommender-systems-research.
54. https://en.wikipedia.org/wiki/DIKW_pyramid.

# Metasearch Engine: A Technology for Information Extraction in Knowledge Computing

**P. Vijaya and Satish Chander**

**Abstract** The increasing number of the Web data due to the increased amount of the digitalized standards, electronic mails, images, multimedia, and Web services, the World Wide Web rises as the cost-effective resource for releasing the data and for discovering the knowledge. Thus, the open challenges for the information retrieval efforts have received much attention among the researchers to deem the browsing as an appropriate searching procedure. To facilitate the most relevant information, it is necessary to develop a unique and extensive structural framework that offers a platter and a navigational proxy to the clients and servers. Consequently, the search engines play a vital role in contributing the users in providing the related Web pages from the Web. A Metasearch engine, in essence, is a search mechanism that sends the user query to a number of modern search engines autonomously and provides the combined outcome through their own unique page ranking technique. This chapter intends to discuss the necessity of metasearch engines, starting with a series of definitions of search engines and its classification. Further, a summary of metasearch engine is provided with the architecture and the result-merging methods. It also states several criteria that validate the stability of metasearch engines and, finally, conclude the chapter explaining the future work.

**Keywords** Knowledge computing · Search engine · Metasearch engine
Page ranking · Result merging

---

P. Vijaya (✉) · S. Chander
Waljat College of Applied Sciences, P.O Box 197, 124 Rusayl,
Muscat, Sultanate of Oman
e-mail: pvvijaya@gmail.com

S. Chander
e-mail: dimrisatish@gmail.com



S. Margret Anouncia and U. K. Wiil (eds.), *Knowledge Computing and its Applications*, https://doi.org/10.1007/978-981-10-8258-0_10

# 1 Introduction

The world, which we live now, has a knowledge repository of information choices—print, spatial, visual, sound, numeric, and so on [1]. Knowledge, which is a facilitator of various capabilities like learning, is transformed into economic value, called knowledge economy, to meet certain organizational goals [2]. Knowledge economy deals with sharing knowledge-based services and products among different users and providers. To facilitate "understanding," which refers to the process of linking new information and knowledge to the existing one, standardization is required for the knowledge obtained from various information sources. For precise computations, knowledge has to be handled or managed effectively. Knowledge management can be considered as a self-organization that adapts to changing environments depending on dynamic evolution, known as knowledge ecosystem. Enhancement of autonomous behavior of knowledge based on Semantic Web standards is known as executable knowledge, and the corresponding process of management of executable knowledge is termed as knowledge computing [3]. The World Wide Web (abbreviated WWW or the Web) is a knowledge management system. The tremendous growth of the World Wide Web in the last decades has promoted searching [4] as one of the most prominent issues in the field of Web research [5]. The World Wide Web consists of a number of Internet servers that protect the formatted documents designed in a Markup Language known as Hyper Text Markup Language (HTML). Moreover, it sustains the link to several documents, with graphics, audio, and video files.

The Users refer the World Wide Web as "the Web," that is a segment of the Internet. In general, Web enables the user to jump from one document to another, through a click on the hot spots. However, all the Internet servers need not be necessarily a part of the www family. The size of the Web has grown exponentially since 1990, and it is estimated to contain not less than 14.27 billion Web documents that are publicly accessible and are distributed all over the world via several thousand Web servers. Searching and obtaining relevant information from a massive collection of Web documents is a complex task, as the Web pages are neither managed as books in a library nor cataloged fully at a central location. Hence, there is a need to design an efficient information retrieval approach for obtaining the relevant information from the Web. An important tool used for accessing online information available on the Web is a search engine. It is a synchronized set of programs, which can access every searchable Web page, by creating an information index that is compared to the query requested by the user, thereby returning the results to the user [6].

According to the studies made by American Life Project, the search engine use has become a regular online activity. The increasing popularity and the influence of search engines generate questions regarding their role in determining the global information order [7]. The word "search" refers to the attempt of finding something, which is either existing or novel [7]. The search engine processes the keyword query received, providing a list of Web pages, called Search Engine Results Page

(SERP) [8]. The results of the search engines are ranked displaying the most relevant results first. In the case of generating irrelevant results, the users must filter their queries to discover the required information [9].

This chapter deals with the basic concepts of knowledge extraction using metasearch engines by discussing the definitions of search engines and its classifications and thereby explains the architecture of metasearch engine [10, 11]. Moreover, it summarizes the ranking concepts that the metasearch engines perform by describing the result-merging techniques. Then, the validation criteria regarding the stability of the metasearch engines are explained with respect to the user queries to show the effectiveness of the metasearch engines. The organization of the chapter is as follows: Sect. 2 provides few definitions of search engines stated by the researchers. The classification of search engine is explained in Sect. 3, along with the architecture of metasearch engine, describing each of its components. In the same section, different merging methods employed in the metasearch engine are demonstrated. Section 4 shows the development criteria, while the conditions to validate the queries are explained in Sect. 5. Common issues to be addressed are discussed in Sect. 6, and Sect. 7 provides the conclusion and the future work.

## 2 Definitions of Search Engines

Search engines are effective in identifying keywords, quotes, phrases, and information that are available in the entire content of Web pages. Search engines permit the user to search the keywords that are stored in a huge database, which can be retrieved when needed. Some of the definitions of search engines explained by the researchers are as follows:

According to [12], a "World Wide Web search engine" is termed as "a retrieval service, consisting of a database (or databases) describing mainly resources available on the World Wide Web (WWW), search software, and a user interface also available via WWW."

According to [13], a search engine is referred as "a program that is accessible by any average user, capable of accepting user input which defines the information it produces as output to this user."

According to [7], a domain-specific search engine is "an information access system that allows access to all the information on the Web that is relevant to a particular domain."

According to [9] a search engine is stated as, "a search tool that allows one to find specific documents through keyword searches and menu choices, in contrast to directories, which are list of Web sites classified by topic."

# 3 Classifications of Search Engines

The classification of search engines can be made in many ways. Herein, the classification is done on the basis of two concepts: working principle and properties. Based on the working principle, the search engines can be categorized into three, as follows.

## *3.1 Based on Working Principle*

**Crawler-based Search Engines**: In crawler-based search engines [6], the listings for the index or catalog are created automatically using a software tool, named "Web crawlers" or "spiders," by sending the spiders to crawl across the Web, following the links on one Web page to the next. Here, the search engine utilizes a computer algorithm for the ranking of all the retrieved pages. These search engines are massive, retrieving a large volume of information from the Web. The search engine is efficient in providing relevant information for the user who has a specific search topic [6]. For general searches, the crawler-based search engines may provide irrelevant responses, where the keyword may found only once in the document [13]. It permits the user to search based on the results of the previous search for complicated searches, enabling the user to filter the search results. This kind of search engines has the complete text of Web pages which are connected so that the user can obtain the pages by matching the words in the desired pages [6]. One of the main challenges in this search engine is the dynamic nature of the Web, where it requires keeping the database up-to-date [9]. Google, Teoma, Altavista, and Lycos are some of the crawler-based search engines.

**Human-powered directories**: These are based on human editors who create listings to construct the directories. Human-powered directories [6] are structured into subject categories, and the pages are classified based on the subjects. They do not hold full text of the Web page to the link they are connected. The Webmasters provide a short description and the URL to the directory for their Web sites. These descriptions edited manually constitute the search base. Hence, a change in individual Web page will not affect the pages listed in the results. The directories are capable of providing much more relevant results than that offered by the search engines. During the search, the directory site focuses only on the matches that are within the descriptions submitted, rather than the information on the Web pages. The user has to submit an online update to the Webmaster of the search engine to update the description of the Web site. Even though the search topic is relevant and accurate, this is not efficient for searching when the user has a specific search topic in his mind. Few human-powered directories are MSNSearch, Yahoo, AskJeeves, Look Smart, and Open Directory.

**Metasearch engines**: Search engines [13] can filter the pages matching explicit queries effectively. However, they require huge memory resources to store the Web index and extreme network bandwidth to construct and refresh the index. As they receive millions of queries a day, the CPU cycles dedicated to satisfy each query are curtailed. This result in information overload that the mandatory intelligence fails to combat at most of the time [1]. To overcome the issues of search engines and to solve the information overload problem over the Web, metasearch engines are introduced [1]. It is likely to attain better results by submitting a query to various search engines, as their databases do not overlap completely. Metasearch engines, such as Mamma, Metacrawler, and Dogpile, perform this automatically. Actually, metasearch engines do not own a database; instead, they submit the query automatically to multiple conventional search engines and obtain the results [14]. Thus, it is possible to acquire more responses, as the combined coverage of multiple search engines is appreciably larger than the result of any single search engine [9]. A searching tool that utilizes the results of other search engines, irrespective of user's preferences, is known as a metasearch engine [1].

A Metasearch engine [15] characterizes an in order retrieval agent which is configured at the apex of parallel search engines. The Metasearch engine receives the query that forwards the queries to a number of individual search engines in parallel. Consequently, the Metasearch engine integrates the results from all the search engines. The numerous models of the Metasearch engines (MSE) are the Metacrawler, Profusion, Savvy search, and the MetaSEEk. There are a lot of challenges in dealing with the Metasearch engine. In contrary, the combined result from various search engines is the hectic challenge in utilizing the Metasearch engine. Upon the arrival of the specific query, the individual principal search engine will re-examine the result, which reasonably leads to a subset of the ultimate post-processed upshot of a Metasearch engine. In case there are a large number of hits of a search engine, the users check only the top-listed documents of the results, and the order of documents in the ultimate outcome assumes added significance.

## *3.2 Based on Properties Included*

The above three categories of search engines have three common properties, such as general, personalization-based, and semantic-based. A hierarchical diagram is presented in Fig. 1, regarding the categorization of the search engine.

**General search engines**: General search engines are those search engines intended to search any general search topic. Google, Bing, and Yahoo are general search engines, where images, maps, news, and much more, can be searched. Specialized keywords and symbols can also be used in general search engines to obtain specific information. These search engines are efficient in finding Web sites with relevant information.

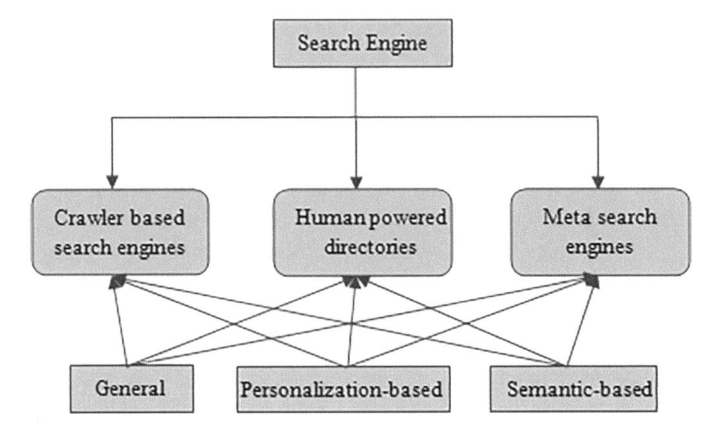


**Fig. 1** Hierarchy of search engine

**Personalization-based search engines**: [16] Personalization-based search engines track the search history of a registered user and then regulate the search results according to the preferences of the users. For example, a person searching for "jaguar" will get the results for the animal if his/her past searches were based on the information regarding animals, and car if the past searches were related to automotives. The results are obtained taking the benefits of information on the Web sites based on the analysis of the behavior of the user in association with the data taken from the Web context. Some of the benefits of personalization-based search engines are as follows: Personalized search can save the user's time by avoiding the repetitive search. The search engine eliminates the tasks that are no longer required from the search results. Thus, the search engine can offer an efficient and faster means to obtain the results as per the requirements of the user. Personalization-based search engines provide the user more satisfactory and accurate result.

**Semantic-based search engines**: The search process of the conventional search engines is improved based on semantic search. Hence, the researchers try to develop metasearch engines based on the idea of including the semantic information. This kind of search includes the context of search, intent, variation of words, specialized queries, location, and so on, to offer relevant results [1]. Major Web search engines, such as Google and Yahoo, incorporate few semantic search elements to enhance the search. LinkedIn, which is a social networking service, provides semantic search for searching the job to recognize and standardize the entities in terms of queries and documents, such as companies, skills. Depending on the entities, several entity-aware features are constructed. This search engine is aware of the context being searched and thereby provides smart and relevant results based on the queries requested. Earlier, the search engines use page ranking approach to provide ranking to the particular link for the relevant search. Conversely, a semantic search engine utilizes ontology for more meaningful and precise search results to be retrieved in minimum time. It ensures that most of the relevant results are returned based on a word's meaning and relations, rather than a specific keyword. The search engine maintains semantic identification regarding Web resources in a way to solve complex queries. By integrating technologies related to Semantic Web into the search engine, semantic offers improved search results, evolving future generation of search engines that are built on Semantic Web.

### 3.3 Architecture of Metasearch Engine

Metasearch engines are constructed on other search engines without collaborating with these search engines explicitly. Even though designing the programs may not be a difficult task for an experienced programmer, preserving their validity can be a serious issue as the connection parameters and the format of result page visualization of the search engines may change. Moreover, in the case of applications, which require connecting several thousand search engines, maintaining these programs is costly and time-consuming. Figure 2 shows the architecture of a typical metasearch engine.

The components of the metasearch engines are explained as follows,

**Database Selector**: Identifying potentially useful databases for a given query is important in metasearch engines. This is the responsibility of the software component database selector. Selecting local databases makes a search engine useless, as it is wastage of effort.

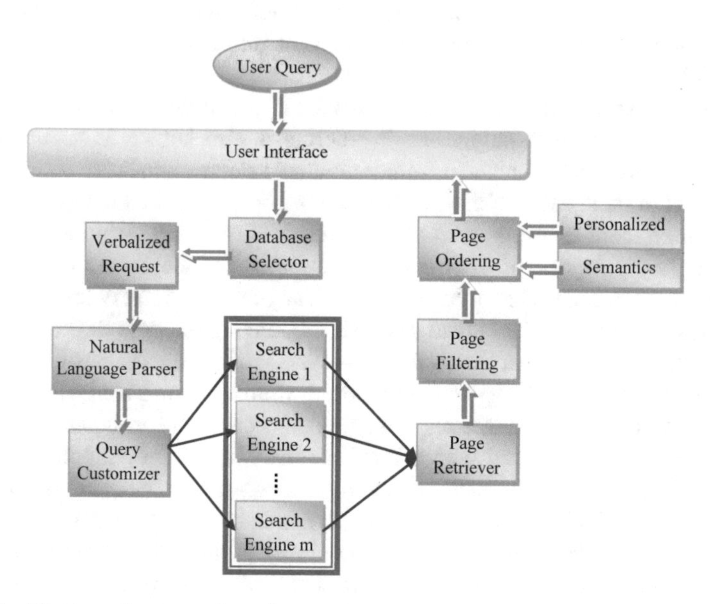


**Fig. 2** Architecture of metasearch engine

**Natural Language Parser**: The users who have information needs create verbalized requests, define what they prefer on the search strategy. They can input their requests either as keywords or a query in a natural language to explain their requirements. The user can select any input format using the user interface. Hence, the user who is unaware of the exact message can access the desired information, with the utilization of natural language parser that has the ability to recognize the natural language query. Thus, the information required will be formatted into a verbalized query.

**Query Customizer**: Query customizer is an agent that progresses the customized query that is determined by the fusing the process of selecting the search engine, query modification settings, and the verbalized query. To facilitate search engine selection process, the information having the document contents of individual component search engine is to be collected. This informative content of a search engine is termed as the search engine representative. The metasearch engine stores the representatives of all the collected search engines in advance. To select a search engine for a particular query, they are ranked according to the matching of the representatives with the query. Various search engine selection methods are being developed using various types of representatives.

**Page Retriever**: The function of a page retriever is to analyze the search results obtained from every search engine and forward the results to the succeeding agent, i.e., page-filtering agent.

**Page Filtering**: The users of search engine must learn few important aspects, such as the selected information resources based on which the results are retrieved, the reliability of the resources, the scope of the resources, information relevancy, and the time when the information is taken. The information gathered by the page-filtering agent is send to the users of the search engine. Moreover, the filtering agent eliminates the out of range and irrelevant information, sending the "good" pages to the following page-ordering agent.

**Page-Ordering**: Ordering the search results is one of the crucial decisions to be made by the Metasearch engine. Page ordering is also known as result merging, where the search results obtained from individual search engines are combined into a single list of rank. Most of the existing search engines used a numerical matching or similarity score in every search result returned, and the techniques for result merging were designed in such a way to "normalize" the scores obtained from various search engines. One of the metasearch engines that employ this technique is Inquirus. Offering a regular way to calculate the ranking scores is the key advantage of this technique such that the ranking provided makes sense. However, it has a limitation due to the delay produced while downloading and analyzing the result instantly, which lead to longer response time. Recent search engines visualize the title of every result returned with a brief summary, known as snippet. These two features of a result can offer better clues suggesting whether the result is relevant to a query or not. Various factors, such as the number of unique terms of the query appearing in the title or snippet and the proximity of those terms appearing in title or snippet, are used to measure a matching score, when the result merging is based on titles and snippets. Even though multiple search engines return same results, they probably tend to be relevant to the user query. This is due to the fact that most of the ranking techniques attempt to offer the same set of relevant results, but various sets of irrelevant results [6]. Adding the ranking scores of the results obtained from several search engines, these results can be ranked higher in the rank-merged list so as to compute the absolute score for the search result.

Different learning methods can be used to enhance the search selection. Moreover, metasearch engines utilize a recommender system to analyze the patterns of the resources of a particular user, and his/her search history, so that the search engine can include user's preferences on their search choices. The process often makes the availability of help information, which is customizable, such that the users can make use of it when desired. Hence, the Metasearch engines are undoubtedly efficient and easily accessible allowing the users to input flexible queries. The customizable preferences like adaptability in using user-desired resources, ability to update the query, providing better ordering policy and the feedback mechanism, add to the benefits of metasearch engines than general search engines.

### *3.4 Merging Methods for Metasearch Engines*

Several merging techniques utilized to order or merge the results of the search engines are summarized in this section, to provide better retrieval of search results for knowledge computing. The process of result merging refers to fusing of the results generated from a number of the search engines into a single ranked list. Following are few commonly used methods for ranking the search results:

(i) Fusing the outcome of the multiple search engines by normalizing the scores into values within a common range with an objective to make them better so that the results obtained from valuable search engines are to be ranked higher.
(ii) Downloading all the retrieved documents from their local servers and utilizing a similarity function to find their matching scores of the search engines used by the metasearch engine.
(iii) Employing voting-based approaches.
(iv) Adopting techniques that depend on titles, snippets and other similar features.

Based on the above concepts, various result-merging techniques [17, 18] employed in metasearch engines are described in detail as follows.

Let $E = \{E_i\}, 1 \leq i \leq n$ represents the search engines present in a metasearch engine, where $n$ denotes the selected number of search engines. Each search engine retrieves a Web page, $W$. Hence, the Web page retrieved by the $i$th search engine is represented as, $W_{ij}; 1 \leq j \leq m$, where $m$ indicates the total number of Web pages recovered by the $i$th search engine. Let $R_k$ be the ranked Web page of a metasearch engine, where $k = 1, 2, \ldots, p$, such that $p = m * n$ corresponds to the number of unique Web pages obtained in the result. Consider $Q = \{q^t\}; 1 \le t \le r$ represents the query, where $q_t$ denotes a query term and $r$ refers to the total number of query terms.

**Simple merge algorithm**: Simple merge algorithm is actually a metasearch engine, which collects the results from other search engines based on a multi-way merge approach. In realistic, users often prefer the first three pages of the search results. The remaining pages could not enhance user experience, rather could only provide completeness to the search results. This simple merge algorithm provides the results of all the search engine by unifying based on the search ranks. Therefore, the merge result provided by the simple merge algorithm is,

$$R^{\mathrm{SM}} = \bigcup_{j=1}^{m} \bigcup_{i=1}^{n} W_{ij} \tag{1}$$

where $R^{\mathrm{SM}}$ is the merge result of the metasearch engine. $W_{ij}$ is the $j$th Web page of $i$th search engine.

**Abstract merge algorithm**: Abstract merge algorithm is based on ranking, where the search results are ranked according to the relevance between the abstract information or snippet and the query of the results. Initially, the key terms are extracted from the query, and the relevance between the terms and the snippet are calculated. Then, the relevance between the query and the page is computed individually for each page. Based on the relevance calculated, the search results are returned to the users. The relevance between the query and the snippet is calculated as follows,

$$R^{\mathrm{AM}}(W_k) = \sum_{t=1}^{r} \ln\left(\frac{l(s)}{\mathrm{loc}(q^t, s)}\right) \tag{2}$$

where $R^{\mathrm{AM}}(W_k)$ is the ranking of $k$th unique page in the search engine results by the metasearch engine. $l(s)$ denotes the length of the snippet $k$th unique page, which is denoted as $s$ and $\mathrm{loc}\left(q^t_{ij}, s\right)$ indicates the location of $i$th query term in the snippet.

**Position merge algorithm**: In position merge algorithm, the objective is to utilize the original information regarding the position entirely from each individual search engine. Few pages may occur in many result lists of various search engines for the same query, but their position may differ for each result list. As a solution to this contradiction, the position of the pages in each search engine is taken into account. For $n$ search engines in the metasearch engine, the rank is obtained based on the position of the Web pages in the result of the search engine as,

$$R^{\mathrm{PM}}(W_k) = \sum_{i=1}^{n} \left[\frac{1}{X(W_{ik}, S_i)} \times n - i + 1\right] \tag{3}$$

where $S_i$ shows the outcome of the $i$th search engine and $X(W_{ik}, S_i)$ means the position of $W_{ik}$ in $S_i$. After the calculation of the relevant results between the query and each result record of the search engine, the outcomes are organized in descending order, returning the results to the users in the HTML format without overlapping.

**Abstract/position merge algorithm**: In abstract/position merge algorithm, the two aforementioned factors, namely abstract and position, are considered synthetically to return an integrated results to satisfy the needs of the user. The algorithm combines the ranking score of both abstract merge and position merge algorithm to obtain the final score, based on which the results are merged. Abstract/position merge algorithm helps in re-ranking the results retrieved from the search engine.

$$R^{\text{AM}}(W_k) = A * R^{\text{AM}}(W_k) + B * R^{\text{PM}}(W_k)$$

where $A$ is the weight value of the abstract merging algorithm, and $B$ is the weight value of the position merging algorithm.

**Take the Best Rank**: This algorithm intends to assign the score values of a Web page by choosing the best rank obtained among the search engine rankings, avoiding the clashes with the aid of search engine popularity.

$$R^{\text{BR}}(W_k) = \min\{R_1(W_{1k}), R_2(W_{2k}), \ldots, R_i(W_{ik}) \ldots, R_n(W_{nk})\} \tag{4}$$

where $R_i(W_{ik})$ indicates the rank of $i$th search engine for the Web page of $W_k$.

**Borda's Positional Method**: Here, the method computes the $L1$-Norm of the ranks for the results obtained in various search engines to find the MetaRank of a Web page. Search engine popularity is adopted to avoid the clashes. The procedure to compute the score of a Web page based on this method is the average of the Web page ranking given by all the search engines.

$$R^{\text{BR}}(W_k) = \frac{R_1(W_{1k}) + R_2(W_{2k}) + \cdots + R_i(W_{ik}) + \cdots + R_n(W_{nk})}{n} \tag{5}$$

**Weighted Borda-Fuse**: In Weighted Borda-Fuse algorithm, the assumption of the search engines is that they are unequal; instead, their votes are computed along with the weights assigned. The vote computation is based on the reliability of the individual search engines. The users can assign the weights that are included to compute the votes, in their profiles. Hence, the votes obtained for the results rendered by $i$th search engine are given by,

$$R^{\text{WB}}(W_k) = A_i * \left(\max_i^n S_i - j + 1\right) \tag{6}$$

where $A_i$ corresponds to the weight of the $i$th search engine and $S_i$ is the outcome of the $i$th search engine.

**The Original KE Algorithm**: KE Algorithm is a score-based approach utilizing the ranking result of the search engines and the number of its occurrences in the listings of the search engine. Since all the component engines are considered to be equal, they are treated as reliable. Equation (7) indicates the formula for ranking the retrieved results.

$$R^{\mathrm{KE}}(W_k) = \frac{R_1(W_{1k}) + R_2(W_{2k}) + \cdots + R_i(W_{ik}) + \cdots + R_n(W_{nk})}{(c * n * (r/10 + 1) * c)} \tag{7}$$

where $c$ represents the total number of search engines retrieved the Web page $W_k$, $n$ denotes the specified number of search engines, and $r$ represents the number of ranked outcome taken from every search engine by the algorithm. Hence, it is obvious that smaller the weight of a result score better is the ranking the result obtains.

**Borda Count**: This is a voting-based technique used for the fusion of the data. In Borda count, the results obtained are assumed as the candidates, while each search engine is considered as a voter. The candidates in the top rank list are assigned "$c$" points; for each voter, the candidates in the second top rank list are given "$c$ – 1" points, and so on. The candidates, which a voter did not rank, are those not returned by the associated search engine. Therefore, the remaining voter points are equally split among the candidates and the ranking of the candidates is done based on the received points.

**Merging Based on Combination Documents Records (SRRs)**: The effective method among all the available merging techniques is the method developed by combining the document evidences, such as search engine usefulness, title, and snippet. In this method, the global similarity is computed by measuring the similarity between the query and its title and its snippet, which is merged linearly for each document. The weight is estimated for each query term in the component search engine using Okapi probabilistic model. The weights of the query term of the search engine are added to get the search engine score. At last, the relative deviation of the score of its source search engine is multiplied with the mean of all the search engine scores obtained, to adjust the global similarity of each result. For a given query, this method also has the possibility of returning the same document from different component search engines. In such case, the method combines their ranking scores. Several linear combination fusion functions have been designed to tackle this issue with the inclusion of min, max, average, sum, and so on.

**TopD—Use Top Document to Compute Search Engine Score**: TopD is an algorithm that merges the result based on the similarity between the query and the top-ranked document retrieved from a search engine. Taking the top-ranked document from the local server introduces a delay, which is negligible, as it requires fetching of only one Web page from the search engine. To estimate the similarity, two functions, namely Cosine function and Okapi function, are used. The similarity estimation using Okapi function [19] is given as follows,

$$\mathrm{ST}(W_k, Q) = \sum_{t=1}^{r} W^o * \frac{(d_1 + 1) * f}{D + f} * \frac{(d_3 + 1) * q_f^t}{d_3 + q_f^t} \tag{8}$$

where $\mathrm{ST}(W_k, Q)$ is the similarity between the Web page and query based on top results, $W^o$ is the Okapi weight, given by, $W^o = \log \frac{b - b^T + 0.5}{b^T + 0.5}$, and $D = d_1 * \left((1 - \alpha) + \alpha * \frac{g}{g_{\mathrm{avg}}}\right)$. $f$ is the frequency of the query term $u^T$, $q_f^t$ is the frequency of $u^T$ within the query $Q$, $b$ is the number of Web pages, $b^T$ is the number of Web pages with $q^t$, $g$ is the length of the Web page, $g_{\mathrm{avg}}$ is the average length of the Web page, and $d_1 = 1.2$, $d_3 = 1000$, and $\alpha = 0.75$ are the constants. This algorithm offers a ranking scores 1 for the top-ranked results obtained from all the search engines. For a Web page obtained from several search engines, the final ranking score is estimated by adding all the ranking scores.

**TopSRR—Use Top Search Result Records (SRRs) to Compute Search Engine Score**: TopSRR algorithm computes the search engine score based on the SRRs of the top results retrieved from each search engine, rather than the top-ranked Web pages. This makes the algorithm sensible as, for a given query, a useful search engine probably provides better results that are revealed in their SRRs. The algorithm merges all the titles of the top SRRs from each search engine to create a title vector and the snippets into a snippet vector. Following equation gives the formula of the search engine score obtained by combining the similarities between a query and a title vector, and between the query and the snippet vector,

$$\mathrm{SRR}(W_k, Q) = P_1 * \mathrm{Sim}\left(Q, V_i^T\right) + (1 - P_1) * \mathrm{Sim}\left(Q, V_i^S\right) \tag{9}$$

where $\mathrm{SRR}(W_k, Q)$ is the similarity between the Web page and query based on title and snippet, $P_1$ is a constant that takes a value 0.5, $V_i^T$ is the title vector and $V_i^S$ is the snippet vector, $\mathrm{Sim}\left(Q, V_i^T\right)$ is a function to compute the similarity between the query and title vector.

**SRRsim—Compute Simple Similarities between SRRs and Query**: As each SRR is treated as the representative of the corresponding document, it is possible to rank SRRs obtained from multiple search engines. SRRsim algorithm [19] measures the similarity between a SRR and a query, defined as a weighted sum of the similarity between the title of SRR and the query and the similarity between the snippet of SRR and the query, as given below,

$$\mathrm{SNT}(W_k, Q) = P_2 * \mathrm{Sim}(Q, T^w) + (1 - P_2) * \mathrm{Sim}(Q, S^w) \tag{10}$$

where $\mathrm{SNT}(W_k, Q)$ is the similarity between the Web page and query based on title and snippet, $P_2 = 0.5$, $T^w$ is the title of $W_k$, and $S^w$ is the snippet of $W_k$. For a document with different SRR returned from several search engines, the similarity is measured between the SRR and the query and the maximum similarity will be taken as the final similarity value for ranking.

**SRRRank—Rank SRRs Using More Features**: The similarity measure designed in the SRRsim algorithm is not suitable in revealing the true matches of the SRRs for a given query. This measure did not consider either the proximity information, like the closeness of the query terms in the title and snippet of a SRR, or the order in which the query terms appear in the title and snippet. SRRRank algorithm overcomes those drawbacks of SRRsim algorithm by considering both the order and the proximity information, as they have a considerable impact on matching the phrases. The consideration of the information is based on five different features [20] defined based on the query terms as follows.

- $N$: The number of distinct query terms appears in the title and the snippet.
- $N^T$: Total number of times the query terms appears in the title and the snippet.
- $X$: The locations of the occurred query terms.
- $A$: Checking the occurred query terms if they are placed either in the same order as specified in the query or in a different order.
- $w^s$: The size of the window with distinct occurred query terms.

The information regarding these five features is obtained for every SRR of the result generated. The procedure involved in the SRRRank algorithm is given in the following steps:

- Initially, the algorithm groups all the SRRs depending on $N$ and the groups that have more distinct terms than the other groups will be ranked higher.
- The SRRs in each group are grouped further to form three subgroups based on the feature regarding the position, denoted as $X$.
- Then, ranking is done by offering the highest rank to the SRRs having large $N^T$ appearing in the title and the snippet within each subgroup. When two SRRs are with the same number of occurrences of query terms, the SRR with distinct query terms placed in the same order as they are given in the query will be ranked higher, after which, the one with least window size will be ranked. In the merged list, the result having the higher local rank is given a higher global rank. When a result is returned from several search engines, the one having the highest global rank will be kept.

**SRRSimMF—Similarities between SRRs and Query Using More Features**: SRRSimMF has a similar procedure of SRRRank, but differ in a way by quantifying the matches depending on the features identified in SRRRank such that the scores are merged into a numeric value. In the specified field of a SRR, $S$, for the given $N$, the matching score is given by the ratio of $N$ to the total number of distinct terms in the query, denoted as $N^D$,

$$C_N = \frac{N}{N^D} \tag{11}$$

For the $N^T$ given, its matching score is computed as the ratio of $N^T$ to the length of title, as

$$C_{N^T} = \frac{N^T}{g^T} \tag{12}$$

where $g^T$ is the length of the title. Based on the order of the query terms and $A$, the matching score, $C_A = 1$, if the distinct query terms are in the same order, adjacently in the title, and $C_A = 0$, for the other case. The score obtained using the $w^S$ of the distinct query terms in the title is $C_{w^S} = \frac{w^S}{g^T}$. By merging all the scores of these features into a single value, the similarity between the title and the query is given by,

$$\mathrm{Sim}\left(T^S, Q\right) = C_N + \left(1/N^D\right) * \left(w_1 * C_A + w_2 * C_{W^S} + w_3 * C_{N^T}\right) \tag{13}$$

where $w_1$, $w_2$, and $w_3$ are the constants, which take values 0, 0.14, and 0.41, respectively. Hence, the final similarity is computed as,

$$\mathrm{Sim} = \frac{N}{g^T} * \left(P_3 * \mathrm{Sim}\left(T^S, Q\right) + (1 - P_3) * \mathrm{Sim}(s, Q)\right) \tag{14}$$

where $N$ is the total number of distinct query terms in title and snippet, $g^T$ is the length of the query terms, $\mathrm{Sim}(s, Q)$ is the similarity between the snippet and the query, and $P_3 = 0.2$.

# 4 Development Criteria of Metasearch Engine

Various criteria that tend to become a part of the development of metasearch engine are explained as follows.

## *4.1 Selection of Search Engine*

Metasearch engine requires multiple search engines and its interfaces to connect with a new Metasearch engine. The result of the Metasearch engine completely depends on the results of the search engines which we have considered. So, the right selection of the multiple search engines plays a major role in the Metasearch engine. An ideal metasearch engine performs database selection, to recognize the most appropriate component search engines such that they access a given user query. A metasearch engine with a huge number of component search engines makes it inefficient, as it requires sending the user query to every component search engine, which in turn inquires additional time to access each search engine and to process the user requests. Hence, for such cases, database selection should be carried out. A perfect database selector must be capable of correctly identifying all the possible databases that are potentially useful in such a way the possibility of identifying a wrong database is considerably reduced. The search engine selection is depending on the value of relevance computed between the user query and the search engine. Generally, database selection is performed based on the collected information that represents the main content of each search engine [20]. On receiving a user query, the usefulness of a search engine is measured using the representative information based on the query [17]. This helps in determining whether a search engine can be selected to process the particular query. The usefulness of the search engine is termed as search engine score [20]. However, there occurs a problem called database selection problem in metasearch engine. The difficulty regarding the identification of potentially useful databases for searching a particular query is termed as the database selection problem [17]. Hence, selecting suitable search engines to process the user query by handling database selection problem is a necessity and a challenging task.

### 4.2 Query Reformation for Search Results

Query applying through the search engine is highly sensitive to the results of the Metasearch engine. So, a user query can be reformulated into meaningful keywords to obtain the relevant information from the search engine. It would enhance the performance of the Metasearch engine. The process of modifying the original query based on its similarity to enhance the effectiveness of the search is known as query reformulation. This process requires the evaluation of a user's input, extending the search query to match similar documents. Add or remove words, acronym expansion, and word substitution are few query reformulation techniques. Earlier query reformation techniques mainly concentrated on the way of extracting useful patterns and ranking of the candidates with the useful patterns, whereas the candidate generation models are simple. This method through the usage of the word-by-word transformation can make the string transformation accurate and efficient. When a user inputs the query, the system returns all possible queries based on the original query. From the generated list of matching queries, the technique selects the top query that the user prefers [18]. Query reformation can also be done based on the keywords and their similarities in the query. The idea behind the usage of keywords in query reformation is to offer better results, instead of sending the original query to the search engine. Moreover, this can avoid the generation of more irrelevant results from the search engine to improve the effectiveness of retrieval [21].

The process of query reformation can be described as follows: For a given input query "*u*," the most probable output queries are generated following a transformation based on a set of operators included in a chain. The input queries can be sentences, strings of words, character, and so on. Every operator in the chain is considered as a transformation rule that replaces a query with another query with similar meaning. Query reformulation aims at solving the issue of maturity mismatch. This can be understood clearly from the given example. If a user submits a query "apjabdulkalam" to the search engine and the document available is only "Famous books by Dr. APJ Abdul Kalam," retrieving results seem to be a difficult task, as the query and the document do not match and the document is not highly classified. In fact, the query reformulation technique tends to transform "apjabdulkalam" into the "Famous books by Dr. APJ Abdul Kalam," and thus, form a better match between the document and the query. This also includes writing the original query or its similar words to match the original query or the similar words in the dictionary so that the search efficiency can be improved [18].

### *4.3 Bringing Semantic Richness*

Before integrating the search results, the chief problem to be addressed is how to obtain the semantic results from the user query or reformulated query even though the intent of keywords is not presented in the user query. In semantic information retrieval applications, keyword similarity is not much useful, because of the difficulty in ensuring the user intention. In addition to keyword similarity measure, semantic similarity measure must be counted in for the information retrieval based on the semantics. Many interesting researches have been developed in semantic search associated with Web mining and ontologies like ODIX platform. ODIX is a metasearch engine offering search results that are structured based on the interests of the specific user and ontologies. The advantages of using ontology in semantic search are many: Ontology-based optimization can locate the Web sites in the search engines, filtering of results using ontology depending on the search concepts extracted from social Web sites to enhance the search, and it improves the relevance of search results based on search-term disambiguation. Semantic richness in the Web mining process that is provided by the ontologies to enable the realization of the outcomes retrieved and the entire performance of the search process. While browsing contents through the Internet using the search engines, such as Bing and Google, the search outcomes are obtained irrespective of the polarities, or they provide the scores depending on the sentiment information generated through the search. Hence, the techniques based on sentiment analysis can be used into a general-purpose sentiment search engine. In general-purpose semantic search engines, the search results are structured by comparing the meaning and not on the popularity of the search terms. Offering functionalities other than sharing the information with the users and search based on the keywords is a complicated task for the search engines, if the semantic information is not specified.

Metasearch engine identifies some new Web pages which are not in the top ten results of any other search engines. These search results are obtained through the semantic matching of keywords obtained through WordNet. Also, the ranking of the Web pages is changed based on the significance and semantic richness. In [22], more user-oriented search results are obtained using the ontological approach in the clustering process. Here, an initial search is done on several search engines and the search results are preprocessed, transforming them into word vectors. WordNet lexical database is utilized to map the word vectors into concept vectors. Recently, with the utilization of text-based search engines, the commercial search engines can make a significant advance with better results for image searching. Even though the research on semantic image retrieval is growing significantly, it is still not widely employed for use due to the rapid index querying, which in turn leads to computational overhead. For the search requests associated with the images, there exist two categories of search requests, such as unique and non-unique queries. The queries that are satisfied based on the retrieval of a unique person, event, or object are known as unique queries, whereas all the other cases come under the category of non-unique queries. Recent search engines like Yahoo and Google utilize text descriptions adopting semantics. Thematically homogeneous groups can be constructed with the help of clustering process from the initial list generated using standard search mechanisms.

## *4.4 Designing of Merging Strategy and Algorithm*

The important challenge to be considered in the Metasearch engine is how to integrate all the results of the multiple search engines and the ranking of those results. This includes the process of selecting the search engine results of important Web pages, removing unnecessary Web pages retrieved from the search engine, the ranking of Web pages. These challenges pose confront of designing aggregate ranking algorithm for Metasearch engine. The merging of multiple criteria is required to generate a score value for every page after the computation of the scores of multiple search engines through different criteria [23]. Convolutional search engines usually performed merging by assigning a numerical matching score to every search result obtained so that the result-merging techniques designed could normalize the scores retrieved from the search engines into values. The ranking is carried out in the search engine to rank the results based on the normalized scores. As mentioned in the architecture of metasearch engine, the usefulness of the search engines selected is estimated to assign weight to the normalized score that helps in ranking the results obtained from useful search engines to be higher. Aggregation of search results can also be performed using voting-based techniques.

Another approach in result merging is downloading all the documents obtained from multiple search engines and then finds their matching scores using a similarity function adopted by the metasearch engine. Thus, the approach computes the ranking score evenly by the score-based ranking. Presently, the search engines visualize the returned result using the title and the snippet, which can determine whether the query is relevant or not [21]. Hence, most of the result-merging techniques developed recently are based on titles and snippets. In such case, it requires a matching score computed relying on the proximity of the query terms in the title and snippet. The results are assumed to be relevant if the same result is obtained from multiple search engines. The ranking score measured helps in ranking the search results in descending order.

### *4.5 Visualization of Merged Results*

The final step is to visualize the ranked Web pages to user-friendly interface to easily read and analyze the retrieved information. The SAVVY is a metasearch engine developed to calculate other search engines through autonomous collaborative decisions and provides the vibrant semantic link design and harmonized multiple view visualization. For illustration, it is considered that a user can have access to a number of resources. The user selects and taps to any one of the links, which are properly created as his operation. The chosen link takes the user to the required resources and benefits the user through providing a number of related links and the user can view all the related resources. Thus, the link provides the contents of the normal Web browsing process and it provides divergence that is harmonized equally. The user clicks the available static link once he adds a keyword into the search engine against the backdrop of modern browsing scenarios. Moreover, this stationary link may not benefit the user, though it is eases the access to view the resources that are linked to the user welfare. This problem is solved by using the traits of the energetic semantic link design and harmonized multiple view visualization. The semantic Web is a Web-based technology that enlarges the XML by providing the way to display the ontologies by describing the objects and relations existing among them. The advantage of SAVVY is that there is no need for the user to click the links of the list and browse the pages frequently.

To check Search Engine Optimization, which is the process of increasing the views of a particular Web site such that the Web site appears top in the result list provided by a search engine, text network visualization of search results can be extremely useful. Text network visualization is useful in identifying the areas between the keyword clusters that attempt to co-occur in the snippets. These empty areas indicate the missing parts in the results so that one can include those parts in the text to appear at the top of the search results of the search engine. The search engines provide snippets of text in their results that are relevant to the search query. Hence, it will be helpful if the words contained in those snippets are known, so that the content that is relevant both for the search engine and the audience can be attained. Therefore, the fundamental step in developing a metasearch engine is to better understand the interests of the people identifying the terms they are actually searching for, i.e., the context. Thus, it is possible to know other keywords to be included in the text to appear in the results of the search engine.

# 5 Stability Validation Criteria of Metasearch Engines Based on Various Queries

This section analyzes the effects of different kinds of metasearch engines on various queries. Here, four kinds of queries are taken for analyzing the retrieval effectiveness of the metasearch engines, as described below.

## *5.1 Contextual Queries*

The contextual retrieval of Web pages is important for the search engines because the Web pages should be different for the same query based on the location. Accordingly, Fathalla et al. [24] had developed the NEC Research Institute (NECI) metasearch engine by downloading and examining each document so that results showing query terms in the context can be displayed. Thus, the efficiency and the precision of Web can be improved. Besides, the users can recognize easily whether the relevant document is retrieved progressively without downloading every page. This technique is simple and more effective, especially when handling with large, poorly organized, and diverse database of the Web [25]. Generally, the metasearch engines can reduce the trouble of returning focused queries if the user interface automatically selects the search engines that are context relevant. When a user who works on an economics paper is determined by the interface, it can return a context description based on this information, and thereby, select a dedicated and context-relevant search engine like CNN financial. According to the discussion made on contextual queries, following three issues are addressed:

(i) Extraction and representation of the context: The way of determining and describing the query context.
(ii) Characterization and Selection of the source: Selecting useful sources such that they support decisions regarding source relevancy to provide an effective access. For the better search, the process of characterization of the sources should be simple, fast, and robust, without depending on the cooperation of the sources.
(iii) Selectivity: Determining when the specialized sources accessible become insufficient.

Eight kinds of context that a search engine used for contextual search are given as follows,

- *Individual*: Considering a person's history and context for an effective search.
- *Demographic Profile*: Including age, gender and occupation, as they can predict the interests of the person.
- *Interest Profile*: Expresses the topics one who is interested in.

- *Location*: Showing accessible shops, hotels, and so on, in the proximity of the town and the country.
- *Device*: Type of device and interface used.
- *Date*: Noticing weekdays, which are in proximity to events like Christmas or New Year.
- *Time*: Time of the day.
- *Weather*: Focusing on current weather to offer a search for local places, such as tourist spots that are popular on summer.
- *Mood*: Search based on the moods like excitement, positivity, and negativity, as they could impact the content.

## 5.2 Personalized Queries

The analysis of the personalized queries is important for the users to understand the benefits and the features of various metasearch engines [16]. More accurate results can be returned for personalized query satisfying the user queries. For an effective personalization, the search engine has to concentrate on the nature of task executed by a user, in addition to the needs, preferences, and interests of the user. These rely on the observed patterns and the resulting probabilities. Few strategies of personalization-based metasearch engines are as follows:

(i) Based on the results retrieved: The average of the results obtained from the database of the search engines is computed. Then, select the useful search engine according to the averages of search results returned.
(ii) Based on the experience of the user: The user selects the search engine based on his/her personal experience. Personalization depending on group-based behavior can personalize the results by analyzing the actions of the users who made similar queries before. The search engine customizes the results based on online behavior, search activity, and other details recorded in an anonymous cookie in the browser of the user. Utilizing these cookies, the search engines collect and store the search history in the databases. Metasearch engines personalize the search query based on certain features, some of which are given below,

- *Device (Operating System and Browser)*: The search results may be customized based on the device used, as the content people search varies depending on the type of device they use. A mobile user may search for more actionable, driving directions, and on-the-go information, while a desktop user may browse completely different information for the same query.
- *Location*: The locations of the searchers have an impact on the search results as people search for the same query at different locations.
- *Previous Searches and Frequent Site Visits*: Metasearch engines like Bing and Google may customize the search results based on previous search behavior.

The techniques used, allow the metasearch engines to consider previous search queries, and the contents taken from previous searches.
- *Bookmarks*: A user may bookmark a site or visit often for specific kinds of searches, such as refilling prescriptions, booking tickets, and so on, as the user has to visit the page for those searches in the future. The search engines utilize all the personal information submitted to show the site when needed.
- *Logged In and Out*: When a user login to an e-mail account or other similar profiles with the search engines, the results are likely to bias heavily to the signals like location and operating system.

## 5.3 Semantic Queries

The query word that has different meanings is a common issue, which is to be considered in the semantic searches. Hence, the metasearch engines designed have to utilize the semantic information so that the results are retrieved focussing on the relevant pages of the user. Following are the features considered by a metasearch engine for semantic queries,

- Language: Understanding the synonymous languages to enhance the search process.
- Linked data: Determine the actual relationships between information to find similar answers.
  - Context: The search engine tool tends to understand the context and meaning of the words in the query to provide accurate search results.
  - Previous Searches: These are analyzed to find semantic connections. The search engines must understand the related topics, to determine the relationship between the items.

## 5.4 Numerical Queries

The effect of numerical queries and its relevancy in retrieving the Web pages by the different metasearch engine is discussed in this part. Even though numbers play a significant role in modern life, most of the search engines assume numbers as strings, neglecting their numeric values. This is because of an unreasonable expectation that the user offers exact numeric values during the searches. In reality, they search for values whose specifications approximately match the values given in the query. As a result, techniques are developed to obtain the specification documents for the extraction of the pairs of attributes and values in a document and are stored in a database. Hence, the queries can be processed with the utilization of nearest neighbor approaches [26]. Numerical queries are carried out based on,

- Economic value of the data;
- Chronological reviews;
- Probability of the value.

## 6 Current Problems and Issues to Be Addressed

This chapter discusses the problems and issues with respect to various kinds of metasearch engines explained. Presently, certain research groups provide search results from the semantics-based search engines but most of which are in their developing stages. One of the large global databases is current Web, which lacks the subsistence of a semantic structure. Another major issue deals with the ranking of the results. Based on the predetermined condition, search engines assign ranks to the generated documents in the descending order of relevance in accordance with the user's preferences. A long list of titles of the documents is the result of the ranking process obtained by a search engine. The major limitation of such method is that the user has to browse through this long list to find the results that the user is actually looking for. Inefficiency in making distinctions between the various concepts available in the query generated in the resultant list of the documents by the search engine is also a drawback, since the list must be ranked sequentially [1].

A common difficulty in metasearch engine is that it relies on the underlying search engines to return a significant set of results. Metasearch engines can only offer a limited number of results to be retrieved to the user. To improve the accuracy of the search results, dealing with the limitations regarding the results due to the issues caused by search engines, metasearch engines modify the query by providing a search range that uses certain options, like sorting by date and time period, specifying the number of items to be retrieved, language constraints, number of items to be displayed. Another issue is the problem regarding the selection of relevant search engines based on the user query [11]. This is a challenging task, which can be solved to an extent by selecting a large number of search engines for the query [25]. Adding more search terms for better results is another problem, as it becomes a burden to the user. Moreover, it is difficult even for the experts to choose the appropriate query terms for the information to be returned by the search engine. Therefore, it is difficult for a machine to recognize the information provided by the user. When the information was circulated in the Web, there exist two types of research problems in search engine; i.e.,

- How a search engine can map a query to the documents having information but are not retrieved in intelligent and useful information?
- The query results provided by the search engines are distributed across various documents that are connected with a hyperlink. How a search engine can identify such a distributed search results efficiently?

## 7 Conclusion and Future Research Paths

In this chapter, a detailed description of various metasearch engines developed for information extraction, solving the common issues of search engines is presented. Different categories of search engines and the architecture of a typical metasearch engine explaining the process of information retrieval are explained in detail. Further, the ranking mechanisms and the important conditions to be satisfied during the development of the metasearch engines are discussed. The chapter has also illustrated several criteria used for the validation of the effectiveness of the metasearch engines, addressing their common issues.

The metasearch engine can support several interesting and specialized applications. The application of metasearch engines to education and e-commerce problems has not been highly studied. In this context, we suggest two main future research directions:

- In a large organization with a number of branches, such as a university having many campuses, a metasearch engine that connects the search engines of all branches turn out to be an organization-wide search engine.
- When a metasearch engine is constructed over different e-commerce search engines that sell the same type of product, it is reasonable to form a comparison-shopping system. Certainly, it requires a different result-merging technique for comparison-shopping applications.

## References


1. Hamdi, M. S. (2011). SOMSE: A semantic map based meta-search engine for the purpose of web information customization. *Applied Soft Computing, 11*(1), 1310–1321.
2. Lawrence, S., & Giles, C. L. (1998). Context and page analysis for improved web search. *IEEE Internet Computing*, 38–45.
3. Terziyan, V., Shevchenko, O., & Golovianko, M. (2014). An introduction to knowledge computing. *Eastern-European Journal of Enterprise Technologies, 1*(2), 27–40.
4. Agrawal, R., & Ramakrishnan, S. (2003). Searching with numbers. *IEEE Transactions on Knowledge and Data Engineering, 15*(4), 855–870.
5. Antoniou, D., Plegas, Y., Tsakalidis, A., Tzimas, G., & Viennas, E. (2012). Dynamic refinement of search engines results utilizing the user intervention. *Journal of Systems and Software, 85*(7), 1577–1587.
6. Kumar, D., & Kumar, A. (2013). Design issues for search engines and web crawlers: a review. *IOSR Journal of Computer Engineering (IOSR-JCE), 15*(6), 34–37.
7. Sharma, S. (2008). Information retrieval in domain specific search engine with machine learning approaches. International Journal of Social, Behavioral, Educational, Economic, Business and Industrial Engineering, 2(6).
8. Meng, W.: Metasearch engines. In *Encyclopedia of database systems* (pp. 1730–1734).
9. Nwosu, O., & Anyira Echezonam, I. (2011). Inquiry into the use of Google and Yahoo search engines in retrieving web resources by internet users in Nigeria. *Indian Journal of Information Sources & Services (IJISS), 1*(2), 61–67.

10. Vijaya, P., Raju, G., & Kumar Ray, S. (2014). S-MSE: Asemantic meta search engine using semantic similarity and reputation measure. *Journal of Theoretical & Applied Information Technology, 60*(2).
11. Ray, S. K., Vijaya, P., & Raju, G. (2013). An ontology based meta-search engine for effective web page retrieval. *International Review of Computers and Software (IRECOS), 8*(2), 533–541.
12. Poulter, A. (1997). The design of world wide web search engines: A critical review. *Program, 31*(2).
13. Weideman, M., & Kritzinger, W. (2003). Search engine information retrieval: Empirical research on the usage of meta-tags to enhance website visibility and ranking of e-commerce websites. In *Proceedings of the 7th World Conference on Systemics, Cybernetics and Informatics* (Vol. 6, pp. 231–236).
14. Vijaya, P., Raju, G., & Ray, S. K. (2016). Artificial neural network-based merging score for Meta search engine. *Journal of Central South University, 23*(10), 2604–2615.
15. Naval, P., & Priyanka, S. (2012). A survey on personalized meta search engine. *International Journal, 2*(3).
16. Jadidoleslamy, H. (2012). Search result merging and ranking strategies in meta-search engines: A survey. *IJCSI International Journal of Computer Science Issues, 9*(3), 239–251.
17. Maloth, B. (2010). *Evaluation of integration algorithms for meta-search engine*. Technical report. http://www.bvicam.ac.in/news/INDIACom%202010%20Proceedings/papers/Group3/INDIACom10_261_Paper%20(3).pdf.
18. Jadidoleslamy, H. (2011). Introduction to metasearch engines and result merging strategies: A survey. *International Journal of Advances in Engineering & Technology, 1*(5), 30–40.
19. Ngu, A. H. H. (2005). Business & economics, In *6th International Conference on Web Information Systems Engineering, WISE 2005*, New York, NY, USA.
20. Ngu, A. H. H., Kitsuregawa, M., Neuhold, E., Chung, J.-Y., & Sheng, Q. Z. (2005). Computers. In *6th International Conference on Web Information Systems Engineering, WISE 2005*, New York, NY, USA.
21. Li, Z., Wang, Y., & Oria, V. (2001). A new architecture for web meta-search engines. In *AMCIS 2001 Proceedings, 31*(84), 415–422.
22. Das, S., & Raghuwanshi, K. S. (2014). Search engine selection approach in meta search using past queries. *Oriental Journal of Computer Science & Technology, 7*(1), 177–183.
23. Rohini, M. (2015). Reformation of query based approach in search engine. *International Research Journal of Engineering and Technology (IRJET), 2*(5).
24. Fathalla, S. M., Hassan, Y. F., & El-Sayed, M. (2012). A hybrid method for user query reformation and classification. In *Proceedings of Computer Theory and Applications (ICCTA), 22nd International Conference on IEEE2012* (pp. 132–138).
25. Pablos, O. (2012). Patricia: Advancing information management through semantic web concepts and ontologies. *IGI Global.*
26. Leake, D. B., & Scherle, R. (2001). Towards context-based search engine selection. In *Proceedings of the 6th International Conference on Intelligent User Interfaces*, ACM (pp. 109–112).

# KEB173—Recommender System

**Subburaj Ramasamy and A. Razia Sulthana**

**Abstract** Trade and commerce are the fundamental necessities of the human beings. International trade existed centuries ago in the most cumbersome manner due to lack of communication facilities, and the lead times required were months and years. Today it happens in seconds owing to the availability of Information and communications technology abundantly at fingertips to everyone. The buyers would always like to consult their peers before taking a purchase decision. The advent of Internet has made it feasible to achieve this painlessly, get the reviews of other customers on the quality of the same or similar products online instantly. Recommender systems (RS) assess the information available on the Web and provide assistance to the users in making an informed selection of products or services. They assist the user in the purchase decision-making process. Research carried out by organizations such as Amazon on recommender systems is noteworthy. In the last few decades, a set of software tools and machine learning methodologies were proposed for implementing recommender systems. The user reviews are analyzed by knowledge discovery or machine learning techniques to extract and understand the pattern of purchase. The recommender systems handle the users online and assist them in choosing products meeting their requirements in consultation with peers worldwide at no cost and instantly. In this chapter, we give an overview about recommender systems and their nuances.

**Keywords** Collaborative filtering · Content based · Ontology based
Context based · Feature extraction · Knowledge based

---

S. Ramasamy (✉) · A. Razia Sulthana
Department of Information Technology, SRM Institute of Science and Technology,
Kattankulathur 603203, Tamil Nadu, India
e-mail: Subburaj.r@ktr.srmuniv.ac.in

A. Razia Sulthana
e-mail: raziasulthana.a@ktr.srmuniv.ac.in


S. Margret Anouncia and U. K. Wiil (eds.), *Knowledge Computing and its Applications*, https://doi.org/10.1007/978-981-10-8258-0_11

# 1 Introduction

A recommender system is a collection of tools and techniques, which provide meaningful suggestions to the users with regard to purchase of items, i.e., products and/or services. Examples of items include automobiles, movies, books, hotel accommodation, or any other item which the user would like to purchase. Growing consumerism and increasing product diversity have necessitated automation of recommender systems. Some examples of recommending items are given below:

- Providing relevant political updates to viewers who regularly browse for political news or visit the political news Web pages.
- Providing movie information to users who frequently book tickets and who regularly browse for movie information or visit the movie-based Web pages.
- Amazon [1], YouTube [2], Netflix [3] are some of the examples of real-world recommender system.

**Objective**: The objective of recommender system is to identify suitable product or service to the users to buy. It identifies a relevance score for each product. The products are ranked in the order based on relevance score and provided to the user.

**Need for Recommender System**: The reasons that caused recommender systems pervasive are given below:

- Driven by business to expand their trade.
- The increasing e-commerce sites and online stores.
- The increasing use of the Internet by the end-users.
- The enlightened users and enhanced quality of life.
- Information overload: The quantity of data is more thereby making the decision-making tedious.

**Uses of Recommender System**: The uses of recommender system are listed below:

**Customer Centric**:

- It aids the users in taking the decision.
- It reduces time taken to search for a suitable item.
- It assists users who lack personal experience in buying online products.
- It assists users in choosing products from its overwhelming alternatives.
- It narrows down the set of choices.

**Producer Centric**:

- Builds trust and assurance in the minds of the users toward buying his product.
- Aids in promotion of a new product.
- Learns the purchase pattern of the users thereby facilitates manufacturing planning of the product.
- Increase sales rate of diverse or new launch product.

In order to get acceptance from the buyers and sellers, the RS should fulfill some requirements. The expectation from RS is given below:

- Transparent: Reveals the true quality of the product.
- Credible: Increases user's trust and confidence in the system, based on the past performance of recommender system.
- Satisfaction: Meets the expectation of the user making him feel happy.
- Scrutinize: Displays the book in the order in which user may like.
- Efficient: Ability to make users feel that their search is productive.
- Effective: Help users take a quick decision.
- Appropriateness: Suggest the relevant product to the right user at the right time.

In this chapter, paradigms of the recommender system are given in Sect. 2, multi-criteria-based recommender system in Sect. 3, performance metrics in Sect. 4, and summary and conclusions in Sect. 5.

## 2 Paradigms of Recommender System

The recommender system introduced in [4] handles a stream of incoming documents in electronic mail [5]. Recommender system remains an active area of research in machine learning [6, 7], data mining [8], and information retrieval [9]. The recommender system twirls around three objects: item [10], user [11], and transaction [12] as represented in Fig. 1.

Recommender system recommends items to the user based on user characteristics, item characteristics, and transaction characteristics.

**Types of Recommender System**: The recommender system is classified into three major categories [13] based on the approach taken to make the recommendation. They are

- Collaborative recommender system: It recommends item by evaluating the preferences of similar users from the past.
- Content-based recommender system: It recommends an item to a user by evaluating his history.

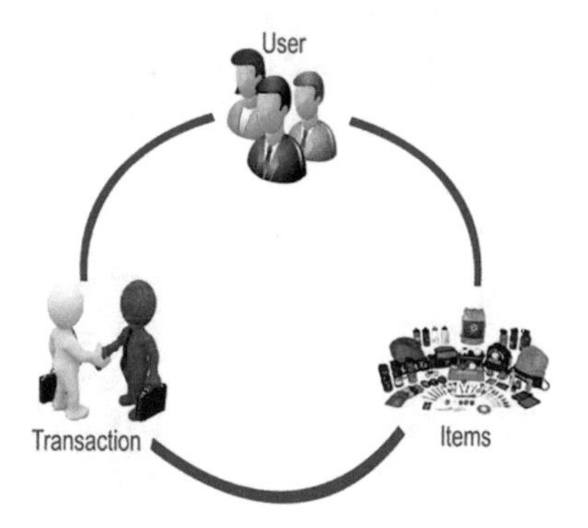


**Fig. 1** Objects of recommender system

- Hybrid recommender system: It recommends item by combining collaborative and content-based approach.

In addition, the recommender systems are classified into generic recommender system [14, 15] and personal recommender system [16–18].

**Generic Recommendation**: It recommends the product to users based on the current trend in the market.

**Personal Recommendation**: It recommends the product to users either based on their past purchase trend or similar users purchase trend. This is carried out based on the personal information provided by the user during the registration process.

## *2.1 Collaborative Filtering Recommender Systems*

Collaborative recommender system [19] recommends an item by evaluating the preferences of similar users in the past. Perhaps, it is one of the well-known and commonly used recommendation methodologies. It applies to all domains, and few algorithms have already been proposed and implemented. Furthermore, the RS is constructed from the behavior of users with similar traits, and it gathers information from groups [20] of similar users. To identify similar users, a utility matrix is constructed. An instance of utility matrix for users against items and their corresponding ratings is given in Table 1.

**Algorithms for Collaborative Recommender Systems**: The two major types of algorithms are used in collaborative recommender system, namely [21, 22]: probabilistic and non-probabilistic. As the name implies, probabilistic algorithms derive a probability model for prediction, whereas non-probabilistic algorithms do not have an underlying probability model. The memory-based collaborative recommender system is non-probabilistic as it measures the similarity between users or items using arithmetic expressions, whereas the model-based recommendation predicts the future by designing a model and hence, it is probabilistic.

**Ratings in Utility Matrix**: The values in the utility matrix are the ratings given by the users against the products. The user can rate the product [23] either implicitly or explicitly. Explicit ratings are given by the users directly, or he is requested to do so by the system. Implicit ratings are obtained from the actions of the user or by

**Table 1** Utility matrix

| | Item 1 | Item 2 | Item 3 | Item 4 | Item 5 |
|---|---|---|---|---|---|
| User 1 | 2 | 5 | 5 | 2 | 2 |
| User 2 | 4 | 3 | ? | 3 | 5 |
| User 3 | 5 | 4 | 5 | 1 | ? |
| User 4 | 3 | ? | 2 | 3 | 3 |
| User 5 | 0 | 1 | 1 | 5 | 4 |

manipulating the reviews given by the users. An implicit rating identifies the polarity [24] of the review as either positive or negative or neutral. These values are then interpreted to know the rating of the products. Hence, explicit ratings can be called as specified ratings, and implicit ratings may be called as observed ratings. Furthermore, the similarity [25] has to be measured between users or between similar items bought by similar users so that we can arrive at more accurate predictions. This led to the classification of collaborative filtering recommender system into memory-based [26–28] and model-based [29].

### 2.1.1 Memory-Based Collaborative Filtering

This approach directly uses the utility matrix to find the similarity information. It finds the similarity of ratings of two different users. Moreover, the current user-item rating is analyzed to predict the user-item rating similar pair in the future. A user $X$ is identified with those products which are preferred or recommended by users who are similar to him. The utility matrix has to be explored to measure similar items. This approach identifies the 'Like users' (neighbors) by one of the following methods [30]. It uses either user-based collaborative filtering or item-based collaborative filtering. Memory-based RS methods are discussed in following subsection, and model-based RS is discussed in subsequent section.

#### User-Based Collaborative Filtering

In this technique, similar users to the target user $X$ are identified based on their purchase pattern. In a movie recommender system, if $A$ and $B$ rate high for a movie: MOVIEX, then they are grouped together and are called neighbors. Subsequently, in the future user $B$ is suggested with those movies that are rated high by user $A$. Almost all the similarity operations have to be done using the ratings in utility matrix. The rows of the utility matrix are worked out to identify the similar users. However, the following issues have to be explored before using this method:

- Which similarity measure has to be used?
- How many users can be grouped at the maximum?
- What prediction method to be used?
- What if the utility matrix is sparse?

Few of the similarity measures include Pearson's correlation-based similarity [31], constrained Pearson's correlation-based similarity [32], and vector similarity [33] and cosine-based similarity [34].

(A) **Pearson's Correlation-Based Similarity**

Pearson's correlation factor identifies the linear relationship between the users. It takes values ranging from −1 to +1. Let us use utility matrix given in Table 1 for

**Table 2** Pearson's similarity

1. Items rated by user 1 ← {1, 2, 3, 4, 5} and items rated by user 2 ← {1, 2, 4, 5}

2. Items that both have rated together ⟹ user 1 ∩ user 2 ⟹ {1, 2, 4, 5} and $N = 4$
3. Pearson's similarity formula is given by

$$\text{Pearson}(\text{user1}, \text{user2}) = \text{sim}(\text{user1}, \text{user2})$$

$$= \frac{\sum_{i=1}^{N} [r(\text{user1}, i) - \mu(\text{user1})] * [r(\text{user2}, i) - \mu(\text{user2})]}{\sqrt{\sum_{i=1}^{N} (\text{rank}(\text{user1}, i) - \mu(\text{user1}))^2} \sqrt{\sum_{i=1}^{N} (\text{rank}(\text{user2}, i) - \mu(\text{user2}))^2}}$$

where $N$ ← Number of items rated by both user1 and user2
$r(\text{user1}, i)$ ← rank given to item $i$ by user1
$\mu(\text{user1})$ ← mean of the ranking given by user1

$$\mu(\text{user1}) = \frac{\sum_{i=1}^{N} \text{rank}(\text{user1}, i)}{N}$$ where $n$ is the number of items rated by user1

understanding this concept. The similarities between the users are calculated by identifying the items that both have rated together and are explained in Table 2.

Pearson's value for those pairs which fall in the high positive zone (0.5–1) is taken to be similar and those which falls in the high negative zone (−0.5 to −1) are taken to be dissimilar. Although Pearson's correlation is easy to calculate; it will not give satisfactory results in the following circumstances:

- When the user's choices are disjoint.
- When the utility matrix is highly sparse.
- It treats the users equally, and emphasis is not given to frequent users.
- It uses the arithmetic mean [35] and hence, outliers can cause errors.
- Users have their scaling range. For example, few users may rate 4 for highest, and 2 for lowest or few other may rate 5 for highest and 1 for lowest.

### (B) Cosine Similarity

It measures the cosine of the angle between two vectors; in our case, it is between two users. The similarity value is 1 if cosine angle is zero degree and for the nonzero angle, it is less than 1. The cosine similarity (Table 3) uses the similarity value calculated by Pearson's. It overcomes the scaling issue of Pearson's similarity measure.

It is analyzed that it filters the neighbors who are far away from the user for whom the cosine similarity is calculated. Also, the scaling errors are eliminated.

**Table 3** Cosine similarity

1. Let the number of neighbors of user 1 be $M$. This is returned by Pearson's similarity measure. The neighbors of user 1 = {user 2, user 3, user 5}, so $M = 3$

2. Cosine similarity formula is given by

$$r(\text{user1}, \text{item}j) = \mu(\text{user1}) + \frac{\sum_{i=1}^{M} \text{sim}(\text{user1}, \text{user}i) * (\text{rank}(\text{user}i, \text{item}j) - \mu(\text{user}i))}{\sum_{i=1}^{M} |\text{sim}(\text{user1}, \text{user}i)|}$$

**Table 4** Constraint Pearson's correlation-based similarity

| Constrained Pearson(user1, user2) = sim(user1, user2) |
|---|
| $= \frac{\sum_{i=1}^{N}[r(\text{user1},i)-\text{med}(\text{user1})]*[r(\text{user2},i)-\text{med}(\text{user2})]}{\sqrt{\sum_{i=1}^{N}(\text{rank}(\text{user1},i)-\text{med}(\text{user1}))^2}\sqrt{\sum_{i=1}^{N}(\text{rank}(\text{user2},i)-\text{med}(\text{user2}))^2}}$ |
| where med(user1) is the median value of the ratings given by user1 |

**Table 5** Vector similarity

| 1. For every user, a record is maintained which is a collection of items rated by him. For instance, the vector of user 3 in Table 1 is given by {item1, item2, item3, item4} |
|---|
| $\text{vecsim}(\text{user1}, \text{user2}) = \sum_{i=1}^{N} \frac{r(\text{user1},i)}{\sqrt{\sum_{j=1}^{M} r(\text{user1},j)^2}} \frac{r(\text{user2},i)}{\sqrt{\sqrt{\sum_{k=1}^{P} r(\text{user2},k)^2}}}$ <br> $N \leftarrow$ Number of items rated together by user1 and user2, $M \leftarrow$ Number of items rated by user1 and $P \leftarrow$ Number of items rated by user2 <br> $r(\text{user1}, i) \leftarrow$ rating given by user1 on item $i$ |

### (C) Constraint Pearson's Correlation-Based Similarity

It is modified version of Pearson's correlation. In Table 4 suggests only those items that are on the same side. In Table 1, user 3 and user 5 are not compared as both are contradicting each other. User 3 gives positive rating for few products for which user 5 has given the negative rating.

More importantly, the mean value in Pearson's is replaced by the median in constrained Pearson's. It reduces the negative coefficients when compared to Pearson's as it uses the median as the measure of central tendency.

### (D) Vector Similarity

The similarity between users is measured by taking each user's record as a vector of items that he has rated. Vector similarity (Table 5) ignores the item which is negatively rated or not rated.

The majority of the similarity measures discussed give preference to users who have rated a large number of items. To overcome this bias, the rating in the denominator is squared. Hence, the ratings are normalized.

### Pros and Cons of User-Based Collaborative Recommender System

As preference is given to user feedback, the recommended items will vary as per the liking of the user. On the contrary, it has many issues:

- The complexity [36] of this method is $O(n^2)$ as it relies on user–item rating matrix.
- Sparse matrices fail to give complete information.

- Scalability problem arises due to increasing number of users. This renders the process challenging.
- Cold-start [37] issue: What to recommend to a new user, who does not have any neighbor?

Item-Based Collaborative Filtering

In this technique, the items bought by other users who buy items similar to target user *X* are identified. These filtered items are recommended to the user *X*. It can be observed from Table 1, that item 2 and item 3 are rated high by almost all the users. The similarity measure has to be calculated between the columns in utility matrix. Few of the similarity measures include cosine similarity, adjusted cosine similarity, Pearson's correlation. The cosine similarity and Pearson's correlation are discussed in Sect. 2.1.1. The only change is that it calculates the similarity between users, and here, we calculate the similarity between the items. Otherwise, the calculations are same.

(A) **Adjusted Cosine Similarity Measure**

The formula for arriving at the similarity measure between two items, say item *i* and item *j*, is given in Table 6.

**Pros and Cons of Item-Based Collaborative Recommender System**

A significant advantage of item-based collaborative approach is that items bought together continued to be bought together. Similar users may not be similar always since their opinion changes with time. As the user group changes dynamically, scalability problem also exists. An increase in the number of items increases the contrary ratings, making the process more tedious. The complexity is $O(n^2)$. Moreover, items which are not co-rated will never be suggested to the user.

Cold-start issue: New products may not be recommended as they do not have similar items.

### 2.1.2 Model-Based Collaborative Filtering

Memory-based collaborative recommender system finds the neighborhoods of user/item and therefore does not create a model to represent the present or predict the

**Table 6** Adjusted cosine similarity measure

| |
|---|
| 1. Let $M$ be the users who have rated item $i$ and $N$ be the users who have rated item $j$ |
| 2. Let $X$ be the users who have rated both items $i$ and item $j$., i.e., $M \cap N$ |
| $\text{sim}_{\text{item}}(\text{item}i, \text{item}j) = \frac{\sum_{u=1}^{X} \text{sim}(u,i) * \text{sim}(u,j)}{\sqrt{\sum_{u=1}^{X} \text{sim}(u,i)^2} \sqrt{\sum_{u=1}^{X} \text{sim}(u,j)^2}}$ |

future, whereas model-based collaborative recommender system [38, 39] builds probabilistic models. It divides the process into the training phase and testing phase. It uses a set of training data, analyzes them and constructs a model during the training phase, and this model is used for future predictions.

Some of the probabilistic models been framed by model-based collaborative recommender system are association rule [40], clustering [41], decision trees [42], regression trees [43], Naïve Bayes classification [44], matrix factorization [45], latent factor analysis [46], etc.

### Association Rules

Most recommender systems give information to the user about the items bought along with the item proposed to be bought for which recommendation is sought. Association rule mining can be used [47, 48] to identify the items that are bought together. Association rules are framed to identify the frequent item pairs. Association rules mining face the scalability problem. The increase in a number of items or transaction increases the association rules generated, thereby making the problem complex.

### Clustering

Clustering is an unsupervised machine learning technique. It is used is in host of applications to gather together, items or users based on similar traits. Popular examples of clustering are text and image processing, etc. It groups objects based on their common characteristics. Using similarity measures, similar items can be clustered or similar users can be clustered. The objects in the clusters are considered to be similar and thus recommended to the users. Some algorithms exists for clustering objects, namely $K$-means clustering [49], hierarchical clustering [50], expectation maximization (EM) [51], self-organizing maps (SOM) [52]. Clustering overcomes scalability problem. It clusters considerably more data compared to other approaches.

### Decision Trees

Decision trees are supervised learning approach and suitable for structured data. Some of the well-known decision tree algorithms are Iterative Dichotomiser (ID.3), C4.5, classification and regression trees (CART). The items classified under a roof are considered to be similar and are recommended. The decision tree acts as prediction model and extracts the necessary attributes from the given test data and provides them as input. A sample decision tree is given in Table 7.

**Table 7** Purchase pattern

| Attributes | Food style | Service | Cost | Score/class label |
|---|---|---|---|---|
| Values | Italian | Self-service | Expensive | Negative |
| | French | Table-service | Expensive | Positive |
| | Italian | Self-service | Affordable | Positive |
| | Italian | Self-service | Expensive | Negative |

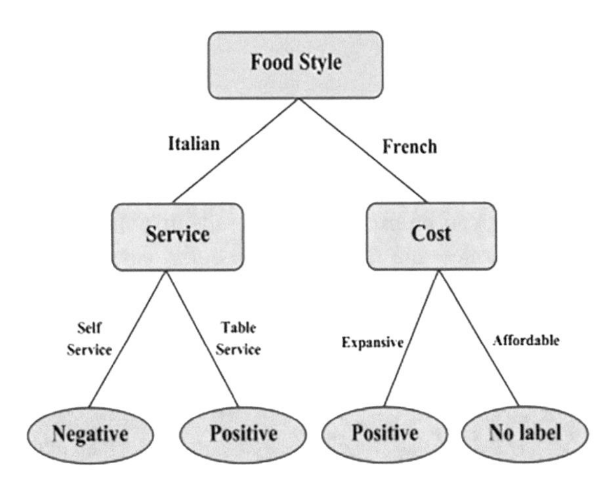


**Fig. 2** Decision tree

When a user likes to receive a recommendation on what food he could order, the system navigates from the root till the leaf node. The constructed decision tree for Table 7 is shown in Fig. 2. Let the given input be ⟨Italian, self-service, Expense⟩. It traverses the tree and returns the result as positive.

Regression Trees

There lies a line of separation between decision trees and regression trees [53]. The type of dependent and independent variables decides to model either decision trees or regression trees. The value of a dependent variable is found or calculated from the known independent variables. The decision trees are preferred when class labels take categorical values, whereas the regression trees are preferred when the class labels take numeric or continuous values. Decision trees contribute to classification and regression trees to prediction.

Naive Bayes

Naïve Bayes [54] uses Bayes theorem for classification. The Bayes theorem is represented by

$$P(A/B) = \frac{P(A)P(B/A)}{P(B)}$$

The job is to assign a class label to an input tuple. Naïve Bayes proceeds by assuming that the attributes are independent of each other. The time complexity of Bayes classification is $O(n)$, where $N$ is the number of training samples, and $A$ is the number of attributes.

Latent Factor Models and Matrix Factorization

Latent factor models replace the missing values using dimensionality reduction methods. Matrix factorization is a type of latent factor model. Dimensionality reduction techniques [55] like principal component analysis (PCA) and singular value decomposition (SVD) add to latent factor models. Matrix factorization: Identifies the latent factors from the utility matrix.

**Pros and Cons of Model-based Collaborative Recommender System**

Building the model takes comparatively more time. And once the model is built; it predicts the rating easily in minimal time executing the model with the test set, thus easing the recommendation process. The recommender system model has to be updated frequently as per the requirement change of the user, to give appropriate suggestions. It faces cold-start and scalability issues.

## 2.2 *Content-Based Recommender System*

Content-based recommender system [56, 57] recommends the item to users based on the user's self-history. This approach filters item that is similar to the items that the user has accessed in the past or by matching with the attribute given by the user. However, it uses the demographic information [58, 59] of the user for predicting his future wish list. It learns the user preferences either from the user profile or the reviews/ratings posted by the user or from his past purchase pattern.

**Algorithms for Content-Based Recommender System**

Content-based recommender system uses many information retrieval algorithms, to extract the information from the text. Some of the methods include weighting schemes like term frequency–inverse document frequency (TF-IDF) [60], entropy, Gini index, and other algorithms like nearest neighbor's, decision trees, neural network, Bayes classification, and regression methods. Learning algorithms have to be formulated to learn the user purchase or navigation pattern. The learning model can be built explicitly by getting the users interest or implicitly using algorithms. Implicitly the content-based recommender system is segregated into attribute based and case based. In case-based reasoning, the environment in which the user has accessed the data is recorded and in an attribute-based key components from the search query posted by the user in the past is recorded. The profile recorded is the mirror reflection of user's preference.

### 2.2.1 Term Frequency and Inverse Document Frequency (TF-IDF)

It returns the weight of a term in a document. It is directly proportional to the frequency of words in the document. Term frequency is the frequency of a word ($w$) in a document ($d$) and is given by:

$$\mathrm{TF}(w,d) = \frac{\text{frequency of wind}}{\text{maximal frequency of atermind}} \quad \&\& \quad \mathrm{IDF}(d) = \log(D/n_\mathrm{d})$$

Term frequency returns words alike 'the' or 'a' or 'an,' which repeatedly occurs in documents. Inverse document frequency minimizes the contribution of such highly repeated words. IDF is the frequency of a word ($w$) in the entire document ($D$) and is given by:

Where $n_\mathrm{d}$ is the number of documents in $D$ containing the word, and TF-IDF is

$$\mathrm{TF\_IDF}(w,d) = \mathrm{TF}(w,d) * \mathrm{IDF}(d)$$

The drawback of TF-IDF is that it never gives weight to new words or words whose frequency is less.

### 2.2.2 Weighting Scheme: Gini Index

These measures are used to extract the features from the item description or user description. Gini index returns the measure of impurity of a feature. Gini index is used for binary ratings and is represented by

$$\mathrm{Gini}(w) = 1 - \sum_{i=1}^{d} p_i^2$$

Let there be two features namely ⟨Quality, Color⟩, and there are 200 users who are rating the product. For every feature, the Gini index needs to be calculated. And each feature has a set of characteristics which is rated say ⟨very good, good, average, poor, very poor⟩. The example in Table 8 illustrates this

Table 8 Raw table

| Feature | Very good | Good | Average | Poor | Very poor |
|---|---|---|---|---|---|
| Quality | 53 | 25 | 45 | 33 | 44 |
| Color | 50 | 43 | 67 | 30 | 10 |

$$\begin{aligned}\text{Gini}(\text{quality}) &= 1 - \left[\frac{53}{200}^2 + \frac{25}{200}^2 + \frac{45}{200}^2 + \frac{33}{200}^2 + \frac{44}{200}^2\right] = 1 - 0.2121 \\ &= 0.7879\end{aligned}$$

The Gini value is thus calculated as shown in Table 8. This process is repeatedly carried out for all the features. The features are ordered based on the Gini index, and these are called as discriminant features. The information retrieval approaches extract information based on either keywords or sentence. Relevance feedback [61, 62] is a technique that automatically upgrades the user's query based on the previous queries. Rocchio's algorithm [63] is adopted in relevance feedback, and it allows the user to rate the documents.

**Pros and Cons of Content-based Recommender system**

This approach overcomes the cold-start issue in contrast to collaborative system. The new items' description is verified against the requirement of the user, and thus, it is recommended. The cold-start problem exists for new users, as there is no implicit information about the new user. It does not recommend items which are beyond the knowledge or scope of the user.

Few references to collaborative system are Ringo [64], Usenet News [65], Amazon [66], Movielens [67], and content-based system are News Weeder [68], Letzia [69], Yournews [70], PSUN [71].

## 2.3 *Hybrid Recommender Systems*

Hybrid recommender system is a combination of both content and collaborative RS. Some articles [72–74] use hybrid to leverage the advantages of both the methods. In paper [75], hybrid recommender system is classified into weighted, switching, mixed, feature combination, cascade, feature augmentation, and meta-level.

### 2.3.1 Types of Hybrid Recommender System

**Weighted**: The hybrid system ranks above all the other recommendation systems as it combines the result obtained from both lower level recommendation and higher level recommendations.

**Switching**: This system switches between different recommendations systems as per the requirement of the user. It can opt for the collaborative system to give wide suggestions or at times content based on given narrow suggestions.

**Mixed**: The results of the different recommender system for a given data are displayed to the user.

**Feature combination**: The selected features from the different recommender systems are combined and given to the hybrid system for further processing.

**Cascade**: The output of one recommender system is given as input to the other. The results are thus refined in cascading stages. This reduces the bias and error.

**Feature Augmentation**: The first recommender system outputs some features which are given as input to the second it goes on. It passes the data, whereas in cascade system the results are passed.

**Meta-level**: The model generated by the first system is used as input to the other. Mostly, the collaborative system model is added to the content-based system.

# 3 Multi-criteria-Based Recommender System

The general recommender system identifies a desirable item for the user either by content or collaborative systems, whereas the multi-criteria-based system identifies a set of possible candidate items. It includes context-based, knowledge-based, constraint-based, ontology-based, trust-based recommender systems. Thus, the recommender system can be extended based on different candidate items chosen. The candidate item set

- May satisfy the complex preferences of the user.
- Contains alternative items, identified based on the recent uses of the recommender system.
- Are chosen by assessing the global preferences of the user.

## *3.1 Context-Based Recommender System*

The implicit feedback given by the user has to be analyzed to know the expectation of the user. The feedback given by the user has to be examined to know the context [76] which he refers. The context refers to the environmental factors of the user and the scope for which he has given his consent. Context may be extracted implicitly or explicitly. Explicit context is the time of posting the review, location of the user, or the social groups in which he is part of. Implicit context is extracted from the reviews such as the feature for which he is posting his opinion. It is used to recommend movie [77, 78], songs, books, products, etc. Moreover, the user's

Table 9 Context-based recommender system's utilities

| Ref. | Additional system | Implementation tools and methods | Tool and datasets | Error | Evaluation metrics |
|---|---|---|---|---|---|
| [79] | Latent factor analysis, matrix factorization, PCA, and unsupervised deep learning | Develops an auto-encoder neural network using backpropagation | Extracts data from data collected from mobile sensors | RMSE | Prediction accuracy |
| [80] | TGSC-PMF, probabilistic matrix factorization, latent dirichlet allocation (LDA) | Does point of interest (POI) recommendation in location aware systems | Extracts data from real-world location-based social network (LBSN) data | RMSE MAE | Precision recall |
| [81] | Latent factors, context-aware ranking, collaborative filtering | Personalized recommendation for multimedia content | Last.fm dataset, uses social tags | MAE | Precision, recall, *F*-measure |
| [82] | Context, latent vectors, principal component analysis (PCA) | COT is compared with SVD, multiverse recommendation, FM, Hetero FM | context operating tensor (COT) captures the context to measure the semantic operation | RMSE and MSE | Precision |

interest can be detected provided the opinion is mapped to its corresponding feature. A product is rated using three-factors, namely user, item, and context.

$$U * I * C \rightarrow \text{rating}$$

Few articles using context-based method, its implementation, and tools are described in Table 9.

## 3.2 Knowledge-Based Recommender System

It is a similar approach to collaborative-based recommender system where the similarity between the users or the items are identified. This information is stored in the database and then reused as knowledge [83] by the user. The users or items knowledge is stored in the repository and acts as metadata, and this is termed as domain knowledge. This ensures to build an organizations knowledge using individual's knowledge. User reasoning is broadly represented as authoritative, motivational, and substantive [84, 85], and these arguments proved a significant impact on decision making of the user. Moreover, when added with collaborative and content-based system shows a significant increase in the quality of the system.

**Constraint-Based Recommender System**:

The knowledge-based system when added with condition on the domain knowledge is termed as constraint-based recommender system [86]. The system [87] arranges the knowledge in the form of taxonomy and uses test cases to handle the constraints.

## 3.3 *Ontology-Based Recommender System*

Ontology is the formal, explicit specification of shared conceptualization [88]. Ontology eases the information overload problems in recommender systems. The knowledge base is searched once, and the extracted knowledge from Web or purchase pattern is stored hierarchically as concepts. This ensures that the knowledge base is not searched multiple time thereby reduces the search time. The constructed ontology can be reused further. A survey of various types of ontologies, the methods for implementing them, the tools that could be used and language used for building them are given in Table 10.

The diversity of ontology justifies its use in all kind of recommender systems. This can be used in combination with machine learning, correlation, statistical tools, and applications. Ontology builds its model using the context of the system.

**Table 10** Ontology-based recommender system utilities

| Ontology type | Implementation tools and methods | Language |
|---|---|---|
| Dynamic domain ontology [89] | Protégé tool, Uses TF-IDF Similarity measures. Framed for job recommender system | Ontology Web Language (OWL) |
| Domain ontology [90] | Protégé tool. Framed for anti-diabetic drug selection | Ontology Web Language (OWL) and Semantic Web Rule Language (SWRL) |
| Domain ontology [91] | Cosine similarity, document identification based on semantic relationship | Ontology Web Language–Description Logics (OWL-DL) |
| Fuzzy-based ontology [92] | Jaccard similarity, Jena Ontology Tool, TF-IDF | Web Service Description Language (WSDL) |
| Domain ontology [93] | Ontology-based tree learner to build the decision tree | J48 algorithm, WEKA tool |
| Domain ontology for context [94] | Recommender System Context (RSC), which represents the context, and Contextual Ontological User Profile (COUP), Compares many ontology models | Contextual pre-filtering using MAE |

### 3.4 Trust-Based Recommender System

The main goal of trust-based recommender system is to increase the level of user's trust in the recommender system. The system [95] qualitatively evaluates the recommender system using Random Walk System (RWS), Majority of Majorities (MoM), and Minimum Cut system (min-cut). The proportion of users trust in the system depends on the performance of the recommender system [96]. It discusses the effect of network density, knowledge sparseness, and user heterogeneity affecting the trust on the system.

## 4 Metrics of Recommender Systems

The performance of a recommender system is assessed based on the level of satisfaction of the user. We will now discuss the various measures used to assess the recommender system.

**Characteristics of recommender system**:

In the following, we discuss some characteristics of recommender system.

- Similarity plays a major role in recommending similar products to the user.
- Diversity ensures not to recommend diverse products of varying characteristics.
- Credibility builds the user's trust toward the system by recommending the products that he knows already.
- The uniqueness of the system is ensured by not repeatedly providing the same results to the user.
- Recommender system surprises the user by recommending products that the user has not seen, thereby providing a new idea to the user.
- A scalable recommender system handles an increase in the number of users or a huge increase in the number of items.
- Confidentiality is ensured so as to hide the personal details of the user specifically in collaborative filtering.
- The recommender system has to provide recommendations to the user instantly and hence has to be updating data continuously. Ontology-based recommender systems may be more suitable since the repository of past data can be updated systematically.

### 4.1 Performance Metrics for Recommender Systems

A set of machine learning methods to measure the performance of recommender system [97, 98] is represented below.

### 4.1.1 Error Metrics

In this method, an initial hypothesis for the system is set following which the system is tested against the hypothesis. During the training phase, the initial hypothesis of the system is set by repeatedly analyzing the system behavior (predicted value). The test phase records the behavior of the system against the hypothesis (observed value). The combinations of items bought by the user and recommended by the system are represented in Table 11.

Let $D$ be the number of tuples and $R_{uj}$ be the rating given by the user ($u$) to the item ($i$). Let the predicted rating be $\mathrm{PR}_i$ and observed ratings are $\mathrm{OR}_i$.

**Root Mean Square Error (RMSE)**: It is used to measure the accuracy and is represented by

$$\mathrm{RMSE} = \sqrt{\frac{\sum_{I=1}^{D} (\mathrm{PR}_i - \mathrm{OR}_i)^2}{D}}$$

**Mean Absolute Error (MAE)**: It formulates the average absolute difference between the predicted value and the observed value. MAE is less sensitive to outliers than MASE.

$$\mathrm{MAE} = \frac{\sum_{i=1}^{D} (\mathrm{PR}_i - \mathrm{OR}_i)}{D}$$

### 4.1.2 Rank Metrics

**Correlation**: Correlation [99] is a statistical measure which finds the dependency between two entities. The correlation coefficient, r, defines the dependence. The correlation coefficient takes values between −1 and +1. A positive value indicates direct relationship, negative value indicates indirect relationship, and a zero value indicates no correlation. A couple of correlation types exist, namely Pearson's correlation and rank correlation.

**Rank Correlation: Spearman's Rank Correlation Coefficient**

Let there be $n$ items. And these items are ranked both by the user and by the prediction system. The individual user rating be $\{\mathrm{ur}_1, \mathrm{ur}_2, \ldots, \mathrm{ur}_n\}$ and the system

**Table 11** Recommendation possibilities

| | Item has been recommended | Item has not been recommended |
|---|---|---|
| Item has been bought | TP (True Positive) | FN (False Negative) |
| Item has not been bought | FP (False Positive) | TN (True Negative) |

generated rating be $\{sr_1, sr_2, \ldots, sr_n\}$. This individual rating of the user and the system generated rating to a product are converted into global rating of the product.

$$\text{spearman}_{\text{rating}} = 1 - \frac{\sum_{i=1}^{n} (\text{ur}_i - \text{sr}_i)^2}{n(n^2 - 1)}$$

The Spearman rating rates the product globally, and the products ranked high are suggested to the user. It is slightly sensitive to data.

**Rank Correlation: Kendall's Rank Correlation Coefficient**

The initial condition is same as in Spearman's coefficient, and moreover, it is insensitive to error as it calculates the concordant and discordant pairs.

- A pair is said to be concordant if $\text{ur}_1 < \text{ur}_2$ and $\text{sr}_1 < \text{sr}_2$ and so on.
- A pair is said to be discordant if (*A*) $\text{ur}_1 > \text{ur}_2$ and $\text{sr}_1 > \text{sr}_2$ (*B*) $\text{ur}_1 < \text{ur}_2$ and $\text{sr}_1 > \text{sr}_2$ (*C*) $\text{ur}_1 > \text{ur}_2$ and $\text{sr}_1 < \text{sr}_2$.
- A pair is said to be neither concordant nor discordant if $\text{ur}_1 = \text{ur}_2$ and $\text{sr}_1 = \text{sr}_2$.

$$\text{Kendall}_{\text{rating}} = \frac{\text{no of concordant pair} - \text{no of discordant pair}}{n(n-1)/2}$$

The items are thus rated globally, and the products ranked high are suggested to the user.

**Utility-based ranking**: Utility-based ranking is a rank metric that is influenced by the position or location of items been displayed. A typical user clicks those items that are displayed in the first, as the user scans from beginning to end. Though correlation ranks the item globally, the bias of these ordering still exists. This bias is overcome in utility-based ranking [100].

### 4.1.3 Information Retrieval Measures

Error metrics are used to measure the error in the model, the difference between the predicted and observed values. Rank metrics help in ranking the items in order based on their values. Information retrieval measure measures the quality of the recommender system. A set of information retrieval measures is listed below. Precision, Recall, Accuracy, and *F*-Score are supporting measures [101, 102] calculated from attributes specified in Table 11.

**Precision and Recall**: These are used in binary classification. The terms retrieved documents, and the relevant documents are used to measure them. Precision answers for 'How useful is the recommended results?' and recall answers for 'How complete is the recommended results?' Precision is known as the positive predicted value, and recall is known as the sensitivity of the value. Also precision says what fraction of the items suggested are suitable to the user and recall says the fraction of suitable items from the total available items.

$$\text{Precision} = \frac{\text{TP}}{\text{TP} + \text{FP}} \text{AND Recall(TruePositiveRate)} = \frac{\text{TP}}{\text{TP} + \text{FN}}$$

$$\text{FalsePositiveRate OR}(1 - \text{Specificity})\text{OR Sensitivity} = \frac{\text{FP}}{\text{FP} + \text{TN}}$$

An information retrieval system having higher precision and accuracy values holds more worth. Higher the recall value, higher is the quality of the system. Higher the precision value, higher is the quality of the system for the given input.

**Accuracy**: Accuracy answers for 'How close is the recommended result to the user expectation?' Accuracy is weighted both by bias and prevalence. Accuracy is a numeral value calculated to certify the system.

$$\text{Accuracy} = \frac{\text{TP} + \text{TN}}{\text{TP} + \text{TN} + \text{FP} + \text{FN}}$$

Higher the accuracy, higher is the likability factor of the recommended item.

***F*-Score**: It is the weighted harmonic mean of precision and recall, conversely the weighted harmonic mean of both is Mathew's correlation coefficient.

$$F_{\text{score}} = 2 * \frac{\text{precision} * \text{recall}}{\text{precision} + \text{recall}}$$

*F*-measure shows better quantification than precision and recall.

## 5 Summary and Conclusions

The e-commerce is pervasive now, and trillions of dollars' worth of goods and services are traded on the Internet. There is an explosive growth of Business to Consumer (B2C) trade owing to the economic development and Internet penetration in the most populous countries like China and India. There has been a dramatic change in the dynamics of online transactions and the way in which the transactions are carried out due to the e-commerce. Due to increasing bandwidth of the Internet in developing countries, the business climate has been phenomenally transformed, thus enhancing a win–win situation for buyers and sellers. One of the important tools used in the e-business is the recommender system, which provides a recommendation to the user on the product or service of interest. Continuous research in recommender system has given benefits to the public at large. In this chapter, we discuss the background, uses of recommender system and the different types of recommender system. The association rules facilitate the users to identify the companion products purchased by similar users, and hence, it can be used in conjunction with any of the recommender system discussed above. Users or items can be clustered using clustering technique, and one popular technique is $K$-means clustering. Similarly, classification can also be used in conjunction with recommender system. Popular classification approaches such as decision trees, regression trees, Naïve Bayes are discussed in this chapter. Ontology is the futuristic approach to recommendation system. We also introduce some performance measures associated with recommender systems.

## References


1. Linden, G., Smith, B., & York, J. (2003). Amazon.com recommendations: Item-to-item collaborative filtering. *IEEE Internet Computing, 7*(1), 76–80.
2. Gomez-Uribe, C. A., & Hunt, N. (2016). The netflix recommender system: Algorithms, business value, and innovation. *ACM Transactions on Management Information Systems (TMIS), 6*(4), 13.
3. Davidson, J., Liebald, B., Liu, J., Nandy, P., Van Vleet, T., Gargi, U., … & Sampath, D. (2010, September). The YouTube video recommender system. In *Proceedings of the fourth ACM conference on Recommender systems* (pp. 293–296). ACM.
4. Goldberg, D., Nichols, D., Oki, B. M., & Terry, D. (1992). Using collaborative filtering to weave an information tapestry. *Communications of the ACM, 35*(12), 61–70.
5. Denning, P. J. (1982). ACM president's letter: Electronic junk. *Communications of the ACM, 25*(3), 163–165.
6. Song, Q., Zhu, X., Wang, G., Sun, H., Jiang, H., Xue, C., et al. (2016). A machine learning based software process model recommendation method. *Journal of Systems and Software, 118,* 85–100.
7. Stone, P., & Veloso, M. (2000). Multiagent systems: A survey from a machine learning perspective. *Autonomous Robots, 8*(3), 345–383.
8. Lin, W. (2000). *Association rule mining for collaborative recommender systems* (Master's thesis, Worcester Polytechnic Institute).
9. Khribi, M. K., Jemni, M., & Nasraoui, O. (2008, July). Automatic recommendations for e-learning personalization based on web usage mining techniques and information retrieval. In *Eighth IEEE International Conference on Advanced Learning Technologies, 2008, ICALT'08* (pp. 241–245). IEEE.
10. Sarwar, B., Karypis, G., Konstan, J., & Riedl, J. (2001, April). Item-based collaborative filtering recommendation algorithms. In *Proceedings of the 10th International Conference on World Wide Web* (pp. 285–295). ACM.
11. Alhoori, H., & Furuta, R. (2017). Recommendation of scholarly venues based on dynamic user interests. *Journal of Informetrics, 11*(2), 553–563.
12. Josang, A. (1999). Trust-based decision making for electronic transactions. In *Proceedings of the Fourth Nordic Workshop on Secure Computer Systems (NORDSEC'99)* (pp. 496–502).
13. Lu, J., Wu, D., Mao, M., Wang, W., & Zhang, G. (2015). Recommender system application developments: A survey. *Decision Support Systems, 74,* 12–32.
14. Mashiach, L. T. (2006). Learning to rank: A machine learning approach to static ranking.

15. Kim, Y. S. (2013). Recommender system based on product taxonomy in e-commerce sites. *Journal of Information Science and Engineering, 29*(1), 63–78.
16. Qian, X., Feng, H., Zhao, G., & Mei, T. (2014). Personalized recommendation combining user interest and social circle. *IEEE Transactions on Knowledge and Data Engineering, 26*(7), 1763–1777.
17. Gao, Y., Feng, Y., & Tan, J. (2017). Exploratory study on cognitive information gain modeling and optimization of personalized recommendations for knowledge reuse. *Journal of Manufacturing Systems, 43,* 400–408.
18. Wang, X., Liu, Y., & Xiong, F. (2016). Improved personalized recommendation based on a similarity network. *Physica A: Statistical Mechanics and its Applications, 456,* 271–280.
19. Zhang, F., Gong, T., Lee, V. E., Zhao, G., Rong, C., & Qu, G. (2016). Fast algorithms to evaluate collaborative filtering recommender systems. *Knowledge-Based Systems, 96,* 96–103.
20. Bok, K., Lim, J., Yang, H., & Yoo, J. (2016). Social group recommendation based on dynamic profiles and collaborative filtering. *Neurocomputing, 209,* 3–13.
21. Zhang, F., & Sun, S. (2014). A robust collaborative recommendation algorithm based on least median squares estimator. *JCP, 9*(2), 308–314.
22. Hernando, A., Bobadilla, J., Ortega, F., & Gutiérrez, A. (2017). A probabilistic model for recommending to new cold-start non-registered users. *Information Sciences, 376,* 216–232.
23. Maes, P. (1994). Agents that reduce work and information overload. *Communications of the ACM, 37*(7), 30–40.
24. Harer, S., & Kadam, S. (2014). Mining and summarizing movie reviews in mobile environment. *International Journal of Computer Science and Information Technologies, 5*(3), 3912–3916.
25. Shimodaira, H. (2014). Similarity and recommender systems. *School of Informatics, The University of Eidenburgh, 21.*
26. Ghazarian, S., & Nematbakhsh, M. A. (2015). Enhancing memory-based collaborative filtering for group recommender systems. *Expert Systems with Applications, 42*(7), 3801–3812.
27. Schafer, J. H. J. B., Frankowski, D., Herlocker, J., & Sen, S. (2007). Collaborative filtering recommender systems. In *The adaptive web* (pp. 291–324).
28. Sánchez-Moreno, D., González, A. B. G., Vicente, M. D. M., Batista, V. F. L., & García, M. N. M. (2016). A collaborative filtering method for music recommendation using playing coefficients for artists and users. *Expert Systems with Applications, 66,* 234–244.
29. Zhao, F., Zhu, Y., Jin, H., & Yang, L. T. (2016). A personalized hashtag recommendation approach using LDA-based topic model in microblog environment. *Future Generation Computer Systems, 65,* 196–206.
30. Zhang, R., Liu, Q. D., Gui, C., Wei, J. X., & Ma, H. (2014, November). Collaborative filtering for recommender systems. In *2014 Second International Conference on Advanced Cloud and Big Data (CBD)* (pp. 301–308). IEEE.
31. Resnick, P., Iacovou, N., Suchak, M., Bergstrom, P., & Riedl, J. (1994, October). GroupLens: An open architecture for collaborative filtering of netnews. In *Proceedings of the 1994 ACM Conference on Computer Supported Cooperative Work* (pp. 175–186). ACM.
32. Ahn, H. J. (2008). A new similarity measure for collaborative filtering to alleviate the new user cold-starting problem. *Information Sciences, 178*(1), 37–51.
33. Breese, J. S., Heckerman, D., & Kadie, C. (1998, July). Empirical analysis of predictive algorithms for collaborative filtering. In *Proceedings of the Fourteenth Conference on Uncertainty in Artificial Intelligence* (pp. 43–52). Morgan Kaufmann Publishers Inc.
34. Rajendra, Q. W., & Raj, J. D. (2015). Recommending news articles using cosine similarity function. *Warwick Business School Journal*, 1–8.
35. Thelwall, M. (2016). The precision of the arithmetic mean, geometric mean and percentiles for citation data: An experimental simulation modelling approach. *Journal of informetrics, 10*(1), 110–123.

36. Shardanand, U., & Maes, P. (1995, May). Social information filtering: Algorithms for automating "word of mouth". In *Proceedings of the SIGCHI Conference on Human Factors in Computing Systems* (pp. 210–217). ACM Press/Addison-Wesley Publishing Co.
37. Schein, A. I., Popescul, A., Ungar, L. H., & Pennock, D. M. (2002, August). Methods and metrics for cold-start recommendations. In *Proceedings of the 25th Annual International ACM SIGIR Conference on Research and Development in Information Retrieval* (pp. 253–260). ACM.
38. Yoon, J., Seo, W., Coh, B. Y., Song, I., & Lee, J. M. (2017). Identifying product opportunities using collaborative filtering-based patent analysis. *Computers & Industrial Engineering, 107,* 376–387.
39. Lozano, E., Gracia, J., Collarana, D., Corcho, O., Gómez-Pérez, B., Villazón, A., ... & Liem, J. (2011). Model-based and memory-based collaborative filtering algorithms for complex knowledge models. *DynaLearn, EC FP7 STREP project, 231526.*
40. Mican, D., & Tomai, N. (2010). Association-rules-based recommender system for personalization in adaptive web-based applications. In *Current Trends in Web Engineering* (pp. 85–90).
41. Pham, M. C., Cao, Y., Klamma, R., & Jarke, M. (2011). A clustering approach for collaborative filtering recommendation using social network analysis. *Journal of UCS, 17*(4), 583–604.
42. Gershman, A., Meisels, A., Lüke, K. H., Rokach, L., Schclar, A., & Sturm, A. (2010, June). A decision tree based recommender system. In *IICS* (pp. 170–179).
43. Gan, M. (2016). COUSIN: A network-based regression model for personalized recommendations. *Decision Support Systems, 82,* 58–68.
44. Engelbert, B., Blanken, M. B., Kruthoff-Brüwer, R., & Morisse, K. (2011, March). A user supporting personal video recorder by implementing a generic bayesian classifier based recommender system. In *2011 IEEE International Conference on Pervasive Computing and Communications Workshops (PERCOM Workshops)* (pp. 567–571). IEEE.
45. Lee, D. D., & Seung, H. S. (2001). Algorithms for non-negative matrix factorization. In *Advances in neural information processing systems* (pp. 556–562).
46. Cheung, K. W., Tsui, K. C., & Liu, J. (2004). Extended latent class models for collaborative recommendation. *IEEE Transactions on Systems, Man, and Cybernetics-Part A: Systems and Humans, 34*(1), 143–148.
47. Slimani, T., & Lazzez, A. (2014). Efficient analysis of pattern and association rule mining approaches. arXiv:1402.2892.
48. Agrawal, R., & Srikant, R. (1994, September). Fast algorithms for mining association rules. In *Proceedings of 20th International Conference on Very Large Data Bases, VLDB* (Vol. 1215, pp. 487–499).
49. Kularbphettong, K., Somngam, S., Tongsiri, C., & Roonrakwit, P. (2014, December). A recommender system using collaborative filtering and K-mean based on android application. In *Proceedings of International Conference Applied Mathematics, Computational Science and Engineering* (pp. 161–166).
50. Zheng, L., Li, L., Hong, W., & Li, T. (2013). PENETRATE: Personalized news recommendation using ensemble hierarchical clustering. *Expert Systems with Applications, 40*(6), 2127–2136.
51. Dempster, A. P., Laird, N. M., & Rubin, D. B. (1977). Maximum likelihood from incomplete data via the EM algorithm. *Journal of the Royal Statistical Society. Series B (Methodological),* 1–38.
52. Roh, T. H., Oh, K. J., & Han, I. (2003). The collaborative filtering recommendation based on SOM cluster-indexing CBR. *Expert Systems with Applications, 25*(3), 413–423.
53. Loh, W. Y. (2011). Classification and regression trees. *Wiley Interdisciplinary Reviews: Data Mining and Knowledge Discovery, 1*(1), 14–23.
54. Jiang, M., Song, D., Liao, L., & Zhu, F. (2015). A Bayesian recommender model for user rating and review profiling. *Tsinghua Science and Technology, 20*(6), 634–643.

55. Yin, C. X., & Peng, Q. K. (2012). A careful assessment of recommendation algorithms related to dimension reduction techniques. *Knowledge-Based Systems, 27,* 407–423.
56. Fong, A. C. M., Zhou, B., Hui, S. C., Hong, G. Y., & Do, T. A. (2011). Web content recommender system based on consumer behavior modeling. *IEEE Transactions on Consumer Electronics, 57*(2).
57. Semeraro, G., Degemmis, M., Lops, P., & Basile, P. (2007, January). Combining learning and word sense disambiguation for intelligent user profiling. In *IJCAI* (Vol. 7, pp. 2856–2861).
58. Solanki, S., & Batra, S. G. (2015). *Recommender system using collaborative filtering and demographic features* (Doctoral dissertation).
59. Qiu, L., & Benbasat, I. (2010). A study of demographic embodiments of product recommendation agents in electronic commerce. *International Journal of Human-Computer Studies, 68*(10), 669–688.
60. Pazzani, M., & Billsus, D. (1997). Learning and revising user profiles: The identification of interesting web sites. *Machine Learning, 27*(3), 313–331.
61. Balabanović, M., & Shoham, Y. (1997). Fab: Content-based, collaborative recommendation. *Communications of the ACM, 40*(3), 66–72.
62. Salton, G., & Buckley, C. (1997). Improving retrieval performance by relevance feedback. *Readings in information retrieval, 24*(5), 355–363.
63. Rocchio, J. J. (1971). Relevance feedback in information retrieval. *The Smart Retrieval System-Experiments in Automatic Document Processing*.
64. Gantner, Z., Drumond, L., Freudenthaler, C., Rendle, S., & Schmidt-Thieme, L. (2010, December). Learning attribute-to-feature mappings for cold-start recommendations. In *2010 IEEE 10th International Conference on Data Mining (ICDM)* (pp. 176–185). IEEE.
65. Konstan, J. A., Miller, B. N., Maltz, D., Herlocker, J. L., Gordon, L. R., & Riedl, J. (1997). GroupLens: Applying collaborative filtering to Usenet news. *Communications of the ACM, 40*(3), 77–87.
66. Wang, L., Fang, L., Wang, L., Li, G., Xie, B., & Yang, F. (2011, November). APIExample: An effective web search based usage example recommender system for Java APIs. In *Proceedings of the 2011 26th IEEE/ACM International Conference on Automated Software Engineering* (pp. 592–595). IEEE Computer Society.
67. Miller, B. N., Albert, I., Lam, S. K., Konstan, J. A., & Riedl, J. (2003, January). MovieLens unplugged: Experiences with an occasionally connected recommender system. In *Proceedings of the 8th International Conference on Intelligent User Interfaces* (pp. 263–266). ACM.
68. Lang, K. (1995, July). Newsweeder: Learning to filter netnews. In *Proceedings of the 12th International Conference on Machine Learning* (Vol. 10, pp. 331–339).
69. Lieberman, H. (1995). Letizia: An agent that assists web browsing. *IJCAI, 1*(1995), 924–929.
70. Ahn, J. W., Brusilovsky, P., Grady, J., He, D., & Syn, S. Y. (2007, May). Open user profiles for adaptive news systems: help or harm? In *Proceedings of the 16th International Conference on World Wide Web* (pp. 11–20). ACM.
71. Sorensen, H., & McElligott, M. (1995, December). PSUN: A profiling system for Usenet news. In *Proceedings of CIKM* (Vol. 95, pp. 1–2).
72. Burke, R. (2007). Hybrid web recommender systems. In *The adaptive web*.
73. Lu, J., Shambour, Q., Xu, Y., Lin, Q., & Zhang, G. (2010). BizSeeker: A hybrid semantic recommender system for personalized government-to-business e-services. *Internet Research, 20*(3), 342–365.
74. De Campos, L. M., Fernández-Luna, J. M., Huete, J. F., & Rueda-Morales, M. A. (2010). Combining content-based and collaborative recommendations: A hybrid approach based on Bayesian networks. *International Journal of Approximate Reasoning, 51*(7), 785–799.
75. Burke, R. (2002). Hybrid recommender systems: Survey and experiments. *User Modeling and User-Adapted Interaction, 12*(4), 331–370.

76. Bazire, M., & Brézillon, P. (2005). Understanding context before using it. *Modeling and using context* (pp. 113–192).
77. Guan, D., Li, Q., Lee, S., & Lee, Y. (2006). A context-aware music recommendation agent in smart office. In *Fuzzy systems and knowledge discovery* (pp. 1201–1204).
78. Park, H. S., Yoo, J. O., & Cho, S. B. (2006, September). A context-aware music recommender system using fuzzy bayesian networks with utility theory. In *International Conference on Fuzzy Systems and Knowledge Discovery* (pp. 970–979). Berlin: Springer.
79. Unger, M., Bar, A., Shapira, B., & Rokach, L. (2016). Towards latent context-aware recommender systems. *Knowledge-Based Systems, 104,* 165–178.
80. Ren, X., Song, M., Haihong, E., & Song, J. (2017). Context-aware probabilistic matrix factorization modeling for point-of-interest recommendation. *Neurocomputing, 241,* 38–55.
81. Alhamid, M. F., Rawashdeh, M., Dong, H., Hossain, M. A., & El Saddik, A. (2016). Exploring latent preferences for context-aware personalized recommender systems. *IEEE Transactions on Human-Machine Systems, 46*(4), 615–623.
82. Wu, S., Liu, Q., Wang, L., & Tan, T. (2016). Contextual operation for recommender systems. *IEEE Transactions on Knowledge and Data Engineering, 28*(8), 2000–2012.
83. Nakagawa, A., & Ito, T. (2002, August). An implementation of a knowledge recommender system based on similarity among users' profiles. In *Proceedings of the 41st SICE Annual Conference, SICE 2002* (Vol. 1, pp. 326–327). IEEE.
84. Giboney, J. S., Brown, S. A., Lowry, P. B., & Nunamaker, J. F. (2015). User acceptance of knowledge-based system recommendations: Explanations, arguments, and fit. *Decision Support Systems, 72,* 1–10.
85. Gregor, S. (2001). Explanations from knowledge-based systems and cooperative problem solving: An empirical study. *International Journal of Human-Computer Studies, 54*(1), 81–105.
86. Zanker, M., Jessenitschnig, M., & Schmid, W. (2010). Preference reasoning with soft constraints in constraint-based recommender systems. *Constraints, 15*(4), 574–595.
87. Felfernig, A., & Burke, R. (2008, August). Constraint-based recommender systems: Technologies and research issues. In *Proceedings of the 10th International Conference on Electronic Commerce* (p. 3). ACM.
88. Moreno, A., Valls, A., Isern, D., Marin, L., & Borràs, J. (2013). Sigtur/e-destination: Ontology-based personalized recommendation of tourism and leisure activities. *Engineering Applications of Artificial Intelligence, 26*(1), 633–651.
89. Kethavarapu, U. P. K., & Saraswathi, S. (2016). Concept based dynamic ontology creation for job recommender system. *Procedia Computer Science, 85,* 915–921.
90. Chen, R. C., Huang, Y. H., Bau, C. T., & Chen, S. M. (2012). A recommender system based on domain ontology and SWRL for anti-diabetic drugs selection. *Expert Systems with Applications, 39*(4), 3995–4006.
91. Kang, J., & Choi, J. (2011, April). An ontology-based recommender system using long-term and short-term preferences. In *2011 International Conference on Information Science and Applications (ICISA)* (pp. 1–8). IEEE.
92. Karthikeyan, N. K., & Raj Kumar, R. K. (2016). Fuzzy service conceptual ontology system for cloud service recommendation. *Computers & Electrical Engineering*.
93. Bouza, A., Reif, G., Bernstein, A., & Gall, H. (2008, October). Semtree: Ontology-based decision tree algorithm for recommender systems. In *Proceedings of the 2007 International Conference on Posters and Demonstrations-Volume 401* (pp. 106–107). CEUR-WS.org.
94. Karpus, A., Vagliano, I., Goczyła, K., & Morisio, M. (2016, September). An ontology-based contextual pre-filtering technique for recommender systems. In *2016 Federated Conference on Computer Science and Information Systems (FedCSIS)* (pp. 411–420). IEEE.
95. Andersen, R., Borgs, C., Chayes, J., Feige, U., Flaxman, A., Kalai, A., … & Tennenholtz, M. (2008, April). Trust-based recommender systems: An axiomatic approach. In *Proceedings of the 17th International Conference on World Wide Web* (pp. 199–208). ACM.
96. Walter, F. E., Battiston, S., & Schweitzer, F. (2008). A model of a trust-based recommender system on a social network. *Autonomous Agents and Multi-Agent Systems, 16*(1), 57–74.

97. Shani, G., & Gunawardana, A. (2011). Evaluating recommender systems. In *Recommender systems handbook* (pp. 257–297).
98. Gunawardana, A., & Shani, G. (2009). A survey of accuracy evaluation metrics of recommendation tasks. *Journal of Machine Learning Research, 10*(Dec), 2935–2962.
99. Kendall, M. G. (1938). A new measure of rank correlation. *Biometrika, 30*(1/2), 81–93.
100. Scholz, M., Dorner, V., Franz, M., & Hinz, O. (2015). Measuring consumers' willingness to pay with utility-based recommender systems. *Decision Support Systems, 72,* 60–71.
101. Raghavan, V., Bollmann, P., & Jung, G. S. (1989). A critical investigation of recall and precision as measures of retrieval system performance. *ACM Transactions on Information Systems (TOIS)., 7*(3), 205–229.
102. Huang, Y. J., Powers, R., & Montelione, G. T. (2005). Protein NMR recall, precision, and F-measure scores (RPF scores): Structure quality assessment measures based on information retrieval statistics. *Journal of the American Chemical Society, 127*(6), 1665–1674.

# Recommender Frameworks Outline System Design and Strategies: A Review

**R. Ponnusamy, Worku Abebe Degife and Tewodros Alemu**

**Abstract** Nowadays, right information and service access are the big challenge in the World Wide Web. There are number of tools available to access the right information in the market. Recommender system is the most valuable tool to provide such service. The applications of recommender systems include recommending movies, music, television programs, books, documents, Web sites, conferences, tourism scenic spots and learning materials, and involve the areas of e-commerce, e-learning, e-library, e-government, and e-business services. These recommender systems can be designed with different objectives, strategies, algorithms, and methods. This article discusses in detail about what is recommender system, needs, benefits, challenges, strategies, algorithms, and measures used for designing the recommender system. It also gives the details about the user personalization and customization.

## 1 Introduction

The explosive growth of variety of information in the Web, different e-services offered, and the large user-access lead to the poor decision making. It is unthinkable for the single client to get to all data with respect to their information/services/products choice by perusing all data accessible to them. More often than not getting to the correct data itself is a major issue for the clients. In spite of the data or items being accessible in the diverse structures like content, video, sound, and movement, the client need a decent strategy to gather information/services/products. Recommender frameworks have ended up being a significant route for online clients to adapt to the data overburden and have turned out to be a standout among the

R. Ponnusamy (✉)
CVR College of Engineering, Hyderabad 501 510, Telangana, India
e-mail: prof.r.ponnusamy@gmail.com

W. A. Degife · T. Alemu
Department of Information Systems, Faculty of Informatics,
University of Gondar, Gondar, Ethiopia



S. Margret Anouncia and U. K. Wiil (eds.), *Knowledge Computing and its Applications*, https://doi.org/10.1007/978-981-10-8258-0_12

most capable and mainstream instruments for data get to or benefits get to [1–7]. There are distinctive strategies for suggestion which have been proposed. Amid the most recent decade, large portions of them have likewise been effectively sent in business situations. Additionally, the recommender frameworks are for the most part used to foresee the recommendation that a client would have for a given thing. These frameworks are generally sent, and we utilize them every day.

When all is said in done, the recommender framework utilizes diverse procedures like client personalization, client conduct tractability and forecast, learning-based proposal, client total suggestion, neighborhood proposal, collective proposal, social labeling, multi-criteria choice, setting-based proposal, requirement-based suggestion, trust-based suggestion, and so on. These procedures are connected autonomously or mutually. There are additionally a few procedures connected to plan the framework to empower the client to choose right information/services/products, for example, neural network systems and soft computing, statistics and machine learning techniques [8], and so on. Personalization is a key component in Web-based social networking and recommender frameworks.

Web site pages can be customized in light of the attributes (interests, social class, setting, and so forth) [4], activities (tap on catch, open a connection, and so on), expectation (make a buy, check status of a substance), or whatever other parameters that can be recognized and connected with an individual, in this way giving them a custom-made client encounter. Take note of that the experience is once in a while just settlement of the client; however, a connection between the client and the yearnings of the site architects in driving particular activities to accomplish goals (e.g., increment deals transformation on a page). The term customization [9, 10] is frequently utilized when the site just uses express information, for example, item evaluations or client recommendations. There are numerous classes of Web personalization including behavioral, relevant, specialized, memorable information, and cooperatively separated. Actually, Web personalization can be accomplished by partner a guest portion with a predefined activity.

Learning-based recommender frameworks [11] (information-based recommenders) is a specific kind of recommender framework that depends on the method of expressing information related to the arrangement of data, client recommendations and suggestion criteria. The information-based recommender framework by and large characterized into three classifications, for example, conversational suggestion, pursuit-based proposal, and route-based proposal framework. In a setting-based recommender framework, the substance suppliers will track every one of the media contents their clients expend with a specific end goal to prescribe the things.

Soft computing (some of the time alluded to as computational knowledge; however, CI does not have a concurred definition) is the utilization of estimated answers for computationally hard assignments, for example, the arrangement of NP-complete issues, for which there is no known calculation that can process a correct arrangement in polynomial time. Soft computing differs from conventional (hard) computing in that, unlike hard computing, it is tolerant of imprecision, uncertainty, partial truth, and approximation. In effect, the role model for soft computing is the human mind. As a result, the good example for delicate processing is the human personality. The principal constituents of soft computing (SC) are fuzzy logic (FL), evolutionary computation (EC), machine learning (ML), and probabilistic reasoning (PR), with the latter subsuming belief networks and parts of learning theory.

Measurements are the customary field that needs to be arranged with the dimensions of evaluation, gathering, examination, translation, and making determinations from information. Recommender framework configuration is an interdisciplinary field that draws on computer sciences (information base, computerized reasoning, machine learning, graphical and representation models), measurements and designing (example acknowledgment, neural systems). Machine learning [7] is a strategy for information investigation that computerizes explanatory model building. Utilizing calculations that iteratively gain from information, machine learning permits computers to discover concealed bits of knowledge without being expressly modified where to look.

The outline of the recommender framework by receiving the distinctive methodologies and strategies is truly a testing one. In this chapter, the author audits the different procedures and methods accessible around the globe for recommender framework plan. The design of the recommender system by adopting the different strategies and techniques is really challenging one [12–14]. In this chapter, the author reviews the various strategies and techniques available around the world for recommender system design. This chapter discusses the details of applications, benefits, types of tools and techniques, algorithms and metrics, and measures of recommender system.

## 2 Application of Recommender System

A recommender framework plans to give clients customized online item or administration proposals to deal with the expanding on the Web data overburden issue and enhances client relationship administration. Different recommender framework procedures have been proposed since the mid-1990s, and many sorts of recommender framework programming have been created as of late for an assortment of uses. Analysts and supervisors perceive that recommender frameworks offer awesome open doors and difficulties for business, government, training, and different spaces, with later effective advancements of recommender frameworks for true applications getting to be noticeably clear. The utilizations of recommender frameworks incorporate prescribing films, music, TV programs, books, records, sites, meetings, tourism beautiful spots, and learning materials and include the ranges of online business, e-learning, e-library, e-government, and e-business administrations. Subsequently, it enables scientists to comprehend the recommender framework advancement and to help engineers to support relevant frameworks improvement in a systematic manner. We bunch recommender framework applications into eight primary spaces: e-government, e-business, Web-based business/e-shopping, e-library, e-learning, e-tourism, e-asset administrations, and e-aggregate exercises.

## 3 Advantage and Disadvantages

Recommendation systems play a vital role in today's Web. Here are some of the benefits and drawbacks of recommendation systems.

### *3.1 Benefits*

The achievement of using recommender system is to suggest products and services. They are being worked for practically every area where we can give proposals, and their focal points are clear. There are a few points of interest exist; these advantages are, for example, (i) one depends on the client genuine past conduct and it foresee the future, (ii) Sometimes couple of items endure on account of non-promotion, yet the recommender framework mitigate this issue since they enable us to find things that are like what we officially like, (iii) It customizes the item and it knows us superior to any other person so they recognize what we like and what we don't care for, (iv) It surrender the to date data about the market and what new items accessible in the market will be prescribed to you (v) It lessen the client perusing time (vi) Most of the hierarchical support of a site is keeping the route framework and scientific categorization in accordance with the clients' evolving needs. With proposal frameworks, a lot of this authoritative upkeep leaves (vii) A suggestion motor can convey movement to your site (viii) It prescribes the Relevant Material in the market (ix) It builds the consumer loyalty (x) It is utilized to control the retailing and stock tenets (xi) it lessens the client work and overhead (xii) it gives different reports about the clients.

### *3.2 Drawbacks*

Proposal frameworks are not great. They have a few disadvantages that outline groups ought to consider in the event that they are pondering executing them. Despite the fact that it gives the different focal points, it has diverse downsides (i) it is exceptionally hard to setup (ii) even however proposal frameworks can diminish hierarchical upkeep, they do not dispose of support itself (iii) sometimes individuals are troubled with suggestions. Infrequently suggestions are not right (iv) it builds the activity to the site, and on occasion, the general framework execution will get influenced.

## 4 Challenges with Recommender Systems

There are different challenges which exist in the recommender system.

(i) Aggregate imitative behavior: Exploiting the "understanding of gathering" has been made less unpredictable with the data amassing openings the Web bears. However, the immense measures of available data in like manner snare this open entryway. For example, but a couple of customers' lead can be illustrated, diverse customers do not demonstrate average direct. These customers can skew the outcomes of a recommender system and reduce its viability. A good recommender system must manage these issues.

(ii) Scalability: One issue that is endemic to vast scale proposal frameworks is adaptability. Conventional recommendations can function admirably with littler measures of information even when the informational collections are increasing based on the customary experiences.

(iii) Privacy protection: Security assurance contemplations are likewise a test. Recommender calculations can distinguish designs people would not know exist. A current case is the situation of an expansive organization that could compute a pregnancy forecast score in light of obtaining propensities. Using focused on advertisements, a father was amazed to discover that his high-school girl was pregnant. The organization's indicator was accurate to the point that it could foresee a planned mother's expected date in view of items she bought.

(iv) Lack of data: Maybe the greatest issue confronting recommender frameworks is that they require a considerable measure of information to successfully make suggestions. It is no occurrence that the organizations most related to having fantastic suggestions are those with a great deal of customer client information: Google, Amazon, Netflix, Last.fm. In general, a recommender framework needs to work with the information in the viewpoint of getting great suggestions of diversified perspectives.

(v) Changing data: The client past conduct [of users] is not a decent instrument on the grounds that the patterns are always evolving. Additionally, thing proposals do not work on the grounds that there are essentially excessively numerous item characteristics in mold and each trait (think fit, value, shading, style, texture, mark, and so on) has an alternate level of significance at various circumstances for a similar shopper.

(vi) Changing user preferences: On user preferences, recommender systems may also incorrectly label users. This may lead the system to recommend the unwanted things.

## 5 Recommender System Strategies

There do different types of strategies used for the design of recommender system exist, and they are classified based on their functionality. These strategies [15, 16] are as follows: (i) collaborative recommendation (ii) content-based recommendation (iii) knowledge-based recommendation system (iv) conversational recommendation (v) search-based recommendation (vi) navigation-based recommendation (vii) social and demographic-based recommendation (viii) hybrid recommendation approaches (ix) constraint-based recommenders (x) neighborhood-based recommendation methods (xi) client-based recommendation. The following section explains one by one in detail.

(i) Collaborative Recommendation

Collaborative recommendation [17] is the most renowned way to deal with the common proposal that has similar synonym derived from multiple users. The primary thought is that you're given a framework of recommendations by clients for things, and these are utilized to foresee missing recommendations and suggest things with high forecasts. One of the key favorable circumstances of this approach is that there has been a colossal measure of research into communitarian separating, making it quite surely knew, with existing libraries that make execution genuinely direct. Another essential preferred standpoint is that community-oriented separating is autonomous of thing properties. All you have to begin is client and thing IDs, and some idea of recommendation by clients for things (evaluations, sees, and so forth).

(ii) Content-based Recommendation

Content-based algorithms [18, 19] are given client recommendations for things and suggest comparative things in view of a space particular idea of thing substance. The fundamental favorable position of substance-based proposal over shared sifting is that it does not require as much client criticism to go ahead. Indeed, even one known client recommendation can yield numerous great proposals (which can prompt the gathering of recommendations to empower synergistic suggestion). In numerous situations, content-based suggestion is the most regular approach. For instance, while suggesting news articles or blog entries, it is normal to look at the printed substance of the things. This approach additionally stretches out normally to situations where thing metadata is accessible (e.g., film stars, book writers, and music kinds).

(iii) Knowledge-based Recommendation

Knowledge-based recommender frameworks [20, 21] (information-based recommenders) are a particular kind of recommender framework that depends on express learning about the thing arrangement, client recommendations, and suggestion criteria (i.e., which thing ought to be prescribed in which setting?). These frameworks are connected in situations where elective methodologies; for example, collaborative separating and content-based sifting cannot be connected. A noteworthy quality of learning-based recommender frameworks is the non-presence of frosty begin (increase) issues. A relating disadvantage is potential learning procurement bottlenecks activated by the need of characterizing proposal information in an express form.

(iv) Conversational Recommendation

Information-based recommender frameworks are regularly conversational; i.e., client necessities and recommendations are evoked inside the extent of an input circle. A noteworthy purpose behind the conversational way of information-based recommender frameworks is the intricacy of the thing space where it is frequently difficult to expressive all client recommendations on the double. Moreover, client recommendations are commonly not known precisely toward the start but rather are developed inside the extent of a suggestion session.

(v) Search-based Recommendation

In a search-based recommender, client criticism is given regarding answers to inquiries which confine the arrangement of important things. A case of such a question is "Which kind of focal point framework do you lean toward: settled or interchangeable focal points?" On the specialized level, look-based proposal situations can be executed on the premise of limitation-based recommender frameworks. Imperative construct recommender frameworks are actualized with respect to the premise of limitation inquiry or diverse sorts of conjunctive question-based methodologies.

(vi) Navigation-based Recommendation

In a route-based recommender, client criticism is normally given as far as "evaluates" which indicate change demands with respect to the thing as of now prescribed to the client. Scrutinizes are then utilized for the proposal of the following "hopeful" thing. A case of a study with regards to an advanced camera proposal situation is "I might want to have a camera like this yet with a lower cost." This is a case of a "unit evaluate" which speaks to a change ask for on a solitary thing characteristic. "Compound evaluates" permit the determination of more than one change ask for at once. "Dynamic investigating" likewise considers going before client studies (the scrutinizing history). Later methodologies also misuse data put away in client connection logs to additionally decrease the association exertion as far as the quantity of required scrutinizing cycles.

(vii) Social and Demographic-based Recommendation

Social and demographic recommenders propose things that are loved by companions, companions of companions, and demographically comparative individuals. Such recommenders need not bother with any recommendations by the client to whom proposals are made, making them effective. As far as I can tell, even inconsequentially executed methodologies can be depressingly exact. For instance, simply summing the quantity of Facebook likes by a man's dear companions can regularly be sufficient to illustrate what that individual preferences.

Given this energy of social and statistic recommenders, it is not amazing that informal communities do not effortlessly give their information away. This implies for some professionals who utilizes social/statistic suggestion and recommendation that are just unimaginable.

(viii) Hybrid Recommendation Approaches

Late research has shown that a mixture approach [22], joining collective separating and substance-based sifting could be more successful sometimes. Half and half methodologies can be executed in a few routes: by making content-based and collaboration oriented-based expectations independently and afterward consolidating them; by adding content-based abilities to collaboration-based approach (and the other way around); or by bringing together the methodologies into one model for an entire audit of recommender frameworks. A few reviews exactly contrast the execution of the half-breed and the unadulterated synergistic and substance-based strategies and show that the hybrid techniques can give more precise proposals than immaculate methodologies. These techniques can likewise be utilized to defeat a portion of the basic issues in recommender frameworks.

Netflix is a decent case of the utilization of hybrid recommender frameworks. The site makes suggestions by looking at the watching and seeking propensities for comparable clients (i.e., community-oriented separating) and by offering motion pictures that offer attributes with movies that a client has evaluated (substance-based sifting).

(ix) Constraint-based Recommenders

Constraint-based methodologies [23, 24] are particularly proper and can settle on the thing decision prepare all the more intense in such spaces. Constraint-based recommendation methodologies offers progressions to the change of databases for prerequisite-based recommenders since reasonable instrument can support essential information. A recommender information base of a limitation-based recommender framework can be characterized through two arrangements of factors and three distinct arrangements of requirements. These factors and requirements are the significant elements of an imperative fulfillment issue. An answer for a limitation fulfillment issue comprises of solid instantiations of the factors with the end goal that all the predetermined requirements are satisfied.

(x) Neighborhood-based Recommendation Methods

As the name proposes, neighborhood-based recommender frameworks consider the recommendations or preferences of the client group or clients of the area of a dynamic client before making proposals or suggestions to the dynamic client. The thought for neighborhood-based recommenders is exceptionally straightforward: given the evaluations of a client, discover every one of the clients like the dynamic client who had comparable recommendations in the past and after that make expectations in regard to every single obscure item that the dynamic client has not appraised but rather are being evaluated in their neighborhood.

(xi) Context-Aware Recommender Systems

Context-aware recommender calculations prescribe things that match the client's present requirements. This enables them to be more adaptable and versatile to current client needs than techniques that overlook setting (basically giving a similar weight to the greater part of the client's history). Thus, logical calculations will probably inspire a reaction than methodologies that are constructing just with respect to recorded information. The key confinements of relevant recommenders are like those of social and statistic recommenders—logical information may not generally be accessible, and there is a danger of crawling out the client. For instance, promotion retargeting can be viewed as a type of logical proposal that pursues clients around the Web and crosswise over gadgets, without having the express assent of the clients to being followed in this way.

(xii) Trust-Based Recommendation

There is a developing need to oversee confide in open frameworks as they may contain deceitful specialist organizations. A trust management framework tries to address the issue of finding a recommendation of the most confided in administrations. A recommender framework decides a level of trust as far as a solitary customized esteem gotten from a few sorts of confirmations, for example, client's criticism, history of client's associations, setting of the submitted ask for, references from outsider clients, and in addition from outsider specialist co-op, and structure of the general public of administrations. The framework can use more trust confirmations towards a more precise estimation of trust. We additionally propose a capacity to make sense of how comparable two clients are in a given setting. The subsequent framework prescribes to the client a rundown of the most trusted Web indexes positioned by the recovery accuracy of archives returned in light of the client's question, and also the level of trust of the Web indexes have picked up by communicating with other related clients inside the setting of the inquiry.

(xiii) Multi-Criteria Recommendation

Multi-criteria recommender frameworks [25] (MCRS) can be characterized as recommender frameworks that consolidate recommendation data upon different criteria. Rather than creating suggestion procedures in light of a solitary paradigm esteems, the general recommendation of client '*a*' for the item '*x*', these frameworks attempt to foresee a rating for unexplored things of client by misusing recommendation data on numerous criteria that influence this general recommendation esteem. A few specialists approach MCRS as a multi-criteria basic leadership (MCDM) issue and apply MCDM strategies and procedures to execute MCRS frameworks.

(xiv) Client-Based Recommendation

It is one sort of recommendation where the server prescribes information in getting to the customer ask for, for instance, clients may have their decision of various customer-side operations and send it back to the server. The framework at that point examines the suggestion, and, when the investigation is finished, transmits its outcomes back to the customer proposal. Operations might be performed in view of the customer ask for in light of the fact that they oblige access to data or usefulness that is accessible on the customer driven, on the grounds that the client needs to watch them or give input, or on the grounds that the server the essential presumption of handling data just in view of the customer ask.

# 6 Personalization

Personalization, now and then known as customization, comprises of fitting an administration or an item to oblige particular people, now and again fixing to gatherings or sections of people. A wide assortment of associations utilizes personalization to enhance consumer loyalty, advanced deals change, promoting results, marking, and enhanced site measurements, and additionally to advertise. Personalization is a key component in online networking and recommender frameworks.

Web site pages can be customized [26] in view of the attributes (interests, social classification, setting, and so on), activities (tap on catch, open a connection, and so on), expectation (make a buy, check status of an element), or some other parameter that can be distinguished and connected with an individual, in this way furnishing them with a custom-made client encounter. Take note of that the experience is once in a while essentially settlement of the client; however, a connection between the client and the goals of the site planners in driving particular activities to accomplish destinations (e.g., increment deals transformation on a page). The term customization is frequently utilized when the site just uses express information, for example, item appraisals or client recommendations. There are numerous classifications of Web personalization including behavioral, contextual, technical, historic information, cooperatively separated.

There are a few camps in characterizing and executing Web personalization. A couple of wide strategies for Web personalization may include: (1) implicit, (2) explicit, (3) hybrid. With certain personalization, the Web personalization is performed in view of the diverse classifications specified previously. It can likewise be gained from direct communications with the client in light of certain information, for example, things bought or pages saw. With unequivocal personalization, the page (or data framework) is changed by the client utilizing the components given by the framework. Crossbreed personalization consolidates a number of ways to deal with use the best among the existing universes.

# 7 Algorithms for Recommender System

There are different algorithms used for recommendations system. In general, these algorithms are classified into two categories either memory-based algorithms or model-based algorithms. Some of these are at times tools dependent. These tools are such as artificial neural models, soft computing models, fuzzy logic, evolutionary computation, machine learning, probabilistic reasoning, and belief theory. The following section discusses the algorithms in detail.

## *7.1 Supervised Latent Dirichlet Allocation (SLDA)*

Supervised latent Dirichlet allocation (SLDA) is a generative statistical model that allows sets of observations to be explained by unobserved groups that explain why some parts of the data are similar. In SLDA, each document may be viewed as a mixture of various topics where each document is considered to have a set of topics that are assigned to it via SLDA. This is identical to probabilistic latent semantic analysis (pLSA), except that in SLDA the topic distribution is assumed to have a sparse Dirichlet prior. The sparse Dirichlet priors encode the intuition that documents cover only a small set of topics and that topics use only a small set of words frequently. In practice, this results in a better disambiguation of words and a more precise assignment of documents to topics. SLDA is a generalization of the pLSA model, which is equivalent to SLDA under a uniform Dirichlet prior distribution. With plate notation, the dependencies among the many variables can be captured concisely.

## *7.2 Support Vector Machines (SVMs)*

In machine learning, bolster vector machines are directed learning models with related learning calculations that investigate information utilized for arrangement and relapse examination. Given an arrangement of preparing cases, each set apart as having a place with either of two classifications, a SVM [23] preparing calculation constructs a model that doles out new cases to one classification or the other, making it a non-probabilistic double straight classifier. A SVM show is a portrayal of the cases as focuses in space, mapped with the goal that the cases of the different classes are isolated by an unmistakable hole that is as wide as could be expected under the circumstances. New illustrations are then mapped into that same space and anticipated to have a place with a classification in view of which side of the hole they fall.

## 7.3 Multinomial Logistic Regression (Maximum Entropy)

In statistics, multinomial logistic regression is a classification method that generalizes logistic regression to multiclass problems, i.e., with more than two possible discrete outcomes. That is, it is a model that is used to predict the probabilities of the different possible outcomes of a categorically distributed dependent variable, given a set of independent variables (which may be real-valued, binary-valued, categorical-valued, etc.).

There are multiple equivalent ways to describe the mathematical model underlying multinomial logistic regression. This can make it difficult to compare different treatments of the subject in different texts. The article on logistic regression presents a number of equivalent formulations of simple logistic regression, and many of these have analogues in the multinomial logic model.

The idea behind all of them, as in many other statistical classification techniques, is to construct a linear predictor function that constructs a score from a set of weights that are linearly combined with the explanatory variables (features) of a given observation using a dot product: where $X_i$ is the vector of explanatory variables describing observation $i$, $\beta_k$ is a vector of weights (or regression coefficients) corresponding to outcome $k$, and score($X_i$, $k$) is the score associated with assigning observation $i$ to category $k$. In discrete choice theory, where observations represent people and outcomes represent choices, the score is considered the utility associated with person $i$ choosing outcome $k$. The predicted outcome is the one with the highest score.

## 7.4 Naive Bayes

In machine learning, credulous Bayes recommenders are a group of straightforward probabilistic classifiers in light of applying Bayes' hypothesis with solid (innocent) autonomy suppositions between the elements. Gullible Bayes is a straightforward system for developing classifiers: models that allot class marks to issue occasions, spoken to as vectors of highlight esteems, where the class names are drawn from some limited set. It is not a solitary calculation for preparing such classifiers, but rather a group of calculations in view of a typical guideline: All credulous Bayes recommenders accept that the estimation of a specific component is free of the estimation of some other element, given the class variable.

## 7.5 Bayesian Belief Nets

It can be envisioned as a coordinated non-cyclic diagram, with curves speaking to the related probabilities among the factors. Guileless Bayes is a basic method for building classifiers: models that dole outclass marks to issue cases, spoken to as vectors of highlight esteems, where the class names are drawn from some limited set. It is not a solitary calculation for preparing such recommenders, but rather a group of calculations in light of a typical guideline: All gullible Bayes recommenders accept that the estimation of a specific component is autonomous of the estimation of some other element, given the class variable. For instance, an organic product might be thought to be an apple on the off chance that it is red, round, and around 10 cm in measurement. Gullible Bayes recommenders consider each of these elements to contribute freely to the likelihood that this natural product is an apple, paying little heed to any conceivable connections between's the shading, roundness, and distance across highlights.

## 7.6 *Neural Networks*

Artificial neural networks [27] are extraordinarily crude digital networks of neurons based on the neural structure of the mind. These method records one by one and learns by comparing their type of the report (i.e., in large part arbitrary) with the acknowledged real category of the record. The errors from the preliminary category of the primary record are fed again into the community and used to adjust the networks algorithm for in addition iterations. A neuron in an artificial neural network is

1. A set of input values ($x_i$) and associated weights ($w_i$).
2. A function ($g$) that sums the weights and maps the results to an output ($y$) as given in Fig. 1.

Neurons are organized into layers: input, hidden, and output. The enter layer is composed not of full neurons, but as a substitute consists simply of the record's values which might be inputs to the next layer of neurons. The subsequent layer is the hidden layer. Numerous hidden layers can exist in one neural community. The very last layer is the output layer, wherein there is one node for each class. An unmarried sweep ahead via the community effects inside the assignment of a value to every output node, and the file is assigned to the class node with the very best fee.

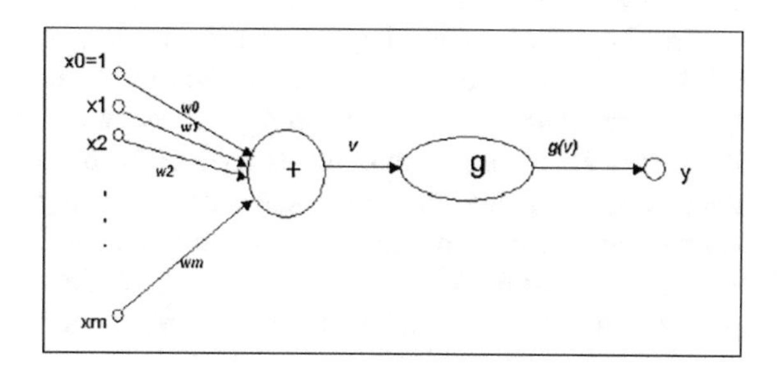


**Fig. 1** Neural network

## 7.7 Decision Trees

A choice tree is a choice help device that uses a tree-like diagram or model of choices and their conceivable results, including chance occasion results, asset expenses, and utility. It is one approach to show a calculation. Choice trees are generally utilized as a part of operations look into, particularly in choice investigation, to help recognize a procedure well on the way to achieve an objective, but on the other hand are a mainstream apparatus in machine learning. A choice tree is a flowchart-like structure in which each inside hub speaks to a "test" on a quality (e.g., regardless of whether a coin flip comes up heads or tails), each branch speaks to the result of the test, and each leaf hub speaks to a class name (choice taken in the wake of processing all qualities). The ways from root to leaf speak to characterization rules.

In choice investigation, a choice tree and the firmly related impact outline are utilized as a visual and scientific choice help apparatus, where the normal esteems (or expected utility) of contending choices are ascertained. A decision tree consists of three types of nodes:

1. Decision nodes—typically represented by squares
2. Chance nodes—typically represented by circles
3. End nodes—typically represented by triangles

Choice trees are ordinarily utilized as a part of operations research and operations administration. On the off chance that, practically speaking, choices must be brought online with no review under fragmented information, a choice tree ought to be paralleled by a likelihood display as a best decision show or online choice model calculation. Another utilization of choice trees is as a graphic means for figuring contingent probabilities.

## 7.8 Boosting and Bagging Algorithms

Bagging and boosting are capable procedures utilized as a part of machine figuring out how to enhance the expectation exactness of classifier learning frameworks. This report will incorporate and prologue to stowing and boosting as a diagram to those new to the strategy and as a refresher to those acquainted with the idea.

Bagging is like subject expects in a workplace. At the point when the leader of the USA needs to settle on approach choices, he depends on the mastery of his bureau individuals to settle on the right choice regarding the strategy. The skill of the bureau individuals compliments each different instead of being repetitive and duplicative. Utilizing this theoretical case of sacking, we can apply it to machine learning ideas commonplace to us. Given some database of preparing information, we can take tests from this database with substitution. Utilizing tests taken from the preparation illustration database, we can prepare our machine learning calculation freely on each of these datasets. After the preparation has finished, we are left with classifiers, these are undifferentiated from the bureau individuals said in the case. At the point when given some obscure illustration, we make an expectation on it utilizing each of the classifiers. The last forecast is made by choosing the most well-known expectation from each of the classifier's. The final order of the test case produced using the objective classifiers is known as a voting plan where the expectation of each objective classifier is a "vote" towards the last forecast.

## 7.9 Hidden Markov Model

Markov chains, which are similar to Bayesian Belief Nets treat the recommendation problem as sequential optimization instead of simply prediction.

A hidden Markov model (HMM) is a triple (Π, *A*, *B*).

| | | |
|---|---|---|
| $\Pi = (\pi_i)$ | the vector of the initial-state probabilities | $\Pr(y_i \vert x_j)\Pr(x_{i_t} \vert x_{j_{x-1}})$ |
| $A = (a_{ij})$ | the state transition matrix | |
| $B = (b_{ij})$ | the confusion matrix | |

Each probability in the state transition matrix and in the confusion matrix is time independent—that is, the matrices do not change in time as the system evolves. In practice, this is one of the most unrealistic assumptions of Markov models about real processes.

## 7.10 k-*Nearest Neighbor Algorithm*

The *k*-nearest neighbor algorithm (k-NN) is a nonparametric technique utilized for order and relapse. In both cases, the information comprises of the *k*-nearest preparing cases in the element space. The yield relies upon whether k-NN is utilized for arrangement or relapse.

The preparation illustrations are vectors in a multidimensional element space, each with a class name. The preparation period of the calculation comprises just of putting away the element vectors and class names of the preparation tests. In the arrangement stage, *k* is a client characterized steady, and an unlabeled vector (an inquiry or test point) is ordered by appointing the mark which is most successive among the *k*-preparing tests closest to that question point.

## 7.11 k-*Means*

*k*-means is one of the least complex unsupervised learning calculations that tackle the outstanding grouping issue. The technique takes after a basic and simple approach to order a given informational index through a specific number of groups (accept *k* bunches) settled apriori. The principle thought is to characterize *k* focuses, one for each bunch. These focuses ought to be set shrewdly in view of various area that causes diverse outcomes. Along these lines, the better decision is to put them however much as could reasonably be expected far from each other. The subsequent stage is to take each direct having a place toward a given informational collection and partner it to the closest focus. At the point when no point is pending, the initial step is finished and an early gathering age is finished. Now we have to re-figure *k* new centroids as barycenter of the bunches coming about because of the past stride. After we have these *k* new centroids, another coupling must be done between similar informational collection focuses and the closest new focus. A circle has been produced. Because of this circle, we may see that the *k* focuses change their area well ordered until the point when no more changes are done or at the end of the day focuses do not move any more. Finally, this algorithm aims at minimizing objective functions know as squared error function given by:

$$J(V) = \sum_{i=1}^{c} \sum_{j=1}^{c_i} \left( \left\| x_i - v_j \right\| \right)^2$$

where '$\left\| x_i - v_j \right\|$' is the Euclidean distance between $x_i$ and $v_j$. '$c_i$' is the number of data points in *i*th cluster. '*c*' is the number of cluster centers.

## 7.12 *Fuzzy* c-*Means*

Fuzzy *c*-means (FCM) is a method of clustering which allows one piece of data to belong to two or more clusters. This method is frequently used in pattern recognition. It is based on minimization of the following objective function:

$$J_m = \sum_{i=1}^{N} \sum_{j=1}^{C} u_{ij}^{m} \left\| x_i - c_j \right\|^2, \quad 1 \le m < \infty$$

where $m$ is any real number greater than 1, $u_{ij}$ is the degree of membership of $x_i$ in the cluster $j$, $x_i$ is the *i*th of *d*-dimensional measured data, $c_j$ is the *d*-dimension center of the cluster, and $\|*\|$ is any norm expressing the similarity between any measured data and the center.

## 7.13 SVD

In this case, it is represented as a conceptual matrix and it also permits one to infer about the relation of expected contextual usage of phrases. A singular value decomposition (SVD) to the matrix is a form of a factor or more properly the mathematical generalization of which factor analysis is a special case. In SVD, a rectangular matrix is decomposed into the product of three other matrices. One component matrix describes the original column entries in the same way, and the third is a diagonal matrix containing scaling values such that when the three components are matrix-multiplied, the original matrix is reconstructed. The reconstructed two-dimensional matrix that approximates the original matrix and a few highest values are selected to reconstruct the original matrix.

Each document in the particular sub-hierarchy represents the rows and each phrase with respect to the document is represented as the columns. Learning human-like knowledge consists in formulating a bivariate frequency table with row $i$ representing the $i$th phrase and column $j$ representing the $j$th document (or between any two entities) and $f_{ij}$ evaluated by the Shannon's measure of information $\sum p \log p$. This together with the dimension reduction will constitute the constraint satisfaction for prediction between the observed and the expected values to make classification.

SVD is a powerful technique employed for solving a linear system of equations **AX** = **B**, in $M$ equations of $N$ unknowns with $M >=< N$ in order to get unique set of solutions; a set of singular solutions, infinite number of solutions; non-trivial solutions, or trivial solutions based upon the nature of the coefficient matrix $A$. Whatever maybe the vectors $X$, $B$ concepts of rank, null space, range space of linear algebra are essential in formulating the computer program for any practical problem in conformity with the decomposition of the matrix $A$

$$[A] = [U][W][V^T]$$

When more equations than the unknowns are given, relevant solutions can be obtained by least squares method.

After the reconstruction of the original matrix, we find the correlation between the new document and the existing document in the sub-hierarchy. If the correlation is high, then it is decided that the new document belongs to the particular sub-hierarchy category.

## 8 Metrics and Measures for Evaluating Recommender System

The last result of a recommender framework [28–34] is a top-rundown of things suggested for the client requested by an assessed score that speaks to the recommendation of that thing for the client. So most noteworthy the esteem, more intrigued the client will be. Be that as it may, to create the correct suggestions is not inconsequential and there are a few reviews and research on assessing the proposals of such motor.

An ideal recommender framework would be the one that could anticipate every one of your recommendations precisely would exhibit an arrangement of things positioned by your future recommendation and be finished. The most recommender system work by attempting different assess rating that are manipulated for each item stored. One conceivable method for assessing recommender's recommendations is to assess the nature of its assessed recommendation esteems, that is, assessing how intently the evaluated recommendations coordinate the real recommendations of the client.

Typically, research on recommender systems is concerned about finding the most accurate recommendation algorithms. However, there are a number of factors that are also important.

- Diversity—Users tend to be more satisfied with recommendations when there is higher intra-list diversity.
- Recommender persistence—In a few circumstances, it is more powerful to re-indicate proposals, or let clients re-rate things, than demonstrating new things. There are a few explanations behind this. Clients may overlook things when they are appeared surprisingly; for example, they had no opportunity to examine the proposals deliberately.
- Privacy—Recommender frameworks typically need to manage security concerns [18] on the grounds that clients need to uncover touchy data. Building client profiles utilizing cooperative separating can be tricky from a security perspective.
- User demographics—The user demographics may influence how satisfied users are with recommendations. In their paper, they show that elderly users tend to be more interested in recommendations than younger users.
- Robustness—When users can participate in the recommender system, the issue of fraud must be addressed.
- Serendipity—Good fortune is a measure of "how shocking the proposals are". For example, a recommender framework that prescribes drain to a client in a market may be superbly exact, yet it is not a decent suggestion since it is a conspicuous thing for the client to purchase. Be that as it may, high scores of luck may negatively affect precision.

- Trust—A recommender system is of little value for a user if the user does not trust the system. Trust can be built by a recommender system by explaining how it generates recommendations, and why it recommends an item.
- Labeling—User satisfaction with recommendations may be influenced by the labeling of the recommendations.

There are different measures used for evaluating the recommender system according the requirements. These measures are such as mean absolute error (MAE), mean squared error (MSE), root mean squared error (RMSE), precision, recall, mean average precision (MAP), normalized discounted cumulative gain (NDCG), intra-list similarity, Lathia's diversity, mean percentage ranking, user-centric evaluation frameworks.

## *8.1 MAE (Mean Absolute Error)*

Measurable exactness measurements assess the precision of a framework by looking at the numerical suggestion scores against the genuine client appraisals for the client thing sets in the test dataset. Mean absolute error (MAE) among evaluations and expectations is a broadly utilized metric. MAE is a measure of the deviation of proposals from their actual client indicated values. For every evaluations forecast, combine $\langle p_i, q_i \rangle$; this metric treats the total blunder between them, i.e., $|p_i - q_i|$ similarly. The MAE is estimated by cumulatively errors of the $N$ related evaluations forecasts. Formally,

$$\text{MAE} = \frac{\sum_{i=1}^{N} |p_i - q_i|}{N}$$

The lower the MAE, the more accurately the recommendation engine predicts user ratings.

## *8.2 MSE (Mean Squared Error)*

Once the assessment for a disconnected examination has been settled, the precision should be measured over the test set. The mean squared mistake (MSE) or mean squared deviation (MSD) of an estimator (of a strategy for evaluating an in secret amount) measures the normal of the squares of the blunders or deviations—that is, the distinction between the estimator and what is watched. In this unique situation, assume that we measure the nature of $\hat{y}$, as a measure of the focal point of the dissemination, regarding the mean square mistake.

$$\mathrm{MSE} = \frac{1}{n}\sum_{i=1}^{n}(y_i - \hat{y}_i)^2$$

MSE is a weighted normal of the squares of the separations among $\hat{y}$ and the class marks with the relative frequencies as the weight elements. Along these lines, the best measure of the middle, with respect to this quantify of mistake, is the estimation of $t$ that limits MSE.

## 8.3 RMSE (Root Mean Squared Error)

Root mean Square Error (RMSE) is the primitive mean/normal of the square of the deviation between the actual and forecasted data. RMSE is usually utilized and makes for a phenomenal universally useful error metric for numerical expectations. Contrasted with the comparable mean absolute error, RMSE increases and extremely rebuffs substantial mistakes.

$$\text{RMS Errors} = \sqrt{\frac{\sum_{i=1}^{n}(\hat{y}_i - y_i)^2}{n}}$$

## 8.4 Precision and Recall

Precision and recall measure are set-based measures. They are computed using unordered sets of documents. We need to extend these measures (or to define new measures) if we are to evaluate the ranked recommendation results that are now standard with recommendation system. We could say that the 'Precision' is the proportion of top results that is relevant, considering some definition of relevant for your problem domain. So if we say "Precision at $n$" would be this proportion judged from the top $n$ results. The 'Recall' would measure the proportion of all relevant results included in the top results.

$$\text{Recall} = \frac{\text{Relevant Retrieved}}{\text{All Relevant}} = \frac{A}{A+C}$$

$$\text{Precision} = \frac{\text{Relevant Retrieved}}{\text{All Retrieved}} = \frac{A}{A+B}$$

## 8.5 MAP (Mean Average Precision)

The two most mainstream positioning measurements are MAP and NDCG. We secured mean normal accuracy a while prior. NDCG remains for normalized discounted cumulative gain. The fundamental distinction between the two is that MAP expect parallel significance (a thing is both of intrigue or not), while NDCG permits pertinence scores in type of genuine numbers. The connection is much the same as with characterization and relapse.

It is hard to upgrade MAP or NDCG straightforwardly, in light of the fact that they are intermittent and therefore non-differentiable. Fortunately, ranking measures and loss functions in learning to rank demonstrate that several misfortune capacities utilized as a part of figuring out how to rank inexact those measurements. Most standard among the TREC people group is mean average precision (MAP), which gives a solitary figure measure of value crosswise over review levels. Among assessment measures, MAP has been appeared to have particularly great separation and solidness. For a solitary data require, average precision is the normal of the exactness esteem acquired for the arrangement of top reports existing after each applicable record is recovered, and this esteem is then found the middle value of over data needs.

## 8.6 NDCG (Normalized Discounted Cumulative Gain)

A approach that has seen increasing adoption, especially when employed with machine learning approaches to ranking SVM-ranking is measures of *cumulative gain*, and in particular *normalized discounted cumulative gain* (NDCG). NDCG is designed for situations of non-binary notions of relevance. Like precision at $k$, it is evaluated over some number $k$ of top search results. For a set of queries $Q$, let $R(j, d)$ be the relevance score assessors gave to document $d$ for query $j$. Then,

$$\mathrm{NDCG}(Q,k) = \frac{1}{|Q|}\sum_{j=1}^{|Q|} Z_{kj} \sum_{m=1}^{k} \frac{2^{R(j,m)} - 1}{\log(1+m)},$$

where $z_{kj}$ is a normalization factor calculated to make it so that a perfect ranking's NDCG at $k$ for query $j$ is 1. For queries for which $k' < k$ documents are retrieved, the last summation is done up to $k'$.

## 8.7 Intra-list Similarity

It is one measure to gauge the differing qualities. Run of the mill target suggestion framework in proposal formalizes the objective of achieving high intra-comparability (archives inside are comparative) and low entomb similitude (reports from various are unique). This is an inner rule for the nature of a grouping. In any case, great scores on an inward standard do not really convert into great viability in an application. A contrasting option to inward criteria is immediate assessment in the utilization of intrigue.

## 8.8 Lathia's Diversity

This is an assessment strategy which ignores the way that clients keep on rating things after some time the worldly attributes of the framework's top proposals are not explored. Specifically, there are no methods for measuring the degree that similar things are being prescribed to clients again and again. In this work, we demonstrate that worldly assorted qualities are a critical feature of recommender frameworks, by indicating how information changes after some time and playing out a client review. We at that point assess the calculations from the perspective of the differing qualities in the arrangement of proposal records they deliver after some time. In looking at how various attributes of client rating designs (counting profile size and time between evaluations) influence differences. At that point, propose and assess set strategies that augment worldly suggestion differences without broadly punishing precision. This is called Lathia assorted qualities.

## 8.9 Mean Percentage Ranking

The basis for calculating a function within the rating is the number of recommendation perspectives generated at the specific question hints. For every question recommendation, we put together a percent distribution of page perspectives made from exceptional systems, after which we calculate the mathematics mean from the formerly counted percentage stocks for each version. The final end result defines the rating position.

## 8.10 User-Centric Evaluation

The term user-centered [35, 36] assessment refers to comparing the software and value of recommended to the meant give up customers. At each step of the consumer-centered recommendation technique, we now not most effective want to acquire facts from the user and take the vital motion based on that information; however, we want to verify that we've got interpreted the facts effectively. So for every step, we gain information from the user, act upon it, and verify that the recommendation became an accurate one or now not.

## 9 Summary

In this chapter, we review the various fundamentals and conventional methods used for design of recommendation systems. These things are such as strategies, algorithms, and measurements. It also analyzed the challenges, benefits, and drawbacks of these recommender systems. The strategies are used independent or combined. With regard to the evaluation, the article discusses the various common measurements available for recommender system. The system designer has to choose the right measurements to ensure the accuracy of the system based on their requirement.

## References

1. Breese, J. S., Heckerman, D., & Kadie, C. (1998, May). *Empirical Analysis of Predictive Algorithms for Collaborative Filtering* (Technical Report, Microsoft Research, WA 98052).
2. Mishra, R., Kumar, P., & Bhasker, B. (2015). A web recommendation system considering sequential information. *Decision Support Systems, 75,* 1–10.
3. Barragáns-Martínez, B., Costa-Montenegro, E., & Juncal-Martínez, J. (2015). Developing a recommender system in a consumer electronic device. *Expert Systems with Applications, 42,* 4216–4228.
4. Liu, Y., & Yang, J. (2015). Improving ranking-based recommendation by social information and negative similarity. *Procedia Computer Science, 55,* 732–740.
5. Chen, H.-H., Ororbia, I. I., Alexander, G., & Giles, C. L. (2008) *ExpertSeer: A keyphrase based expert recommender for digital libraries*. http://expertseer.ist.psu.edu/.
6. Felfernig, A., Isak, K., Szabo, K., & Zachar, P. (2012). *The VITA financial services sales support environment*. www.aaai.org.
7. Takacs, G. (2009). Scalable collaborative filtering approaches for large recommender systems. *Journal of Machine Learning Research, 10,* 623–656.
8. Melville, P., & Sindhwani, V. (2011). Recommender systems. In C. Sammut & G. Webb (Eds.), *Encyclopedia of machine learning*. Berlin: Springer.
9. Lee, J., Lee, K., Kim, J. G., & Kim, S. (2015). Personalized academic paper recommendation system. In *SRS'15*, August 10, 2015, Sydney, NSW, Australia.
10. Rubens, N., Elahi, M., Sugiyama, M., & Kaplan, D. (2015). Active learning in recommender systems. In F. Ricci, L. Rokach, & B. Shapira (Eds.), *Recommender systems handbook*. Boston: Springer. https://doi.org/10.1007/978-1-4899-7637-6_24.
11. Beel, J., Gipp, B., Langer, S., & Breitinger, C. (2015). Research paper recommender systems: A literature survey. *International Journal on Digital Libraries,* 1–34. https://doi.org/10.1007/s00799-15-0156-0.

12. Beel, J., Genzmehr, M., Langer, S., Nürnberger, A., & Gipp, B. (2013). A comparative analysis of offline and online evaluations and discussion of research paper recommender system evaluation. In *Repsys'13*, October 12, 2013, Hong Kong China. ACM. ISBN: 978-1-4503-2465-6/13/10.
13. Song, Y., Dixon, S., & Pearce, M. (2012). A survey of music recommendation systems and future perspectives. In *9th International Symposium on Computer Music Modelling and Retrieval* (CMMR 2012), June 19–22, 2012, Queen Mary University of London.
14. Adomavicius, G., & Tuzhilin, A. (2005). Toward the next generation of recommender systems: A survey of the state-of-the-art and possible extensions. *IEEE Transactions on Knowledge and Data Engineering, 17*(6), 734–749.
15. Ricci, F., Rokach, L., & Shapira, B. (2015). Introduction to recommender systems handbook. In F. Ricci, L. Rokach, B. Shapira, & P. Kantor (Eds.), *Recommender systems handbook*. Berlin: Springer. https://doi.org/10.1007/978-1-4899-7637-6_24.
16. Ge, Y., Xiong, H., Tuzhilin, A., Xiao, K., Gruteser, M., & Pazzani, M. (2010). An energy-efficient mobile recommender system. In *KDD'10*, July 25–28, 2010, Washington, DC, USA. ACM. ISBN: 978-1-4503-0055-110/07.
17. Rennie, J. D. M., & Srebro, N. (2005). Fast maximum margin matrix factorization for collaborative prediction. In *Proceedings of the 22nd International Conference on Machine Learning*, Bonn, Germany.
18. Puglisi, S., Parra-Arnau, J., Forné, J., & Rebollo-Monedero, D. (2015). On content-based recommendation and user privacy in social-tagging systems. *Computer Standards & Interfaces, 41,* 17–27.
19. Melville, P., Mooney, R. J., & Nagarajan, R. (2002). Content-boosted collaborative filtering for improved recommendations. In *Proceedings of the Eighteenth National Conference on Artificial Intelligence* (AAAI-2002) (pp. 187-192), Edmonton, Canada, July 2002.
20. Bobadilla, J. (2013). Recommender systems survey. *Knowledge-Based Systems, 46,* 109–132.
21. Ricci, F. (2002, November/December). Travel recommender systems. *IEEE Intelligent System, 17*, 54–57.
22. Burke, R. (2007). Hybrid web recommender systems. In P. Brusilovsky, A. Kobsa, & W. Nejdl (Eds.), *The adaptive web* (pp. 377–408)., LNCS 4321 Berlin: Springer.
23. Kothari, A. A., & Patel, W. D. (2015). A novel approach towards context based recommendations using support vector machine methodology. *Procedia Computer Science, 57,* 1171–1178.
24. Felfernig, A., & Burke, R. (2008). Constraint-based recommender systems: Technologies and research issues. In *10th International Conference on Electronic Commerce*, (ICEC)'08 Innsbruck, Austria. ACM. ISBN: 978-1-60558-075-3/08/08.
25. Adomavicius, G., & Kwon, Y. (2015). Multi-criteria recommender systems. In F. Ricci, L. Rokach, & B. Shapira (Eds.), *Recommender systems handbook*. Berlin: Springer. https://doi.org/10.1007/978-1-4899-7637-6_24.
26. Beel, J., Langer, S., Genzmehr, M., & Nürnberger, A. (2013). Persistence in recommender systems: Giving the same recommendations to the same users multiple times. In *Proceedings of the 17th International Conference on Theory and practice of Digital Libraries*. LNCS. Berlin: Springer.
27. Reed, R., & Marks II, R. J. (1999). Neural smithing: Supervised learning in feedforward artificial neural networks. MIT Press. ISBN-13: 978-0262181907.
28. Waila, P., Singh, V. K., & Singh, M. K. (2016). A scientometric analysis of research in recommender systems. *Journal of Scientometric Research, 5*(1).
29. Beel, J., Langer, S., & Genzmehr, M. (2013). Sponsored vs. organic (research paper) recommendations and the impact of labeling. In T. Aalberg, C. Papatheodorou, M. Dobreva, G. Tsakonas, & C. J. Farrugia (Eds.), *Research and advanced technology for digital libraries*. TPDL 2013, LNCS 8092 (pp. 391–395). Berlin: Springer.
30. Parra, D., & Sahebi, S. (2013). Recommender systems: Sources of knowledge and evaluation metrics. In J. D. Velasquez, V. Palade, & L. Jain (Eds.), *Advanced techniques in web intelligence-2*. SCI 452 (pp. 149–175). Berlin: Springer.

31. Isinkaye, F. O., Folajimi, Y. O., & Ojokoh, B. A. (2015). Recommendation systems: Principles, methods and evaluation. *Egyptian Informatics Journal, 16,* 261–273.
32. Marung, Ukrit, Theera-Umpon, N., & Auephanwiriyakul, S. (2016). Top-N recommender systems using genetic algorithm-based visual-clustering methods. *Symmetry, 8,* 54. https://doi.org/10.3390/sym8070054.
33. Kaminskas, M., & Bridge, D. (2014). Measuring surprise in recommender systems. In *Workshop on Recommender Systems Evaluation: Dimensions and Design* (REDD 2014), held in conjunction with RecSys 2014.
34. Murakami, T., Mori, K., & Orihara, R. (2008). Metrics for evaluating the serendipity of recommendation lists. K. Satoh, A. Inokuchi, K. Nagao, & T. Kawamura (Eds.), *New frontiers in artificial intelligence*. JSAI 2007, LNAI 4914 (pp. 40–46). Berlin: Springer.
35. Beel, J., Langer, S., Nürnberger, A., & Genzmehr, M. (2013). The impact of demographics (age and gender) and other user-characteristics on evaluating recommender systems. In *Proceedings of the 17th International Conference on Theory and Practice of Digital Libraries*. Berlin: Springer.
36. Cremonesi, P., Garzotto, F., & Turrin, R. (2013). User-centric vs. system-centric evaluation of recommender systems. In P. Kotzé, G. Marsden, G. Lindgaard, J. Wesson, & M. Winckler (Eds.), *INTERACT 2013, Part III*. LNCS 8119 (pp. 334–351). 2013 © IFIP International Federation for Information Processing.

# Ontology Design and Use for Subjective Evaluation of Computer Science Examinations

**Himani Mittal**

**Abstract** Ontology is the study of groups or classes of things. It is a classification of concepts in a domain. Domain Ontology represents the concepts which belong to part of the world. Ontology along with set of individual instances of classes constitutes a knowledge base. In this paper, Domain Ontology of Computer Graphics is prepared using subject–predicate–object representation, where subject is a class or instance, predicate is property of class or instance and object is values of properties. This Ontology is then used for subjective evaluation of answers submitted by students in examinations. The statistical techniques, namely Latent Semantic Analysis (LSA), BiLingual Evaluation Understudy (BLEU), Generalized Latent Semantic Analysis (GLSA), Maximum Entropy, and Hybrid Technique based on LSA and BLEU are used for performing evaluation of student answer along with Ontology. When the students' answers are to be evaluated, the details related to the concept are fetched from the Ontology. The point of access is provided by human examiner. When short questions are to be answered, then directly related information is fetched. When longer questions are to be answered, then more details are fetched using Ontology property, instances, and subclasses. After performing preprocessing, the students' answers are given as input along with subject-specific Ontology to statistical techniques. The sentences in each student answer are classified as belonging to an Ontology concept using statistical techniques. Then, total number of concepts found in each answer is divided by total number of concepts in concept map to generate the similarity score between concept map and student answer. The development and implementation of Ontology-based evaluation involve: development of Ontology, extraction of Ontology from RDF file, and implementation of statistical techniques in Java Programming Language. The Ontology is implemented with the help of Protégé tool. The developed application is tested using 10 questions of Computer Graphics. The techniques are tested both with and without Ontology. The OHYBRID technique shows the best correlation as compared to all other techniques. Maximum Entropy has correlation varying from 0.61 to 0.96. There is a lot of improvement in Maximum Entropy with Ontology.

H. Mittal (✉)
GGDSD College, Chandigarh, India
e-mail: research.himani@gmail.com



S. Margret Anouncia and U. K. Wiil (eds.), *Knowledge Computing and its Applications*, https://doi.org/10.1007/978-981-10-8258-0_13

GLSA Ngram 1 and 2 do not show much improvement with use of Ontology as correlation varies from 0.44 to 0.87 for NGRAM1 and 0.41 to 0.84 from NGRAM2. The performance of the Ontology-based evaluation is compared with evaluation without Ontology. It is found that with use of Ontology the performance is more streamlined as individual feedback can be given to students. The feedback includes the concepts that are missing in student answer which is more near to human-like evaluation. It is concluded that use of Ontology makes the evaluation more thorough and near to accurate.

# 1 Introduction

An Ontology is conceptualization of knowledge in a domain [1]. Ontology defines the concepts, properties of concepts, and relation among concepts. Ontology develops common vocabulary for researchers who need to share information in a domain. It includes machine interpretable definitions of basic concepts in the domain and relations among them. Ontology can be used to classify documents discussed in [2]. Use of Ontology takes advantage of semantic relationships among concepts for finding relevant documents and eliminates irrelevant documents by identifying mismatch in concepts. Dridi [3] used Ontology for information retrieval. It prepared a bag of concepts by placing the words in the documents under the Ontology concepts. Then, this bag of concepts along with vector space model was used to classify the documents. The results were compared with those of traditional classification methods, and it was found that use of Ontology to create bag of concepts instead of bag of words improved the classification of document. Oleiwi and Yasin [4] and Wu et al. [5] used Ontology for classification of documents.

Bulakh [6] developed system for evaluation of short answers using Ontology. It proposed that answer model as well as student answer should be converted into Ontology and compared using XML. Accuracy of results was found 78%. Converting the student's answers and model answers into Ontology is a complex job for long answers. Several tools for converting running text into Ontology exist, but they have limited success rate [7–10].

# 2 Design of Ontology

Ontology is the study of groups or classes of things [1, 4, 11]. It is a classification of concepts in a domain. Domain Ontology represents the concepts which belong to part of the world. Particular meanings of terms are applied to that Domain Ontology. For example, the word 'card' has many meanings. Ontology about domain of 'poker' has meaning playing card, while Ontology of computer hardware has a meaning of 'punch cards' and 'video cards'. Ontology is a formal explicit description of concepts in a domain of discourse (classes or concepts), properties of each concept describing features and attributes of concepts (slots), and restrictions on slots (facets). Ontology along with set of individual instances of classes constitutes a knowledge base. The Domain Ontology of Computer Graphics is prepared using subject–predicate–object representation, where subject is a class or instance, predicate is property of class or instance, and object is values of properties. The following components of Ontology are defined for Computer Graphics subject:

(1) Classes: Sets, collections, concepts, and types of objects. These represent the concepts in a domain that cannot be initialized. These are names of categories, events, algorithms, etc. For example, the main class is 'thing'. Under 'thing', the starting class is 'Computer_graphics'. 'Computer_graphics' has subclasses including 'computer_graphics_applications', 'Computer_graphics_systems', and 'types_of_media'. The 'Computer_graphics_systems' has subclasses 'hardware' and 'software'. Further classification of each is performed as depicted in Fig. 1a–c. The algorithms and processes are designed as classes with steps as individuals and properties.
(2) Individuals: The classes have instances or objects of classes. For example, 'image_processing' class has several applications. These applications are designed as individuals of 'image_processing', namely 'color_coding', 'improve_picture_quality', 'improve_quality_of_shading', 'machine_perception', and 'rearrange_picture_parts'. The instances further have properties. Instances cannot have children or subclasses or instances but have attributes like graphics packages implementing these applications—'hasPackage'. The algorithms implementing these applications are represented as classes and linked to individuals of respective 'image_processing' applications.
(3) Attributes: The classes and individuals have properties. Some of the attributes identified are: 'uses' with values depicting application of device, 'technique' depicting name of technique used in the device, 'purpose' depicting the purpose like input or output or pointing device, 'standard' depicting the universal standards followed and adopted by the devices. Each class has separate attributes. The 'Events' class has subclasses depicting working of devices. These classes have attributes like 'actor' depicting the part of device performing the action, 'target' depicting the part of device effected by event, 'output_of_event' depicting the result of action or event.
(4) Relations among Classes: Apart from subclass relation, Ontology uses disjoint and equivalence relations for linking classes and subclasses of different hierarchies. For example, 'Working_of_CRT' is a subclass of 'video_display_devices'. Former is also a subclass of 'Events'. So equivalence is created between the child nodes of the 'Events' and 'video_display_devices' classes.
(5) Process representation: There are many processes and algorithms in graphics for example, 'working_of_CRT', 'working_of_plasma_panel', etc. These are represented as event class individuals. These have properties like 'actor', 'target', 'output_of_event', 'input_to_event', 'part_of_event', 'predecessor', and 'successor'.

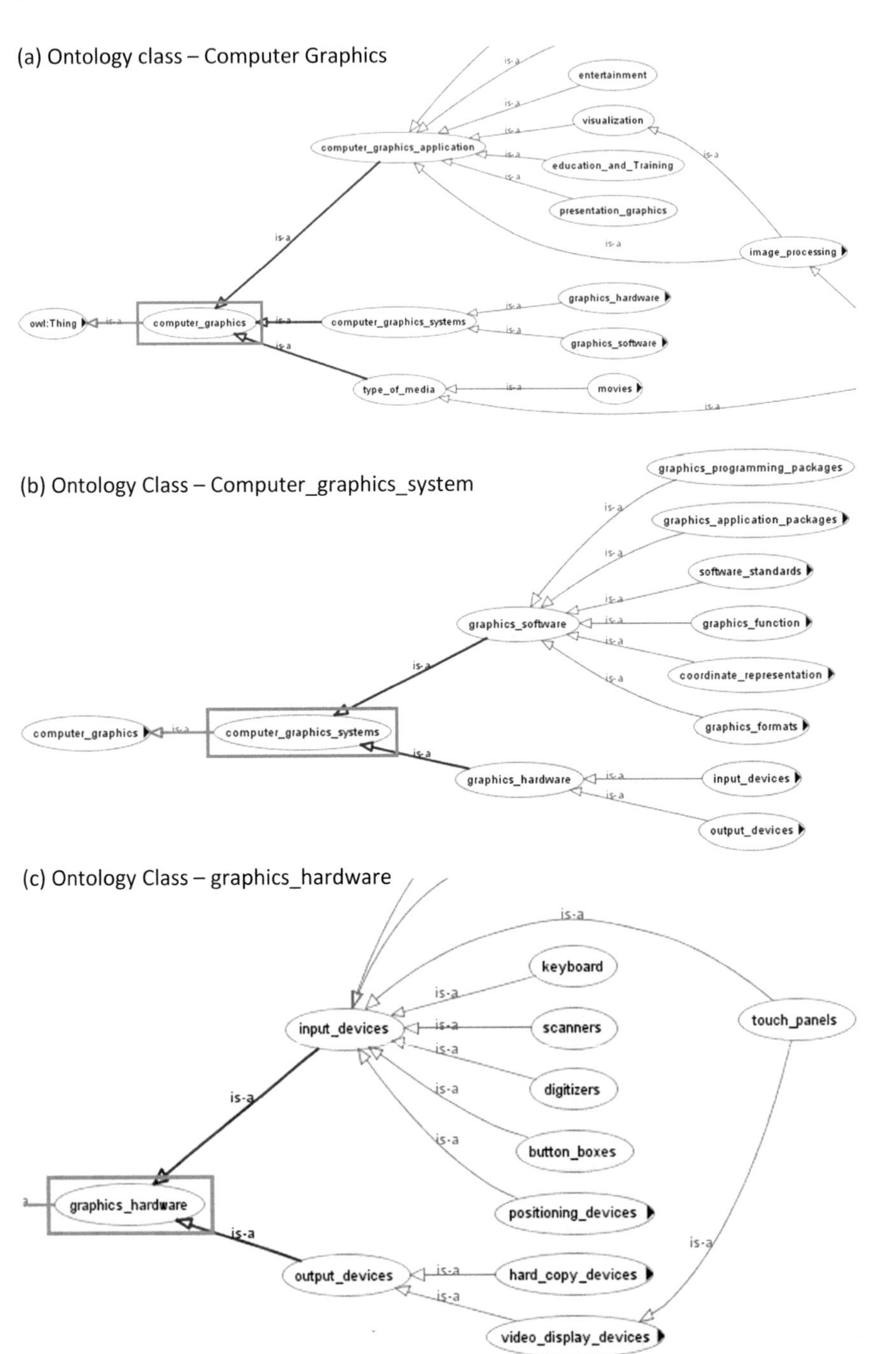


**Fig. 1** Hierarchy of classes defined in Computer Graphics Ontology

The designed Ontology is a dummy Ontology and does not cover exhaustively all the concepts of Computer Graphics.

## 3 Information Retrieval Statistical Techniques

The input to all the techniques is students' answers. The output is a similarity measure in the range of [0, 1], where a value of 0 indicates no similarity and 1 indicates high similarity. The input is preprocessed before applying the techniques. The steps of preprocessing are: tokenization (find all words in all students' answer), stop word removal (removing common words like a, the, as, an), and synonym search (for each word left after stop word removal, find its synonyms) as already discussed in previous section. Stemming is not performed as Ontology concept classes cannot be stemmed, so matching will become difficult.

**Latent Semantic Analysis:** Latent Semantic Analysis (LSA) [12] is a technique in natural language processing for analyzing relationships between a set of documents and the terms they contain. The basic assumption is that there exists a hidden semantic space in each text that is the accumulation of all words' meaning. It usually takes three steps to compress the semantic space—filtering, selection, and feature extraction. First, the stop words are filtered. Second, word frequency matrix is constructed by selecting reference texts. Third, singular value decomposition is done to extract features by factorizing the feature matrix. Last, cosine similarity is found between feature vectors. The cosine similarity of these vectors (correlation value) signifies the degree of relation between the student answer and the keywords. The semantic presence of the keywords in student answers is indicated by higher cosine correlation value. The standard LSA technique is modified in two ways: firstly, preprocessing the input to find synonyms and performing stemming. This makes the output of LSA more precise as all the forms of the words are provided for. It gives the students freedom to use any similar word or form of same word depending on tense and sentence structure being used. Secondly, in this work the selection of words (second step above) in LSA is modified. The vocabulary required for calculating word frequency is originally prepared by counting unique terms in model answer or reference text. This is modified by counting unique terms in model answer and unique terms in all student answers, thus building a complete model of all the student answers and model answer.

The detailed steps of modified LSA technique are given in Fig. 2. First, the preprocessing steps of stop word removal, tokenization, synonym search, and stemming are performed (discussed above). Then, word frequency matrix is calculated. Singular valued decomposition (eigen decomposition of non-invertible matrices) is performed on term-frequency matrix. This generates vectors of terms and answers. These term and answer vectors are multiplied to get cosine similarity.

Algorithm Latent Semantic Analysis (model answer, Student Answer)

Variables Used in Algorithm

- Terms matrix of order 1XM. It has all the keywords from model answer and unique words from students' answer.
- Stu matrix of order NX1. It contains all the students' answer.
- Matrix 'A' of order MXN. It is the frequency value a[i][j] of number of times term[i] appera in stu[j].
- S is the left Eigen vectors of $A^TA$.
- E is the square roots of Eigen values of A.
- U is right Eigen vectors of $AA^T$.
- SE matrix of order based on rank of E is product of S and E.
- D matrix is product of E and U.
- Q is sum of all rows of SE matrix where, se[i][j] is added if the ith element in Term is included in model answer.

1. Find out all the unique keywords in model answer as Term 1XM matrix. Add the list of unique words in student answers to this matrix. Note that no words are repeated in Term matrix.

2. Keep all student answers in an NX1 matrix, $Stu_{NX1}$.

3. Perform stop word removal, tokenization, synonym search and stemming on model answer and all student answers.

4. Construct the term to document frequency (tdf) matrix- $A_{mXn}$. Here m is number of terms and n is number of answers. The rows of A correspond to entries in Term matrix and columns correspond to Stu matrix. ... continued on next page

a[i][j] = 0 if ith word in Term is not present in jth answer in Stu.

a[i][j] = x, where x is the number of times the word appears in answer j in Stu.

5. Calculate Singular Value Decomposition (SVD) => $SVD(A_{mXn}) = S * E * U^T$

Where, S = Left Eigen Vectors of $A^TA$. Terms in the concept space are represented by row vectors of S;

E = Identity of square roots of Eigen values of $A^TA$. It gives the degree of relationship between the S and $U^T$ matrix.

**Fig. 2** Steps in Latent Semantic Analysis technique

$U^T$ = Right Eigen Vectors of $AA^T$. Documents are represented by column vectors of $U^T$.

6. Reduce the dimension of S,E and U by finding Rank of E. This step removes noise in the data.

7. Calculate SE = S*E . It gives the participation of term in the model answer (concept).

8. Calculate d= $E*U^T$ . It describes the participation of student answer in the model (concept).

9. Compute the weight vector of the keywords in model answer as q using following method:

for j=1to N q[j] = $\sum_{i=1 \text{ to } n}$ se[i][j]    where, se[i][j] is an element of SE.

Each row of SE, se[i] signifies a word vector. By adding these individual vectors for the words included in list of keywords in model answer, the term vector weight is found. The result in q is a column matrix of order 1XN.

**Fig. 2** (continued)

**BiLingual Evaluation Understudy:** The BLEU technique [13] is used to overcome the drawback of LSA technique. LSA overrates the answers that repeat the keywords many times. It means if a student repeats the same sentence a number of times then LSA evaluates it as good answer. However, in reality it is a bad answer. To overcome this problem, BLEU is used. BLEU generates a metric value ranging between 0 and 1. The value indicates how similar student answers are to the standard answer. The frequency of a keyword in student answer and total number of words in student answer are calculated. These values cannot be more than frequency of each keyword in model answer and total number of keywords in model answer, respectively. If any value is more, then the corresponding value calculated from model answer is used. Divide the frequency of each keyword in student answer by the total number of words. The algorithm used is given in Fig. 3.

A modified version of BiLingual Evaluation Understudy (BLEU) algorithm is used.

Firstly, the original BLEU algorithm makes use of n-grams (phrases of words). In this work, individual words are used because stop words are removed and phrases cannot be constructed.

Secondly, the original method calculates brevity factor and multiplies the bleu value with it. Brevity factor (BF) is exponent value of *r*, where *r* is number of words in student answer divided by total words in model answer. The significance is not to penalize a short response. In this work, BF is not calculated. BLEU is used to clip the maximum usage of keywords so that unnecessary repetition of keywords does not fetch more marks to students.

**Hybrid technique:** The outputs of BLEU and LSA are mapped using Fuzzy Logic. Fuzzy Logic is an extension of two-valued logic to handle the concept of

```
Algorithm BLEU ( model answer, students' answer)

    Variables Used

Term matrix of order 1XM – unique words in model answer.
masterFrequencyMatrix – frequency of words in Term in model answer.
Stu matrix of order NX1 – all student answers
bleuRepOfKeywords matrix of order NXM – repetition of each model answer keyword in
    students' answer. N is the number of Answers in Stu and M is the number of keywords
    in Term.
num- stores the numerator for word average for bleuVal
den – stores the denominator of word average for bleuVal
bleuVal- the final bleu score
 1. The Preprocessing steps- tokenization, stop word removal and synonym search are
    applied to input.
 2. Find frequency of all keywords in model answer as masterFrequencyMatrix by
    counting the number of times each unique word appears in model answer. Keep
    record of all unique words in Term matrix of order 1XM.
 3. Loop for each word in student answer i in Stu matrix of order NX1.
 4. Loop for j in Term
 5. Count the frequency of each jth keyword in ith student answer and store it in
    bleuRepOfKeywords[i][j]. Note: do not count frequency of words not present in Term
    matrix.
 6. End loop in step 4.
 7. End loop in step 3.
 8. Loop for i -  each student answer in Stu
 9. Loop for each keyword j in Term
10. If number of times keyword appears in student answer is less than equal to number of
    times it occurs in model answer (bleuRepOfKeywords[i][j] <
    masterFrequencyMatrix[j]), then set num = bleuRepOfKeywords[i][j]. Else set num =
    masterFrequencyMatrix.
11. den = total number of words in model answer.
12. bleuVal = bleuVal + num/den. This ratio finds the significance of each word in student
    answer with respect to model answer clipping away over use of the keywords.
13. End loop in step 9.
14. End loop in step 8.
```

**Fig. 3** Steps in BiLingual Evaluation Understudy algorithm

**Table 1** Rules for inference engine of Fuzzy Logic

| Input1 (LSA) | Input2 (BLEU) | Output (final) |
|---|---|---|
| Bad | Bad | Bad |
| Bad | Average | Ok |
| Bad | Excellent | Ok |
| Average | Bad | Ok |
| Average | Average | Average |
| Average | Excellent | Average |
| Excellent | Bad | Ok |
| Excellent | Average | Average |
| Excellent | Excellent | Excellent |

partial truth. Compared to traditional crisp variables, a fuzzy variable has a truth value varying between 0 and 1 showing there degree of membership. LSA and BLEU both generate correlation value between 0 and 1. In sharp set, zero correlation is interpreted as no relationship. A correlation value less than or equal to 0.30 represents weak relationship, 0.50 represents moderate relationship, and 0.70 represents strong relationship. Value less than or equal to one represents a perfect relationship. However, this interpretation does not deal well with boundary cases like correlation value of 0.51 is in moderate relationship (0.50) or strong relationship (0.70). Therefore, Fuzzy variables are used to define the correlation values. Both LSA and BLEU give output as degree of correlation, so they are used as fuzzy input variables. These two independent variables are pointing toward the different aspects of level of similarity between model answer and students' answer. This work defines two input variables named LSA and BLEU with three membership functions (bad, average, and excellent) and one output variable Final with four membership functions (bad, ok, average, and excellent). Trapezoidal class is used for all membership functions. The inference engine rules of Fuzzy Logic are shown in Table 1. Whenever the BLEU and LSA techniques both give bad correlation value, the output is bad. When any one is average or excellent and the other is bad, the result is ok. When both are average, result is average. Lastly, when both give excellent correlation, result is excellent.

**Generalized LSA (GLSA):** GLSA [14] finds the Term matrix as used in LSA algorithm (Fig. 2) by generating phrases (ngrams) from model answers and students' answers. Ngrams are formed by placing the n adjacent words in model answer together and moving right word by word. The phrase length used is 2, 3, and 4 neighboring words. For example, 'computer_graphics_system' is a 3 g. These phrases constitute the Term matrix. The frequency of each phrase in student answer is calculated to generate tdf ($A_{MXN}$ matrix). The remaining steps are same as in LSA.

**Maximum Entropy (MAXENT):** In MaxEnt [15], the input is training essays (multiple model answers) and student answers. The training essays are labeled with category to which they belong like good, bad, etc. No preprocessing steps are performed. The training data is used to study the word context—that is words that mostly follow and precede the word under consideration. The entropy is calculated for the current word to appear in a given context. This entropy is calculated for each word in model answer. The entropy for all word pairs in the model answers and corresponding target category are given to Perceptron for training. This builds a model using which the student answer evaluation is performed.

# 4 Design of Subjective Evaluation Technique Based on Ontology

The answers submitted by students on the online system are textual information. The model answer consisting of keywords expected in the students' answer is provided as input. The problem is to evaluate the text-based answers. The evaluation of subjective answers is viewed as the task to find the correlation between each student answer and model answer. The more similar meaning words are used, the more is the correlation. The keywords in an answer are the main criteria for scoring a technical answer. Therefore, the more domain-specific keywords are present in the answer, the more accurate it is. However, we cannot mark the answers by just counting the number of keywords. A more wholesome approach is required, which can evaluate based on semantic relationship between words and concepts. From the literature review, two areas of research were identified—improving the existing techniques by creating a hybrid technique and use of domain Ontology along with existing techniques. Both these areas are explored in this work. A hybrid technique is proposed in this thesis for subjective evaluation. The hybrid technique is based on clustering technique Latent Semantic Analysis (LSA), BiLingual Evaluation Understudy (BLEU), and Fuzzy Logic. It provides the student freedom of word choice and helps in overcoming problems inherent in LSA technique. Several Machine Learning (ML) techniques exist that try to capture the latent relationship between words. The techniques explored in this work are—Latent Semantic Analysis (LSA), Generalized Latent Semantic Analysis (GLSA), Maximum Entropy (MaxEnt), and BiLingual Evaluation Understudy (BLEU). These techniques are applied to subjective evaluation in previous work of other authors. The existing techniques show an accuracy of results from 59 to 93%. However, the evaluation is purely based on keyword presence. An evaluation process is required which measures the wholesome relationship between words, words and concepts and among concepts. Ontology is a concept map of domain knowledge. This work explores the effect of using Ontology with ML techniques. All the above-mentioned techniques are implemented both with and without Ontology and tested on common input data. The use of Ontology with these techniques ensures proper semantic-based evaluation as the answers are matched against an exhaustive knowledge base. This ensures categorization of answers based on concept category to which the answers belong. Ontology makes the evaluation process holistic as presence of keywords, synonyms, the right word combination, and coverage of concepts can be checked.

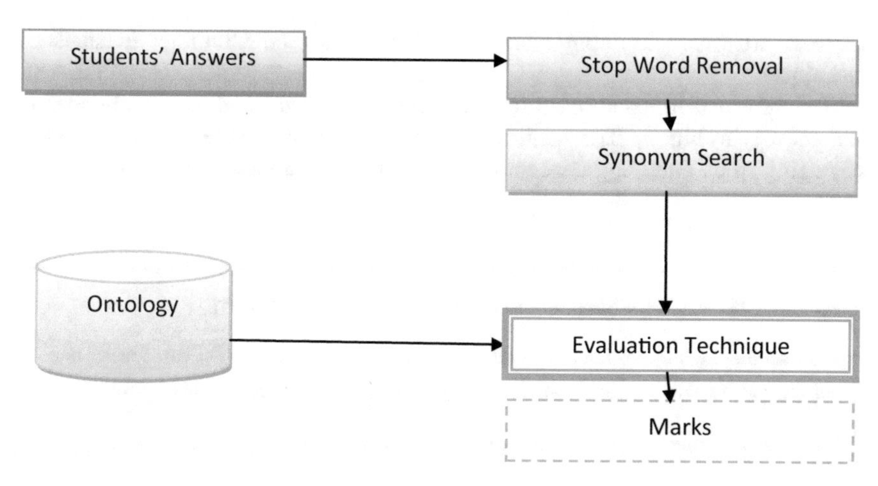


**Fig. 4** Design of subjective evaluation methodology with Ontology

**Table 2** Criteria to fetch information from Ontology

| Type of question | Level of detail fetched from Ontology |
|---|---|
| One-line questions | Concepts direct properties, instances, and subclasses |
| Short-length questions | All the concepts under the main concept |
| Long answer questions, stating facts and phenomenon | All the concepts under the main concept along with equivalent and inverse classes |
| Essay length questions, reflective and open ended | All the concepts directly under the main concept and related concepts with inverse, equivalence, part-of, steps in phenomenon, etc., are fetched. All the nodes directly related to main concept and indirectly related are fetched in the form of triples |

The scheme of evaluation is given in Fig. 4. First, the preprocessing steps are performed. The processed answers are given as input to statistical techniques along with subject-specific Ontology. The statistical techniques used are—Latent Semantic Analysis, Generalized Latent Semantic Analysis, BiLingual Evaluation Understudy, proposed hybrid technique, and Maximum Entropy. These techniques are discussed below.

When the students' answers are to be evaluated, the details related to the concept are fetched from the Ontology. The point of access is provided by human examiner along with model answer. When the students' answers for the question are to be evaluated, the details related to the concept are fetched from the Ontology. The level of detail will depend on the type of question as shown in Table 2. When short questions are to be answered, then directly related information is fetched. When longer questions are to be answered, then more details are fetched.

The combination of Ontology with the statistical techniques is performed using the design depicted in Fig. 4 [16]. After performing preprocessing, the students' answers are given as input along with subject-specific Ontology to statistical techniques.

The steps performed for each statistical technique (discussed in previous section) are shown in Fig. 5. The sentences in each student answer are classified as belonging to an Ontology concept using statistical techniques. Then, total number

Algorithm OntologyIREvaluation( Ontology, model answer, student answer)

1. Extract the Ontology related with question for which answers are to be evaluated taking the ontology access point mentioned at time of adding the question to database of MASMEE.

2. Represent all students' answers (i) and sentence in student answer (j) in matrix form.

3. Classify each sentence in students' answer as belonging to one of the ontology concept using following steps.

a) Split the student answer into constituent sentences.

b) Then perform the following steps.

BLEU- count the frequency of each word in each answer and all words in each concept. Find bleuVal for each sentence with each concept. The concept with which bleuVal is highest, is the concept to which sentence belongs.

LSA – perform SVD for each sentence with the ontology concepts replacing the Term matrix of order 1XM (Figure 4.6).

GLSA- form phrases of length 2,3, and 4 from ontology concepts and students' answer and then perform LSA.

Hybrid technique – Output of Onto-LSA and Onto-BLEU are applied to fuzzy.

Maximum Entropy – classify each sentence as belonging to one concept using concept name as category.

4. The total number of concepts pertaining to ontology access point and those found in each student answer separately are used to find similarity score for each answer.

correlation for ith student = no. of concepts in each student answer[i] divided by no. of concepts in ontology for that question.

5. This value is multiplied by maximum marks for the answer.

6. The explanation for the marks assigned includes the concept not found in student answer.

**Fig. 5** Algorithm for Ontology-based evaluation

of concepts found in each answer is divided by total number of concepts in concept map to generate the similarity score between concept map and student answer.

## 5 Implementation of Proposed Ontology-Based Evaluation

The development and implementation of Ontology-based evaluation involve: development of Ontology, extraction of Ontology from RDF file, and implementation of Machine learning techniques in Java Programming Language.

The Ontology is implemented with the help of Protégé tool. The Sparql queries were written to access information from the Ontology. Jena API was used to access the Ontology from Java code. The machine learning techniques LSA and BLEU are used as implemented for Hybrid technique with some customization. Generalized LSA is implemented in Java Programming Language. Open-source package is used for Maximum Entropy technique, available from SourceForge Web site.

(1) Development of Ontology using Protégé: Protégé was used to develop the Ontology for 'Computer Graphics' subject. Protégé is a free, open-source platform that provides tools to construct domain Ontology. It supports a variety of formats for Ontology like XML, RDF, and OWL. RDF format is used in this work, as it is fully developed and supported by a number of applications. The Resource Description Framework (RDF) is a family of World Wide Web Consortium (W3C) specifications originally designed as a format for Ontology and is similar to entity–relationship diagrams, as it is based upon the idea of making statements about resources (in particular Web resources) in the form of subject–predicate–object expressions. These expressions are known as triples in RDF terminology. The subject denotes the resource, and the predicate denotes traits or aspects of the resource and expresses a relationship between the subject and the object. For example, the Ontology subject—'Video_device_CRT_Monitor'—in Computer Graphics has predicate/property 'hasResolution'. The value of 'hasResolution' is 640X480, becoming the object of this subject. Figure 6a–b shows the Ontology developed using Protégé. In Fig. 6a, the interface shown is Protégé tool. The classes start from 'thing' which is main class of all the classes, beginning of everything. Under 'thing', all classes appear. The main class is Computer Graphics. Under Computer Graphics, all other classes are arranged. In the middle box, individuals belonging to each class are shown. In the third box, properties of individuals are shown. Figure 6a shows the individuals designed for Computer Graphics Ontology. The 'cross_sectional_view_design' individual belongs to classes 'computer_aided_design' and 'three_dimensional'. Figure 6b shows the properties of individuals. The class 'working_of_CRT' has individual 'Step_1_generation_of_electron_beam'. This individual has properties- 'target', 'actor', and

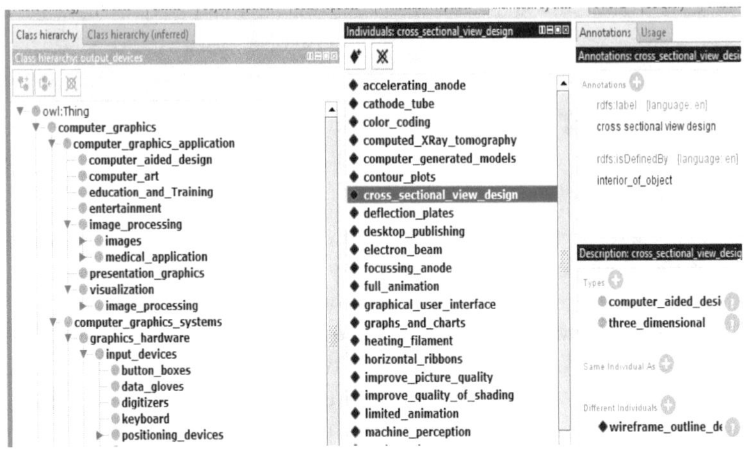


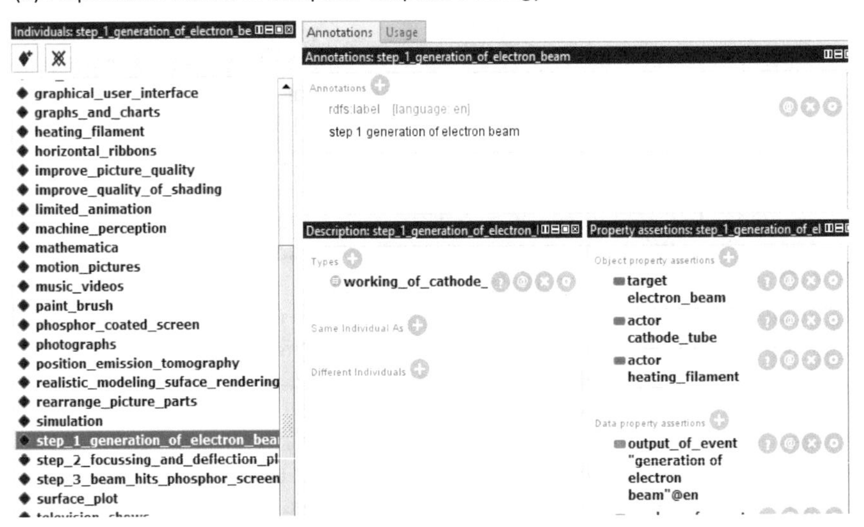


**Fig. 6** **a** Individuals of classes in Computer Graphics Ontology and **b** properties of classes in Computer Graphics Ontology

'output_of_events'. The 'target' property symbolizes the need for the step in working of CRT. The 'output_of_event' is expected result of performing step 1. 'Actor' property is used to identify the elements involved in performing step 1.

(2) Extraction of Ontology from Ontology file: After the Ontology is developed using Protégé, it is stored in a file on the hard disk. Java Programming Language along with Apache Jena Library and Sparql queries are used to extract the Ontology and use it for evaluation. Sparql 1.1 is a query language to

```
String prefix=   "PREFIX rdf: <http://www.w3.org/1999/02/22-rdf-syntax-ns#>\n" +
     "PREFIX owl: <http://www.w3.org/2002/07/owl#>\n" +
     "PREFIX rdfs: <http://www.w3.org/2000/01/rdf-schema#>\n" +
     "PREFIX xsd: <http://www.w3.org/2001/XMLSchema#>\n"
     + ontoPrefix;
String query1=prefix+"\r\n"+
        "SELECT distinct ?x ?p ?o  WHERE {{?x rdfs:subClassOf* my:"+ontologyConcept+
     " . ?x ?p ?o} UNION {?class rdfs:subClassOf* my:"+ontologyConcept+
     ". ?x rdf:type ?class .  ?x ?p ?o  } "
     + "UNION {"+"?class owl:unionOf my:"+ontologyConcept+
     ". ?x rdf:type ?class .  ?x ?p ?o}"
     + "UNION {"+"?class owl:intersectionOf my:"+ontologyConcept+
     ". ?x rdf:type ?class .  ?x ?p ?o}"
     + "UNION {"+"?class owl:complementOf my:"+ontologyConcept+
     ". ?x rdf:type ?class .  ?x ?p ?o}"
     +"UNION {"+"?class owl:disjointWith my:"+ontologyConcept+
     ". ?x rdf:type ?class .  ?x ?p ?o}"
     + "} order by ?x";

model=ModelFactory.createDefaultModel();

modelOnt=ModelFactory.createOntologyModel();

try
{
  File file=new File(ontologyFilename);
  FileInputStream reader=new FileInputStream(file);
  model.read(reader,null);
  modelOnt.add(model);

  Query query=QueryFactory.create(query1);

  QueryExecution exe;
  exe = QueryExecutionFactory.create(query, modelOnt);
   result = exe.execSelect() ;
```

**Fig. 7** Jena API and Sparql usage

access the knowledge saved in RDF triples of Ontology. It is similar to SQL. Apache Jena Library 2.13.0 is an open-source Semantic Web Framework for Java Programming Language that helps to create an in-memory model of Ontology where Sparql can be executed on it. It provides API to extract from and write to RDF graphs. The graphs are represented as an abstract model. A model can be sourced with data from files, databases, URLs, or combination of these. A model can be queried through Sparql 1.1. Figure 7 shows Jena and Sparql used for accessing information from designed and developed Computer Graphics Ontology. In Fig. 7, the Sparql query used to extract Ontology detail is shown. As can be seen, this query is like Select statement in SQL. Apache Jena package org.apache.jena.rdf.* is used. Model class is used to read the Ontology from file. Query class is used to create the Sparql query. ExecSelect function is used to execute the query and fetch results.

(3) Statistical Techniques used with and without Ontology:

(a) Latent Semantic Analysis (LSA): The modified LSA algorithm is implemented with and without Ontology. The use with Ontology is shown in Fig. 8. In the figure, all the words in Ontology that belong to the concept are used to create submodel answers. Then, LSA is used to find presence of each submodel in student answer by changing the value of q vector. This is done by adding only those terms' weight in q that is present in the submodel concept.

(b) BiLingual Evaluation Understudy (BLEU): The modified BLEU algorithm is tested both with and without Ontology using a similar scheme as shown for LSA.

(c) Generalized LSA (GLSA): Generalized Latent Semantic Analysis is implemented in Java Programming Language. The difference between GLSA and LSA is that of n-grams. None of the preprocessing steps are performed in GLSA. The word phrases of length n are formed using sliding window as shown in Fig. 9. The ngram size is varied between 2 and 4 in the first code snippet. This value of ngram is passed to callGLSA function. In callGLSA function, *i*th student's answer is tokenized. Then, sliding window is implemented to generate ngrams of specified size. The LSA is performed with phrases by document matrix. Theoretically, the GLSA method is improvement over LSA as it looks for phrases and maintains the word order. However, the presence of exact phrases in student answer is difficult to find. It does not consider the synonyms of words. The size of matrix generated becomes even larger than term by document matrix. The GLSA algorithm is implemented with and without Ontology.

(d) Maximum Entropy (MaxEnt): The Maximum Entropy package is downloaded from SourceForge Web site. MaxEnt is used both with and without Ontology. Without Ontology, the model answers are given to maxent, and based on these model answers, it classifies the new student answers. With Ontology, instead of classifying the complete answer, individual sentences of student answer are classified and matched to Ontology concepts. The scheme is similar as that in LSA. The code snippet showing use of Maximum Entropy with Ontology is shown in Fig. 10. In Fig. 10, the program snippet for creating category file is shown. First, the concepts are extracted from Ontology. Second, these concepts are written to a category file.

Domain Ontology is a concept map of subject knowledge and contains detailed information about the field. It is tested whether the domain Ontology can be used to replace the model answers completely with statistical techniques. The testing is performed using a database of ten questions of Computer Graphics with several answers. The same answers are evaluated by human examiner. Pearson correlation

```
String smallModelAnswers1[] = setMasterTermVector(smallModel1,"model",null); // create small model from Ontology concept for each concept
String []masterTermVector1=setMasterTermVector(Answer,"concept",smallModelAnswers1); // count the term vector
double answerTermFrequency1[][];
System.out.println("frequency calculation");
answerTermFrequency1 = setAnswerTermFrequency(masterTermVector1,line1);
System.out.println("Calling LSA");
//System.out.println(model);
double dd1=0.0;
int index1=0;
double [][]cor_LSA1=new double[keys1.size()][line1.length]; // keys is a multimap holding the ontology map (multimap class is available in guava
// corLSA1 variable is used for holding correlation of each answer with each concept.
int answerNo = 0;
// for all concepts run the following loop
for(String k1:keys1)
{
  //System.out.println(k+ " = "+multiMap.get(k));
  String smallModelQuery1[]= multiMap.get(k1).toString().split("[ _,]");
  if(index1==0)
  cor_LSA1[index1] =calculateSVD(answerTermFrequency1, masterTermVector1, proxy, processor,smallModelQuery1,1);
  else
  cor_LSA1[index1] =calculateSVD(answerTermFrequency1, masterTermVector1, proxy, processor,smallModelQuery1,0);
  //category1[index1]=k1;
  index1++;
}
```

**Fig. 8** Code to implement LSA with Ontology

```
System.out.println("Calling GLSA with ontology");
        GLSA g1=new GLSA();

        double[][] value_SYN_LSA1=new double[4][Answer.length()];
        for(int ngram=2;ngram<4;ngram++)
        {
          System.out.println("ngram = "+ngram);
          value_SYN_LSA1[ngram-2]=g1.callGLSA(model2, (float)marks, Answer, myAgent,ngram);
        }


String[]terms = answers[i].split(" ");
answers[i]= " ";
int end=2;
for (int start=0;start<terms.length ; start++)
{ end=start;
   String sub=terms[start];
   for(int j=1;j<ngram;j++)
   {
   while(end <start+j)
   {   end=end+1;
   if(end <terms.length)
   sub= sub+"_"+terms[end];
   }
   answers[i]=answers[i]+" "+sub;
   }
}
```

**Fig. 9** Code snippet for Ngram generation for GLSA

```
// fetch words from ontology
SubjectiveEvaluation.OntoEvaluation.sparqll s=new SubjectiveEvaluation.OntoEvaluation.sparqll();
String []ontowords=s.fetchResultFromOntology(prefix, fileWithPath, concept);
if(ontowords==null)
{  System.out.println("no words could be fetched from ontology");
   System.out.println(prefix+" "+ fileWithPath+" "+ concept);
   return;
}
// count words in ontology are not blank
String model2=""; System.out.println("this is after fetch"); int count=0;
for(String a:ontowords)
{  if((a!="")||(a!=null)) {model2 = model2+"| |"+a;  count++; }}
if(count <= 4) { System.out.println("not enough words in ontology ... so quitting from ontoMax");return;}
String [] modelSplit=model2.split("\\| \\|");
// write ontology to category file
String examiner=System.getenv("EXAMINER_HOME");if(examiner.contains(" ")) examiner="\""+examiner+"\"";
String CategoryFileContents=""; FileWriter fw;
      try {
        fw = new FileWriter(examiner+"\\maxOntoFile.txt");
        Multimap<String, String> multiMap = ArrayListMultimap.create();
        for(String  x: modelSplit) {
          String keyval[]=x.split(":");
          if((keyval!=null)&&(keyval.length == 2))
         {   String ss= keyval[0] + " "+ keyval[0].replace("_"," ")+" "+keyval[1]; ss=ss.concat("\r\n");
             CategoryFileContents = CategoryFileContents+ss; multiMap.put(keyval[0],keyval[1]);
         }
        }
 CategoryFileContents = CategoryFileContents.substring(0,CategoryFileContents.lastIndexOf("\r\n"));
 fw.write(CategoryFileContents);
 fw.close();
```

**Fig. 10** Code snippet to create category file with Ontology

is calculated between computer-generated and human-assigned scores. The proposed technique OHYBRID is tested, and results are analyzed to verify its performance. The OHYBRID technique is further analyzed for effect of change in level of detail in Ontology and change in number of answers evaluated together. Finally, the OHYBRID technique is compared on common data with other statistical techniques like MAXENT and GLSA, using Ontology in all the techniques.

# 6 Testing of Ontology-Based Technique

The statistical techniques, namely BLEU, LSA, and HYBRID are tested using a common database consisting of ten different questions with several answers for each question from the domain of Computer Graphics. The marks are generated for each student answer using the techniques with Ontology. The variation of individual marks generated by computer and human-assigned marks is shown in Table 3.

In Table 3, question numbers and number of answers evaluated simultaneously are included in first two columns. The first column of the table also includes headings ±1 marks, ±2 marks, and ±3 marks or higher which means difference of one mark, difference of 2 marks, and difference of three or higher marks, respectively. Comparing the performance of techniques with Ontology, it is observed that in Q1, Q4, Q6, Q7, Q8, and Q9 the Ontology-based HYBRID technique, OHYBRID, shows a better performance as compared to OLSA and OBLEU with lesser number of answers in ±3 marks or higher category. In Q1, Q3, Q4, Q5, Q6, and Q9 OHYBRID has minimum number of answers in ±3 marks or higher as compared to number of answers in ±1 marks and ±2 marks. The performance of HYBRID with Ontology is also acceptable. The maximum and minimum correlations between human-assigned and computer-generated scores for OLSA, OBLEU, and OHYBRID techniques are shown in Table 4. The performance of OHYBRID is better than OLSA and OBLEU.

## *6.1 Analysis of OHYBRID Technique*

OHYBRID technique is further analyzed for effect of change in Ontology and number of answers evaluated simultaneously using nine questions of Computer Graphics.

### 6.1.1 Effect of Change in Ontology

Two levels of Ontology are considered one with less details and other with more details. The performance of OHYBRID changes with change in Ontology as shown

**Table 3** Variability in marks with Ontology

| | Number of answers | OBLEU | OLSA | OHYBRID |
|---|---|---|---|---|
| Q1 | 48 | | | |
| ±1 marks | | 31 | 30 | 24 |
| ±2 marks | | 9 | 10 | 19 |
| ±3 marks and higher | | 8 | 8 | 5 |
| Q2 | 34 | | | |
| ±1 marks | | 10 | 14 | 12 |
| ±2 marks | | 10 | 14 | 8 |
| ±3 marks and higher | | 14 | 6 | 14 |
| Q3 | 18 | | | |
| ±1 marks | | 8 | 11 | 5 |
| ±2 marks | | 6 | 4 | 8 |
| ±3 marks and higher | | 4 | 4 | 5 |
| Q4 | 34 | | | |
| ±1 marks | | 8 | 6 | 18 |
| ±2 marks | | 6 | 11 | 9 |
| ±3 marks and higher | | 20 | 17 | 7 |
| Q5 | 54 | | | |
| ±1 marks | | 36 | 30 | 20 |
| ±2 marks | | 6 | 14 | 20 |
| ±3 marks and higher | | 12 | 10 | 14 |
| Q6 | 22 | | | |
| ±1 marks | | 12 | 4 | 12 |
| ±2 marks | | 8 | 6 | 6 |
| ±3 marks and higher | | 2 | 12 | 4 |
| Q7 | 28 | | | |
| ±1 marks | | 8 | 14 | 12 |
| ±2 marks | | 6 | 2 | 4 |
| ±3 marks and higher | | 14 | 12 | 12 |
| Q8 | 34 | | | |
| ±1 marks | | 18 | 12 | 18 |
| ±2 marks | | 6 | 12 | 12 |
| ±3 marks and higher | | 10 | 12 | 4 |
| Q9 | 54 | | | |
| ±1 marks | | 42 | 30 | 41 |
| ±2 marks | | 9 | 11 | 5 |
| ±3 marks and higher | | 3 | 13 | 8 |
| Q10 | 68 | | | |
| ±1 marks | | 49 | 39 | 41 |
| ±2 marks | | 15 | 15 | 13 |
| ±3 marks and higher | | 4 | 14 | 14 |

**Table 4** Correlations between human-assigned and computer-generated scores

| | OBLEU | OLSA | OHYBRID |
|---|---|---|---|
| Maximum | 0.96 | 0.90 | 0.93 |
| Minimum | 0.32 | 0.58 | 0.63 |

**Table 5** Effect of change in Ontology on OHYBRID technique

| OHYBRID technique with change in Ontology | | |
|---|---|---|
| Questions | Basic Ontology | Enhanced Ontology |
| Q1 | 0.93 | 0.94 |
| Q2 | 0.67 | 0.76 |
| Q3 | 0.71 | 0.74 |
| Q5 | 0.36 | 0.75 |
| Q4 | 0.85 | 0.86 |
| Q6 | 0.83 | 0.91 |
| Q7 | 0.71 | 0.74 |
| Q8 | 0.86 | 0.94 |
| Q9 | 0.77 | 0.83 |

in Table 5. The correlation is improving for all the questions. Therefore, change in level of detail in Ontology improves the quality of evaluation.

### 6.1.2 Effect of Change in Number of Answers

OHYBRID technique is used with 10, 50, and 100 answers for evaluating simultaneously to check the effect of change in number of answers on evaluation results. It is observed from Table 6 that correlations are consistent for question Q2, Q3, Q4, Q5, and Q7. Out of nine questions, five questions show consistent performance. Therefore, change in number of answers evaluated simultaneously does not affect the accuracy of results.

## *6.2 Comparison of OHYBRID Technique with Other Statistical Techniques*

The Ontology is further used with MAXENT and GLSA along with LSA, BLEU, and HYBRID, and the results are compared. These techniques are implemented and tested using ten questions of Computer Graphics. Table 7 shows the correlations of all the techniques. The OHYBRID technique shows the best correlation as compared to all other techniques. Maximum Entropy has correlation varying from 0.61 to 0.96. There is a lot of improvement in Maximum Entropy with Ontology. GLSA

**Table 6** OHYBRID with change in number of answers evaluated simultaneously

| | 10 answers | 50 answers | 100 answers |
|---|---|---|---|
| Q1 | 0.72 | 0.59 | 0.58 |
| Q2 | 0.66 | 0.60 | 0.60 |
| Q3 | 0.64 | 0.69 | 0.69 |
| Q4 | 0.76 | 0.70 | 0.70 |
| Q5 | 0.82 | 0.86 | 0.86 |
| Q6 | 0.77 | 0.64 | 0.64 |
| Q7 | 0.75 | 0.71 | 0.71 |
| Q8 | 0.97 | 0.87 | 0.87 |
| Q9 | 0.75 | 0.83 | 0.83 |

**Table 7** Correlation between human-assigned and computer-generated scores using statistical techniques with and without Ontology

| | OBLEU | OLSA | OHYBRID | OMAX | ONG1 | ONG2 |
|---|---|---|---|---|---|---|
| Q1 | 0.96 | 0.82 | 0.93 | 0.91 | 0.77 | 0.76 |
| Q2 | 0.66 | 0.58 | 0.63 | 0.72 | 0.61 | 0.57 |
| Q3 | 0.32 | 0.65 | 0.88 | 0.61 | 0.44 | 0.41 |
| Q4 | 0.80 | 0.86 | 0.83 | 0.66 | 0.87 | 0.84 |
| Q5 | 0.71 | 0.83 | 0.85 | 0.71 | 0.71 | 0.71 |
| Q6 | 0.79 | 0.76 | 0.87 | 0.96 | 0.65 | 0.57 |
| Q7 | 0.93 | 0.87 | 0.93 | 0.72 | 0.82 | 0.82 |
| Q8 | 0.96 | 0.90 | 0.93 | 0.91 | 0.64 | 0.63 |
| Q9 | 0.96 | 0.73 | 0.79 | 0.93 | 0.54 | 0.48 |
| Q10 | 0.87 | 0.64 | 0.65 | 0.70 | 0.62 | 0.59 |
| Maximum | 0.96 | 0.90 | 0.93 | 0.96 | 0.87 | 0.84 |
| Minimum | 0.32 | 0.58 | 0.63 | 0.61 | 0.44 | 0.41 |

ngram 1 and 2 do not show much improvement with use of Ontology as correlation varies from 0.44 to 0.87 for NGRAM1 and 0.41 to 0.84 from NGRAM2.

## 7 Conclusion

The hybrid technique for subjective evaluation is capable of handling irrelevant answers. Sometimes students, who do not know the answers, write invalid content like repeating some keywords available in question paper itself in the answer. Such invalid answers are not given marks. The hybrid technique combines the best features of LSA and BLEU as observed from results and performs better than LSA and BLEU. It has advantages like semantic study and irrelevant answer identification both inherited from LSA and BLEU, respectively.

For hybrid technique, better results are achieved with use of best model answer. However, statistical techniques like LSA and MAXENT give good results with use of Ontology instead of model answer.

## References


1. Noy, N. F., & McGuinness, D. L. (2001). What is an ontology and why we need it.
2. Ozcan, R., & Aslangdogan, Y. A. (2004). Concept based information access using ontologies and latent semantic analysis. *Department of Computer Science and Engineering, 8,* 2004.
3. Dridi, O. (2008). Ontology-based information retrieval: Overview and new proposition. In *Second International Conference on Research Challenges in Information Science, 2008. RCIS 2008,* pp. 421–426.
4. Oleiwi, S. S., & Yasin, A. (2013). Classify the scientific paper to multi categories using ontology.
5. Wu, S. -H., Tsai, T. -H., & Hsu, W. -L. (2003). Text categorization using automatically acquired domain ontology. In *Proceedings of the sixth international workshop on Information retrieval with Asian languages-Volume 11* (pp. 138–145).
6. Bulakh, P. (2013). Automation of subjective answer evaluation, *4*(1), 1328–1332.
7. Youn, S., Arora, A., Chandrasekhar, P., Jayanty, P., Mestry, A., & Sethi, S. (2009). Survey about ontology development tools for ontology-based knowledge management (pp. 1–26). *University of Southern California.*
8. Mylopoulos, J., Buitelaar, P., Olejnik, D., Sintek, M., Della Valle, E., Castagna, P., & Brioschi, M. (2003). OntoLT: A protégé plug-in for ontology extraction from text (pp. 31–44).
9. D'Avanzo, E., Lieto, A., & Kuflik, T. (2008). Manually vs semiautomatic domain specific ontology building (Doctoral dissertation, Thesis in Information and Commercial Electronics).
10. Sarda, N. L. (2007). Ontology-enabled database management systems. In *Ontologies* (pp. 563–584). Springer, Boston, MA.
11. Zouaq, A., & Nkambou, R. (2008). Building domain ontologies from text for educational purposes. *IEEE Transactions on Learning Technologies, 1*(1), 49–62.
12. Foltz, P. W., Kintsch, W., & Landauer, T. K. (1998). The measurement of textual coherence with latent semantic analysis. *Discourse Processes, 25*(2–3), 285–307.
13. Pérez, D., Pérez, D., Gliozzo, A., Gliozzo, A., Strapparava, C., Strapparava, C., Alfonseca, E., Alfonseca, E., Rodriguez, P., Rodriguez, P., Magnini, B., & Magnini, B. (2005). Automatic assessment of students' free-text answers underpinned by the combination of a BLEU-inspired algorithm and latent semantic analysis. In *Proceedings of the Eighteenth International Florida Artificial Intelligence Research Society Conference, FLAIRS 2005—Recent Advances in Artificial Intelligence* (pp. 358–363).
14. Islam, M. (2010). Automated essay scoring using generalized. In *Proceedings of 13th International Conference on Computer and Information Technology (ICCIT 2010).*
15. Sukkarieh, J. Z., & Blackmore, J. (2009). C-rater: Automatic content scoring for short constructed responses. *FLAIRS Conference* (pp. 290–295).
16. Devi, M. S., & Mittal, H. (2016). Machine learning techniques with ontology for subjective answer evaluation. *International Journal on Natural Language Computing, 5*(2), 01–11.

# A Survey on Feature Selection and Extraction Techniques for High-Dimensional Microarray Datasets

**G. Manikandan and S. Abirami**

**Abstract** In recent years, lots of data are generated and stored in the field of information technology, bioinformatics, text mining, face recognition, microarray data analysis, image processing, etc. From this microarray gene expression, data analysis gained the more importance due to role of disease diagnosis and prognoses to choose the appropriate treatment to the patients. Generally, gene expression data are a sort of high-dimensional data with small number of observation and large number of attributes. Interpreting the results from the gene expression data are difficult one due to the "curse of dimensionality." For this issue, dimensionality reduction plays an important role, since it reduces the number of variables by using the techniques such as feature selection and feature extraction. The main aim of these approaches is to reduce/downscale the high-dimensional feature space to low-dimensional representation with an affection of classification accuracy. For this concern, the objective of this chapter is to gather and provide the up-to-date knowledge in the field of feature selection methods applied in the microarray data analysis as possible for the readers. In this chapter, a brief introduction about feature selection methods in the DNA microarray analysis was presented in Sect. 1. The taxonomy of the dimensionality reduction methods is represented in the diagrammatic way. Five feature selection methods such as filter, wrapper, embedded, hybrid, and ensemble methods have been discussed in detailed manner and tabulated the recent proposed algorithms, datasets used and accuracy achieved in the respective methods, and also the advantages and disadvantages of each method are discussed in Sect. 2. In Sect. 3, supervised, unsupervised, semi-supervised gene selection methods, with advantages and disadvantages were discussed. Section 4 discusses the feature extraction techniques applied in the microarray data analysis. Finally, Sect. 5 provides intrinsic characteristics of microarray data with respect feature selection.

---

G. Manikandan (✉) · S. Abirami
Department of IST, Anna University, Chennai 600025, India
e-mail: manitamilm@gmail.com

S. Abirami
e-mail: abirami@auist.net



S. Margret Anouncia and U. K. Wiil (eds.), *Knowledge Computing and its Applications*, https://doi.org/10.1007/978-981-10-8258-0_14

# 1 Introduction

Over the last few decades, due to the advent of microarray datasets creates the new way of research in both bioinformatics and machine learning. Among many active researches in the DNA microarray technology, gene expression classification is the emerging topic in recent years and attracted to pay attention of many researchers from various fields such as data mining, machine learning, statistics. The main aim of gene expression data classification is to build the efficient classification model to classify the historical gene data and to use the classifier to classify the future incoming data or to predict the trend of the data. Microarray data is usually in very high dimension (large number genes) with small number of samples. One of the most applications of the microarray data is to classify the microarray data tissue samples into normal or cancerous. In the microarray data, some number genes are irrelevant or insignificant to clinical applications and also they are not going use in classification task for further analysis. So, interpreting the significant genes from the large set of genes is the most important challenges in microarray data analysis domains. Usually, gene expression data contain very small samples for training and testing set but the numbers of genes ranges from 6000 to 60,000. The task is to classify the normal healthy patients to the abnormal patients based on their gene expression data; it is referred as binary classification. There is also microarray datasets in which the task is to classify the samples into different types of tumors referred as multiclass classification. For these reasons, microarray data classification leads lots of attention and poses serious challenges to the machine learning researchers. To deal with the 'curse of dimensionality' problem, feature (gene) selection plays the important role in the field of bioinformatics to analyze the DNA microarray data. The main aim of machine learning algorithm is to focus and identify the relevant genes from the large set of genes by removing irrelevant genes.

Generally, these datasets are high dimensional which contains irrelevant and redundant data too. Structuring, predicting, and classifying these high dimensionally data are highly expensive. To overcome these problems, data preprocessing mechanisms such as aggregation, discretization, sampling, data cleansing, normalization, and dimensionality reduction can be used. Among these techniques, dimensionality reduction plays an important role, since it reduces the number of variables by using the techniques such as feature selection/feature ranking and feature extraction. The main aim of these approaches is to reduce/downscale the dimensional feature space to low-dimensional representation with an affection of classification. In feature extraction techniques, the new features are generated from the original set of feature by merging and transforming the original features into reduced features. As a result, sometimes the newly constructed feature may contain irrelevant and redundant features also and may affect the performance of the learning process too. The main target of the feature selection is to select the subset of features by eliminating redundant and irrelevant features, because these features will not give unique information for analyzing and modeling the data.

In general, the feature selection method is broadly classified into five categories such as filter, wrapper, embedded, hybrid, and ensemble method [1]. Each method performs significantly to select the subset of features with various learning methods such as supervised, unsupervised, and semi-supervised approaches. Filter or open-loop methods select the features based on their intrinsic characters using four evaluation metrics such as dependency, consistency, distance, and information. This method is independent of learning algorithm; hence, it does not affect the classifier performance, and also it does not correlate with the bias and gives considerable generalization property. The second one, wrapper method which is also called as closed-loop method, selects the predominant features with the help of learning algorithm, thereby it minimizing the prediction error. Embedded method is an inbuilt feature selection method which is embedded with learning algorithm and guides to feature evaluation. This method ignores repetitive implementation of learning algorithm and examination of every individual subset which leads to the less risk of overfitting. A hybrid method performs the feature selection by combing two different feature selection methods either filter or wrapper based on same criterion. Ensemble method groups the features into subset, and then it aggregates the results of the individual groups. It gives the better approximation results by selecting the optimal features based on the ranking mechanisms. This method uses different subsampling techniques to run on ample number of subsamples to achieve the highly stable subset which has been explained in Sect. 2.1.5. In addition to the feature extraction and feature selection techniques, feature ranking methods are available to evaluate the individual relevance of the features from the available set of features. The main objective of the feature ranking is to group the highly correlated features based on some scoring function. In order to find the relevant features, it eliminates the least significant features which affect the performance of the classifier considerably. The main advantage of the feature ranking mechanism is that it involves a low computational cost. Next section deals with the brief introduction and explanation about the dimensionality reduction techniques.

## 2 Dimensionality Reduction

Dimensionality reduction is one of the powerful techniques to remove the irrelevant and redundant features. The taxonomy of the feature selection technique is shown in Fig. 1. It can be categorized into feature selection and feature selection. Feature extraction techniques produce the new features from the original features with low dimensionality, and the newly created features are the linear combinations of the original features. Some of the examples of the feature extraction techniques are linear discriminant analysis (LDA), principal component analysis (PCA), and canonical correlation analysis (CCA). In the contrast to the feature extraction, feature selection selects the subset of the features from the original set with minimal redundancy and maximum relevance to the target class to gain the classification accuracy. Some of the examples of the feature selection techniques are information

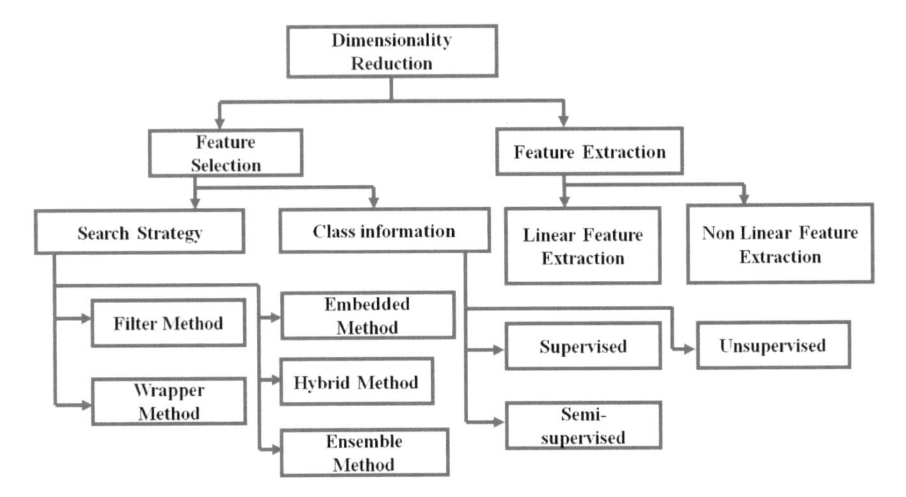


**Fig. 1** Taxonomy of dimensionality reduction techniques

gain, symmetrical uncertainty, relief, Lasso, etc. The main motivations of both the techniques are to improve the classification performance with the low computational cost with high classification accuracy and also less consumption of the storage space. In feature extraction, the original features space is mapped into new feature space with minimal dimension by combining the original features. In this perspective of further analysis, the newly constructed features are problematic because there is no physical meaning in the transformed features. But in feature selection technique, it selects only subset of features without any transformation of original features. In this, sense feature selection techniques are superior to the feature extraction techniques by in terms of better interpretability and readability. In microarray data classification, the main difficulty of the machine learning technique is to get trained with the more number of genes. The upcoming section explains the feature selection techniques applied in the microarray data analysis.

## 2.1 Feature Selection in DNA Microarray Data

In the field of machine learning, feature selection is also called as attribute selection, variable selection, and feature subset selection. Based on the availability of the class information, the feature selection technique can be broadly classified into three machine learning families such as supervised method, unsupervised method, and semi-supervised method. In supervised feature selection technique, the discriminative features are selected based on the importance and relevancy of the features with respect to the class. Some of the literatures of supervised feature selection techniques are discussed in [2]. In unsupervised feature selection, there is no prior among the class; it evaluates the feature relevance by developing the internal structure of the data with help separability, data variance, and data distribution. Semi-supervised feature selection can take the account of both labeled and unlabeled data for finding the discriminative features. Most of the semi-supervised algorithms rely on constructing the similarity matrix, and the features are selected from the similarity matrix. According to [3], the features fall into four categories: (a) completely irrelevant with noisy features, (b) weekly relevant and redundant features, (c) weakly relevant and non-redundant, and (d) strongly relevant features. From these categories, the optimal subset falls in the categories (c) and (d). Strongly relevant features are those features which have the main role in the target classification and represents high information about the problem. Redundant features are modeled as the features which will not provide any information about the discriminative features, but it is correlated with relevant features. An irrelevant feature does not contribute to the predictive accuracy, so it is to be eliminated to build good prediction model. Hence, all the strongly relevant features and few weakly relevant features should be considered as good subset, whereas irrelevant, redundant, and noisy features should be eliminated. As a result, to achieve the objective of the feature selection, several criteria have been framed to evaluate the feature relevance. Based on the various search strategies and mechanisms, the feature selection is broadly classified into five categories, namely filter, wrapper, embedded, hybrid, and embedded.

### 2.1.1 Filter Method

Filter method is also called as the open-loop method. It selects the features based on the internal characteristics of the features with respect to the classification task. In general, filter methods measure the internal characters of the features by using four types of evolution criteria, namely dependency, consistency, distance, and information. In most of the cases, it uses the variable ranking methods as the standard criteria for ordering and selecting the features. Statistical ranking methods are used here for simplicity and goodness of the feature selection in practical application domains. Various statistical and multivariate methods are used to score the features, and based on the threshold value, discriminative features are selected from the large set.

Filter method evaluates the features without using any classification algorithm. It undergoes two stages; in the first stage it ranks the feature according to the certain evaluation criteria, it would be either univariate method or multivariate method. In univariate method, each and individual features are ranked independently in feature space, whereas multivariate method evaluates the feature in the batch perspective. So the multivariate methods are capable of handling the duplicate and redundant features. In second stage, the features with the highest ranking are selected to induce the machine learning classification models. Some of the well-known classical methods applied to the microarray data are chi-square, information gain, consistency-based filter, correlation-based feature selection (CFS), fast correlation-based feature selection (FCBF), consistency-based feature selection, INTERACT, relief, and minimum redundancy maximum relevance (mRMR).

Filter method is further classified into two approaches as: ranking and subset search approach. The former approach selects the best features based on the rank assigned to each features, such rank is calculated based on the independent measures such as consistency, information, and distance between the features [4]. The gene with higher value is selected while the gene with smaller value is ignored. The drawback of rank approach is that they compute only the relevance between the feature and not the dependency between them. The latter approach includes dependency as they evaluate the data as a whole. Filter method is computationally more efficient, fast, and scalable, and provides more generalized result. The limitation of filter includes [5]: fails to detect the indirect relationship between the features leading to a varying predictability, ignore the classifier algorithm, and difficult to set the threshold for feature ranking. For example, Hoque et al. [6] implemented a fuzzy mutual information-based feature selection with non-dominated solution (FMIFS-ND) based on the fuzzy mutual information computing feature-class fuzzy mutual information and select the feature that has the highest mutual information.

Raza et al. [7] incorporated the proposed incremental dependency calculation (IDC) with the GA, QuickReduct, ReverseReduct, supervised PSO-based quick reduct, and fish swarm as IDC calculates attributes dependency by avoiding the positive region and based on the movement from one attribute to another. IDC could be a replacement for dependency measure and enhanced the performance accuracy with reduced execution time and runtime memory. The method proposed by Guo et al. [8] called regularized logistic regression (RLR) with SVM as the selection mechanism provided global optimal solution showed a linear complexity for the number of gene and samples and outperformed the other considered feature selection algorithm.

### 2.1.2 Wrapper Method

Filter method selects the features independently without involving of any classifiers. So the major disadvantage of the filter method is that it totally ignores the importance of the features with respect to the induction algorithm, because some features would purely depend on the biases and heuristics of the induction algorithm. Based on this issue, the wrapper methods utilize the classifiers to evaluate the quality of the selected subset. It offers a simple and powerful way to overcome the feature selection problem. It undergoes the following stages for selecting the features:

1. Searching the subset of features
2. Evaluating the obtained subsets of features by the performance validation of the classifier
3. Repeat the step1 and step2 until finding the good quality of subset with respect to the classifier accuracy.

In wrapper method, the classifier algorithms work as the black box manner, the feature search component will generate the set of features, and then the feature evaluation component will estimate the performance of the selected subset with help of classifier. The feature set with highest classifier accuracy is selected as the best subset. Here the search space of the '$n$' features is the $O(2^n)$, so the exhaustive search is impractical unless the size of '$n$' is small, i.e., the NP-hard problem. For solving these issues, various search strategies such as genetic algorithms hill climbing, best first, and branch-and-bound has been used.

Wrapper methods assess the subsets of variables according to the usefulness to the given predictor. Wrapper method considers the interaction between feature subsets searches, and thus it has the ability to take into account feature dependencies. The wrapper approach evaluates the values of features by applying a predetermined learning algorithm to the data for finding the quality of selected subsets. In wrapper method, the induction algorithm is used. The induction algorithm is used for selecting optimal feature set. Training examples are provided to the induction algorithm where each example may be feature attributes and values. This method is followed by evaluating subsets and searching the subset space. An $n$-fold cross validation is adopted to find the accuracy of obtained structure by selected features. First, the initial data is divided into $n$ equally sized partitions. By keeping $n - 1$ partitions as initial data sets and other partitions as a test data, the induction algorithms run $n$ times. The results obtained in each run are combined together to obtain the accuracy. Feature space is searched for finding better features. For finding the better features, algorithm like forward selection and backward elimination are used. The working of forward and backward algorithms is similar with only difference in the selection of features. The backward algorithm begins with all features while the forward algorithm begins with no features. The wrappers use cross-validation measures of predictive accuracy to avoid overfitting and hence acquire the ability of generality [5, 9–11].

The wrapper method uses the rank search algorithm which evaluates exactly $n$ subsets, containing the first-ranked variable. Therefore, it is linear in the number of wrapper evaluations. The wrapper method also uses best agglomerative-ranked subsets (BARS) alternates between the construction of the ranking available subset and growing heuristic process that obtain all the combinations of the first three subset in the ranking with each one of the remaining ones. The wrapper method also uses the linear forward selection method in that linear forward selection algorithm speed up the wrapper search by limiting the number of available attributes to those ranked in the top $k$ positions. The wrapper method uses the algorithm named re-ranking-based feature selection. It tries to overcome the static ranking by iteratively using the blocks of variables taken from the beginning of ranking. The re-ranking algorithm involves four steps. They are selection algorithm, stop criterion, block size, and re-ranking algorithm [12].

Successive feature selection (SFS) implemented by Sharma et al. [12] allows a size of $h \leq 10$ features to be processed at a time. Based on the rank assigned, at each stage, one feature is dropped until the stopping criteria are reached and yield a top-r feature set. Kim et al. [13] eliminated the redundant feature if it had a Markov blanket. *F*- test or *t* test can also be applied to remove the irrelevant features. Kang et al. [14] selected relevant feature by performing random forward search (RFS) by incorporating the randomness to achieve global optimum. It also uses other sequential selection techniques such as sequential forward selection (SFS), sequential backward elimination (SBE) until a final model is built. The individual use of SFS and SBS suffers from nesting effect [5]. In SBS (top-down), once a feature is ignored, it cannot be reselected, and in SFS (bottom-up), once a feature is selected, it cannot be ignored. To alleviate these effects, sequential floating forward selection (SFFS) and its modified extension algorithm [13, 15] are used to improve the goodness of the selected feature. The result of subset generation should lead to selection without transformation.

There is also called an evolutionary wrapper approach. The evolutionary wrapper approach uses a nonparametric density estimation method and a Bayesian classifier. The nonparametric methods are a good alternative for scarce and sparse data, since they do not make any assumptions about its structure and all the information comes from the data itself. The another wrapper-based algorithm is multiple-filter and multiple wrapper method. The rationale behind this proposal is that filters are fast but their predictions are not accurate while wrapper maximizes the classification accuracy at the expense of a formidable computation burden. The wrapper algorithm has the advantage achieving greater accuracy than filters.

### 2.1.3 Embedded Method

In the embedded approach, feature selection intrinsic to the learning algorithm which simultaneously learns the classifier and chooses the subset of features. These methods typically work by including in the objective function of the learning algorithm a sparsity-inducing regularizer or prior, which encourages the weights assigned to some feature assigned to zero. The algorithms used in embedded approach are recursive feature elimination for support vector machines (SVM-RFE), sparse multinomial logistic regression (SMLR), sparse logistic regression (SLogReg), joint classifier, and feature optimization (JCFO), Bayesian logistic regression (BLogReg) [efficient].

During the classification process, the algorithm decides which attributes to use and which attribute to ignore. Just like wrapper approach depends on specific learning algorithm. Decision trees are the example of embedded approaches. This algorithm evaluates feature subset with the classification algorithm to measure efficiency according to incorrect classification rate. The algorithm used is known as sequential forward selection (SFS) [5]. Probably the most famous embedded method is support vector machine based on recursive feature elimination (SVM-RFE). This embedded method use the core of the classifier to establish criteria to rank the features. This embedded method performs feature selection by iteratively training the SVM classifier with the current set of features and removing the least important feature indicated by the SVM. After that, new embedded method is introduced. It simultaneously selects the relevant features during classifier construction by penalizing each feature's use in the dual formulation of support vector machine. This approach is called kernel-penalized SVM, and it optimizes the shape of an anisotropic RBF kernel eliminating features that have low relevance for the classifier. Another technique used in embedded approach is the embedded gene selector, and it is introduced and applied to four microarray datasets. This algorithm uses the backward elimination approach and the criterion to determine which features are the least important, which relies on the classification performance impact that each feature has when perturbed by noise. The disadvantage of this backward elimination algorithm is that it results in the imbalance of some microarray data sets. To overcome the problem, a new embedded method based on the random forest algorithm is presented. First, the algorithm finds the best training error for each class in run; in order to deal with the imbalance of microarray dataset. random forest is run to select relevant features [2].

The embedded method can be viewed as an intermediate position between wrappers and filters. It works with the classifier, but it relies on the error. Based on the literature review, various proposed techniques, dataset used, classifier used, and accuracy achieved in various feature selection methods are tabulated in Table 1.

**Table 1** Review on feature selection methods

| Method | Proposed technique | Classifier used | Accuracy |
|---|---|---|---|
| Filter | Centroid-based gene selection [8] | SVM (linear kernel function) | 76–79% |
| | IDGA [7] | SVM, Naïve Bayes, k-NN | 100% |
| | Kernel-based clustering method for gene selection (KBCGS) [9] | SVM, k-NN | 85–95% |
| | Fuzzy mutual information-based [6] | k-NN-ND, SVM, k-NN | |
| | CAM, DSCFS, CFS, IG, relief [10] | CART, C4.5, NB, SVM, k-NN, SL | 75–85% |
| Wrapper | Gravitational search, modified gravitational search, and sequential quadratic programming [42] | k-NN | 93.49 ± 1.46% |
| | CSMSVM [43] | SVM | 91–97% |
| | Cat swarm, modified cat swarm optimization [5] | RR, OSRR, KRR, SVM, RFSVM | 95% |

(continued)

**Table 1** (continued)

| Method | Proposed technique | Classifier used | Accuracy |
|---|---|---|---|
| Hybrid | Wrapper-BDE, filter-ranking based [16] | Naive Bayes (NB), SVM, k-NN, C4.5-based decision trees | 99% |
| | Filter—correlation-based feature weight, wrapper—Taguchi genetic algorithm (TGA) [17] | k-NN | 99% on average |
| | GADP, $X^2$ test [18] | SVM | 100% except GCM-87.04% |
| | iSFSM [19] | SVM, NN | 98.6% |
| | ABC and DE [20] | SVM | |
| | Cellular learning automata—ant colony optimization feature selection (CLACOFS) [44] | K-nearest neighbor (k-NN), support vector machine (SVM), and Naïve Bayes (NB) | 82.035% |
| | Laplacian score-WNCH-simple ranking (LS-WNCH-SR) and Laplacian score-WNCH-backward elimination (LS-WNCH-BE) [45] | K-means clustering | – |
| | MIMAGA-selection algorithm [46] | Backpropagation neural network (BP), support vector machine (SVM), ELM, and regularized extreme learning machine (RELM) | 78.73% |
| | Genetic algorithm with dynamic parameter setting (GADP) and $X^2$-test [33] | Support vector machine (SVM) | 97.84% |
| | BDE $X_{\text{rank}}$ and BDE $X_{\text{rank}_{\text{fi}}}$ [16] | K-nearest neighbor (k-NN), support vector machine (SVM), Naïve Bayes (NB) and C4.5 | 91.16% |
| Embedded | Embedded backward feature selection method [47] | MCLP | 79.25% |
| | Support vector machine-bayesian $T$ test-recursive feature elimination (SVM-BT-RFE) [48] | Support vector machine (SVM) | 97.805% |
| | EGSEE and EGSIEE [49] | Easy ensemble classifier | 89.28% |
| | Multiple-filters (MF), tabu search, genetic algorithm [50] | Support vector machine (SVM), K-nearest neighbor (k-NN), linear discriminant analysis (LDA) | 94.83% |
| | Block diagonal linear discriminant analysis with embedded feature selection [51] | Linear discriminant analysis (LDA) | – |

(continued)

**Table 1** (continued)

| Method | Proposed technique | Classifier used | Accuracy |
|---|---|---|---|
| Ensemble | Hierarchical ensemble of error-correcting output codes (HE-ECOC), maximizing local diversity (MLD) and maximizing global diversity (MGD) [52] | K-nearest neighbor (k-NN) and support vector machine (SVM) | 92.43% |
| | Ensemble of filters and classifiers [53] | C4.5, Naive Bayes (NB) and IB1 | – |
| | PSO-diCA algorithm [54] | Support vector machine (SVM) | 92% |
| | Homogeneous distributed ensemble and heterogeneous centralized ensemble [55] | Support vector machine (SVM) | 80 ± 15.63% |
| | Bi-objective genetic algorithm based feature selection (FSGA) [56] | Naïve Bayes (NB), SVM, k-NN, AdaBoost (BOOST), MLP, SMO, and DT | 92.36% |
| | Ensemble gene selection by grouping (EGSG) [57] | three-Nearest Neighbor (3-NN) | – |
| | MRMR and MRMR-HFS [58] | C4.5, Naïve Bayes and support vector machine (SVM) | 86% |

#### 2.1.4 Hybrid Method

Hybrid combines filter and wrapper approach and takes the advantage of these two approaches to achieve the best performance. It alleviates the computational complexity of wrapper and less prone to overfitting. To select the most relevant feature, an efficient hybrid algorithm called binary differential evolution (BDE) algorithm was proposed by Apolloni et al. [16]. Filter method fails to detect the simultaneous correlation between the features. To address this, wrapper approach—is used to optimize the subset. BDE computes the fitness function based on the frequency of the features present and remarked that subset feature obtained had an accuracy close to 100%. Chuang et al. [17] combined correlation-based feature selection with the Taguchi genetic algorithm. CFS implements the ranking approach, while TGA utilizes SNR to improve the efficiency at each level.

During analysis of feature selection, simple genetic algorithm fails to consider the number of genes needed. To overcome this problem, in [18], Chien et al. ranked the genes according to the frequency of occurrences. To select the minimal subset, $X^2$ test is performed to determine the threshold of occurrence frequencies. *F*-score computes the discrimination ability for each feature and not for multiple features, whereas the information gain selects only the feature with the more information. Hsu et al. [19] effectively combined these two feature subsets. For fine-tuning, reverse concept of sequential floating search method (SFSM) is used with SBS performed before SFS. To maintain a balance between the exploration and exploitation of searching, Zorarpacı et al. [20] proposed hybrid method of ant bee colony (ABC) and differential evolution (DE) for feature selection, reducing the higher CPU time required by applying parallelization and using only DE for exploration. The experimental analysis paved the way to solve other search and optimization problem such as 0–1 knapsack problem, job scheduling, etc.

### 2.1.5 Ensemble Method

The goal of the ensemble filter method is to introduce the diversity and increase the regularity of the feature selection process, so as to take the advantage of the strength of the individual selectors and overcome their weakness. There are two different approaches to evaluate the features of the dataset: individual evaluation and subset evaluation. In individual evaluation, the rank of the feature is calculated by assigning a level of relevance to each feature. But individual evaluation is incapable of removing redundant features because of these are likely to have similar ranking. In subset evaluation, successive subsets of features generated according to a predefined search strategy are evaluated iteratively according to an optimality criterion until a final subsets of selected features is obtained. The subset evaluation removes the redundant features, and they are not computationally efficient. To avoid the problem of these two approaches, the ensemble method has been introduced. The ensemble method can be achieved by combining different machine learning methods so as to solve the problem.

The most widely used ensemble learning methods applied to classification are bagging and boosting. Bagging creates an ensemble by training individual classifiers generated by random selection with the replacement of $n$ instances. In boosting approach, sampling is proportional to the weight of an instance. The ensemble method uses four search strategies; they are genetic search, hill climbing, ensemble forward, and ensemble backward search. The ways to design the ensemble are homogeneous distributed ensemble and heterogeneous centralized ensemble. In homogeneous distributed ensemble, $N$ models are generated using the same feature selection method but different training data. An important issue with this method is the computation time required is high. In heterogeneous centralized method, $N$ models are generated using different feature selection methods, but same training data; here, the computational time is comparatively low than homogeneous methods. The advantages and disadvantages of each technique are given in Table 2.

**Table 2** Examples, advantages, and disadvantages of filter, wrapper, embedded, and hybrid methods

| Method | Advantages | Disadvantages |
|---|---|---|
| Filter | Fast and scalable | It ignores the classifier interaction |
| | Independent to classifier | It ignores the dependency among the features/ variables |
| | Faster than wrapper method | Redundancy may be included |
| | Better computation complexity than wrapper | |
| Wrapper | It interacts with the classifier | Risk of overfitting |
| | It interacts among the features | Expensive operation |
| | Prone to the local optima | High computational cost |
| | Higher performance than filter method | |
| Embedded | Higher accuracy, performance than filter method | Classifier dependent |
| | Less prone to the overfitting<br>It considers the dependency between variables | It considers the feature dependency between the features<br>High computational cost |
| Hybrid | Higher performance than filter<br>Less prone to the overfitting<br>Better computational cost | Classifier dependent<br>Dependents among the combination of various feature selection algorithms |
| Ensemble | Less prone to overfitting<br>More scalable for high dimensional datasets | Understanding the combination of classifiers is difficult |

# 3 Machine Learning Techniques for Gene Selection

Data mining which is also referred as knowledge discovery, involves preprocessing of large datasets, selection and transformation of important features followed by evaluation and presentation with the knowledge visualization. The goal of data mining is converting the raw data into an useful information. Discovering knowledge or making prediction from large datasets or for a dataset with no explicit algorithm is often difficult. To serve this purpose, machine learning algorithm is used. Machine learning algorithms are often categorized as

1. Supervised learning method
2. Unsupervised learning method and
3. Semi-supervised learning.

These algorithms lend themselves for predictions, thereby revealing the hidden insights. Learning algorithms stop when an acceptable level of performance is achieved. Recent review on supervised, unsupervised, and semi-supervised gene selection algorithms can be clearly mentioned in [2].

### 3.1 Gene Selection Methods Using Supervised Learning

Supervised learning algorithm involves analyzing the input (training dataset) and generates function for mapping any valid new data. It also allows prediction of class label for new instances. This learning technique is commonly used for training decision trees and neural networks because both require information from predetermined classifications, and also to minimize the error rate. The main drawback of supervised learning technique is overfitting, as the learning obtained from general classification is not robust enough to train complex data; hence this technique produces only generalizable results. The steps involved in supervised learning are as follows:

1. Determine the data type of the dataset to be trained and gather input object of the specified data type
2. Determine the representation of input object and the corresponding learning algorithm.

Supervised learning technique is further categorized based on the target variables as classification and regression. Classification technique is used, if the target variable is categorical and could be classified based on the categories (two or multiple categories), and the latter is used, if the target variable is numeric and if the prediction is based on the linear or nonlinear relationships between the predictors and the target. Local-learning-based feature selection, Canonical-depended degree-based feature selection approach [21], fuzzy preference-based roughset (FPRS) [22], locality sensitive Laplacian score (LSLS) [23], maximum relevance minimum multicollinearity (MRmMC) [24], improved particle swarm optimization algorithm, mutual information-based feature selection algorithm (MIFS) are some of the filter approach-based supervised features selection technique. Sample selection method (FS-SSM) [25], fuzzy interactive self-organizing data algorithm (ISODATA) [26], sample weighting SVM-RFE [10], null LDA technique, top-r feature selection called successive FS (SFS) and block reduction, incremental wrapper-based subset selection (IWSS) [27], incremental wrapper-based subset selection with Marko blanket are some of the wrapper-based supervised feature selection technique.

### 3.2 Gene Selection Methods Using Unsupervised Learning

Unsupervised learning technique uses only the input data and has no target variable. This builds the underlying structure for the input data and learns more from that structure. It is further categorized as clustering and association. Clustering technique allows discovering the inherent groupings in the data, while association allows discovering the rules among the data. Compared to the supervised and semi-supervised feature selection technique, unsupervised technique lacks the class information. The main aim of the unsupervised learning is to how to generate the pseudolabels from the DNA microarray data.

Many unsupervised learning [28] methods have been proposed in the literature to generate the pseudolabels; some of the important methods to consider are matrix factorization, spectral clustering, spectral embedding, consensus clustering, etc. Recently authors from [28] proposed new unsupervised feature selection method called feature-level self-representation framework. It first uses a feature-level, self-representation loss function to represent the features sparsely by representing each features by other features; then it uses $l_2$, $p$—norm regularization to gain row-sparsely on the coefficient matrix for doing feature selection. Their proposed method yields better results better accuracy when compared to the state-of-the-art algorithms. Unsupervised probabilistic ACO-based method for feature selection (UPFS) proposed in [29], which reduces the redundant and irrelevant features by taking the less amount of time. This algorithm is generally based on the graph approach; the ants traverse the fully connected graph in an iterative strategy. Then each ant selects the features iteratively using the previous states of the selected features and the heuristic information about the features until traverse of fully connected graph. Their proposed UPFS algorithm does not need learning step and presence of class information to select the features.

Consensus learning [30] is the class of unsupervised learning is one of the frequently adopted methods to address the issues of patient stratification discrepancies from multiple set of datasets. In consensus clustering, the patients are clustered based on the individual modality, then these clusters are aggregated into the set of consensus clusters based on some optimization criteria. In general, there are two main approaches which exist to achieve the consensus solution and to assess the quality, namely probabilistic approach and similarity approach. In probabilistic approach, from the given distribution of labeling, the maximum likelihood formulation is used to return the consensus. In similarity approach, it finds directly once the consensus clustering agrees the input base clustering. Molecular-regularized consensus patient stratification (MRCPS) [30] is novel consensus clustering algorithm based on the optimization strategy with regularization. Usually, the traditional consensus clustering methods such as Bayesian consensus ensemble (BCE), cluster-based similarity partitioning algorithm (CSPA), hypergraph partitioning algorithm, and HGPA can take either soft or hard base method to cluster the data. But MRCPS method can automatically clusters both numerical and categorical data with any criteria of the similarity metrics.

## 3.3 Gene Selection Methods Using Semi-supervised Learning

Semi-supervised learning is used when there is a dataset with large number of input data, where only few input data are categorized based on the target variable.

Labeled data is used to maximize the margin between input data of different categories, while the unlabeled data is used to discover the structure of the dataset. A survey about semi-supervised feature selection can be found in the literature [31]. The microarray data or the gene expression data can either be fully labeled or unlabeled, or partially labeled and the dataset may contain many redundant, irrelevant, and noisy features along with the informative features. To discover the patterns and class prediction from such data marked, the use of supervised, unsupervised, and semi-supervised feature selection techniques. Semi–supervised learning learns from the large number of unlabeled data with small number of labeled data [31]. Cluster assumption and manifold assumption are the two smoothness assumption must meet. Cluster assumption defines that if the instances are in the same cluster, they are likely to be the same class. Manifold assumption assumes that the high-dimensional data on to the low-dimensional manifold to extract the features.

Based on the literature review from [31, 32] these methods can be classified into generative methods, discriminative methods, co-training, self-training, and semi-supervised support vector machines, and graph-based methods. Generative semi-supervised model purely depends on the model for the distribution of the data, and it can fail when the selected model is wrong or correlation among the data is not correlated with the classification task. Discriminative semi-supervised methods inclusive of probabilistic and non-probabilistic techniques [32] such as transductive or semi-supervised support vector machine (TSVMs, S3SVMs) and other graph-based methods assumes low-density between the class and high density within class, and this method can fails strongly overlapped classes. Self-training is one of the commonly used semi-supervised learning techniques. Initially, the classifier is trained with the small amount of labeled data then classifier used to classify the unlabeled data. Co-training method assumes that, the whole feature set is split into two set, each subset is enough to train the classifier and two subsets are conditionally independent to given class. Initially, the two individual classifiers are trained with labeled data based on the two selected subset, then the classifier classifies the unlabeled data and supervises (teaches) the other classifier to classify the unlabeled data [32]. Semi-supervised support vector machines (S3VMs) are also called as transductive support vector machines (TSVMs). The main aim of the S3VMs is to find the linear boundary with maximum margin from labeled and unlabeled data. Finally, the graph-based method constructs the graph from the training samples. The vertices in the graph represent the labeled and unlabeled training samples, whereas the undirected edge between the two vertices represents the similarity between the samples [33, 34]. The advantages and disadvantages of each supervised, semi-supervised, and unsupervised techniques are given in Table 3.

**Table 3** Advantages and disadvantages of supervised, semi-supervised and unsupervised techniques

| Learning method | Advantages | Disadvantages |
|---|---|---|
| Supervised learning | Works well on labeled data<br>Provides general classification | Require prior knowledge<br>Limited in terms of scalability of the target variable risk of overfitting<br>Heterogeneity of data |
| Unsupervised learning | Unbiased<br>Easier to bridge the gap between input and output<br>Performs well even when no prior knowledge is available and provides hidden meaningful insights | Ignores the correlation between different feature, might not result in the most optimal subset<br>Ignore dependency |
| Semi-supervised learning | Works well on both labeled and unlabeled data | Output is provided only for a subset of the training set |

## 4 Feature Extraction in Gene Selection

Feature extraction is alternative method to feature selection to reduce the high-dimensional data. It is also called as “feature construction” and projection onto a low-dimensional subspace. It creates the new features from the original set of features by the combination of the other features to reduce the dimensionality. There are two types of feature extraction techniques: linear extraction and nonlinear extraction technique [35, 36].

### *4.1 Linear Extraction Technique*

Linear feature extraction technique assumes that the input data are linearly separable in the low-dimensional subspace; thereby, it transforms the features by using any of the matrix factorization method. The popular well-known technique principal component analysis (PCA) [35, 37] is the best example of the feature extraction technique. PCA finds it principal components in the given data by using covariance matrix, and its eigenvalues and eigenvectors. PCA and several techniques related to the PCA are applied to microarray data for predicting the cancer. But these methods are highly effective to identify the important features of the data. It cannot find the nonlinear relationship between the features and the class in this way it researchers concluded that PCA cannot find the variables related to the target variables. For this drawback, supervised principal component (SPCA) analysis based on the target has been proposed by Hira and Gillies [35] and Khalid et al. [36]. SPCA performs well when comparing to the unsupervised method. Another major problem in PCA is that the mean square error is dominated because of large number of errors. For solving these issues, PCA-based on L2—norm is proposed, but this approach is also sensitive to the outliers. Later PCA-based on L1—norm is proposed by applying the maximum likelihood estimation. PCA based on maximum correntropy criterion (MCC), rotational PCA (R1-PCA), robust two-dimensional PCA (RTDPCA) are some of the best techniques proposed to solve the problem of outliers.

Linear discriminant analysis (LDA) is the supervised feature extraction technique. Based on the global structure of the training samples, it determines the linear discriminate vectors and these vectors are assumed as global. Extracting the global linear discriminant features from the test samples may lead classification error because of global data structure is not completely inconsistent. To overcome this, local linear discriminant analysis is proposed to extract the features from the local structure of samples. It is more powerful than LDA. Another major issues in the LDA are it does not work for small size samples (SSS) problem (dimension of the data exceeds the number of samples) and common mean problem. Orthogonal centroid method (OCM), null space-based LDA (NLDA), kernel discriminative common vector (kDCV) [35, 38], LDA with generalized singular value decomposition are some the techniques used to solve the small sample size problem.

## 4.2 Non-Linear Extraction Technique

Most of the data in the real word are in the form nonlinear data. Handling these types of the data for the further analysis is very difficult. In nonlinear dimensionality reduction, the low-dimensional feature space is mapped on a high-dimensional space by doing this nonlinear relationship among the features can be determined. Here, mapping the features onto the high-dimensional space is achieved by lifting $f(x)$, from the high-dimensional space the relation among the features will be viewed as linear. Therefore, the relationship among the features can be easily determined. After this, the features are mapped back to the lower-dimensional space and the relationship among the feature can be viewed as nonlinear [39].

Manifold approach is another nonlinear dimensionality reduction technique which is based on the assumption that the genes lie on the embedded nonlinear method which lowers the dimensions than original dimension. Plethora's of the methods are applied on the microarray data analysis. Isomap [38] is one of the widely used techniques. It constructs the manifold by merging each point based on the nearest neighbors. Isomap has able to extract more goodness features than PCA methods. Kohonen self-organizing map is one of the neural methods used to create the low-dimensional mapping on the features. Independent component analysis (ICA) [35] is also nonlinear feature extraction technique which determines the correlation among the data by minimizing and maximizing the contrast information which called whitening.

# 5 Intrinsic Characteristics of Microarray Data

In literature, there are two types of microarray datasets available, and the first one is binary datasets, and here the task is to classify the normal healthy patients from the normal patients. Second one is multiclass dataset; the goal is to classify the different types of tumors; here the classification is difficult task. Due to the large dimensionality with the small number of genes, the classification is difficult task and also it requires high computational time. Based on the literature from [40, 41], some internal characteristics of the microarray data is explained below.

## 5.1 *Small Size Problem*

The very first problem with the microarray data is usually it contains small sample size (less than 100). Due to this, the error estimation is very difficult, without correct error estimation the final conclusion from those dataset may lead improper results. To overcome this problem, it is important to select the correct feature selection method as well as the correct validation methods for estimating the errors.

## 5.2 *Class Imbalance*

It is another common problem in microarray dataset. It occurs when a dataset is dominated by the majority class or the classes which have more number of instances than other class. For example, usually the cancer class patients tend to be less when comparing the non-cancer patients because of more number of healthy patients. In this situation, the learning algorithms preselect (bias) toward the more number of instances because of weightage of the healthy patients. For this issue, more number of rules is required to classify correctly from the class imbalance dataset. Some traditional preprocessing techniques such as undersampling (created the original subset by eliminating the instances) and oversampling (by replicating the instances) are used to resample the data. SMOTE, tree splitting criterion, Cost-sensitive learning methods are some well-known methods used to deal with this issue.

## 5.3 *Data Complexity*

Data complexity measures are the measures to represent the characteristics of the data. It measures separability, linear decision boundaries, overlapping among the classes to achieve better classification. Maximum Fisher's discriminative ratio (F1) measures the class overlapping which focuses the effectiveness of the single feature dimension in classifying the labels.

### 5.4 *Data Shift*

In general, the datasets were originally divided into the training and testing set; it arose when there is a change in the distribution between the training and testing sets. This problem was first addressed in the book [40]. Concept shift or concept drift, changes of classification, changing environments, contrast mining in classification learning, fracture points, and fractures between data is the different terms used to represent the word data shift. Data shift [41] appears when the training data and testing data distribution are in different, i.e., $P_{\mathrm{train}}(y, x) \neq P_{\mathrm{test}}(x, y)$. Covariate shift, prior probability shift, concept shift are some types of data shift.

### 5.5 *Outliers*

An outliers are a data point in the dataset which significantly different from the others. Due to very nature of the real data, the outliers may exist. Outliers are also called as anomalies, abnormalities, discordant, etc. The main causes of outliers are poor data quality, low-quality measurement, equipment malfunction, manual error, etc. Statistical-based detection, deviation-based method, distance-based detection are some of the outlier detection techniques.

## 6 Summary

Feature selection method plays a crucial role to segregating important features as well as it provides the direction to remove the irrelevant features without affecting the final performance. Further, it improves the classification accuracy and faster induction of final prediction. In order to identify/extract optimized features, different algorithms and techniques in various application fields have been proposed earlier. This chapter has reviewed the up-to-date contributions in the topic of feature selection applied to the field of microarray data analysis in bioinformatics domain. The objective of this chapter is to gather and provide the up-to-date knowledge in the field of feature selection methods applied in the microarray data analysis as possible for the readers. Bearing this scope in mind, the recent literature has been analyzed in order to give best representation to the interested readers. In this chapter, a brief introduction about feature selection methods in the DNA microarray analysis was presented. With this, the taxonomy of the dimensionality reduction methods is represented in the diagrammatic way. Five feature selection methods such as filter, wrapper, embedded, hybrid, and ensemble have been discussed in detailed manner, and also tabulated the recent proposed algorithms, datasets used and accuracy achieved in the respective methods for easy understanding of readers. In order to decide best method, the advantages and disadvantages of each method are discussed. In machine learning perspective, supervised, unsupervised, semi-supervised gene selection methods, with advantages and disadvantages were discussed. Finally, this chapter provides how the feature selection methods differ from the feature extraction methods and few intrinsic characteristics of microarray data.

## References

1. James, A. P., & Dimitrijev, S. (2012). Ranked selection of nearest discriminating features. *Human-Centric Computing and Information Sciences, 2,* 12.
2. Ang, J. C., et al. (2016). Supervised, unsupervised and semi-supervised feature selection: A review on gene selection. *IEEE/ACM Transactions on Computational Biology and Bioinformatics, 13*(5), 971–989.
3. Yu, L., & Liu, H. (2004). Redundancy based feature selection for microarray data. In *Proceedings of the Tenth ACM SIGKDD Conference on Knowledge Discovery and Data Mining* (pp. 737–742).
4. Ambusaidi, M. A., et al. (2016). Building an intrusion detection system using a filter-based feature selection algorithm. *IEEE Transactions on Computers, 65*(10), 2986–2998.
5. Mohapatra, P., Chakravarty, S., & Dash, P. K. (2016). Microarray medical data classification using kernel ridge regression and modified cat swarm optimization based gene selection system. *Swarm and Evolutionary Computation, 28,* 144–160.
6. Hoque, N., et al. (2016). A fuzzy mutual information-based feature selection method for classification. *Fuzzy Information and Engineering, 8*(3), 355–384.
7. Raza, M. S., & Qamar, U. (2016). An incremental dependency calculation technique for feature selection using rough sets. *Information Sciences, 343,* 41–65.
8. Guo, S., et al. (2016). A centroid-based gene selection method for microarray data classification. *Journal of Theoretical Biology, 400,* 32–41.
9. Chen, H., Zhang, Y., & Gutman, I. (2016). A kernel-based clustering method for gene selection with gene expression data. *Journal of Biomedical Informatics, 62,* 12–20.
10. Wang, S., & Wei, J. (2017). Feature selection based on measurement of ability to classify subproblems. *Neurocomputing, 224,* 155–165.
11. Liu, H., Lui, L., & Zhang, H. (2008). Feature selection using mutual information: An experimental study. In *PRICAI 2008: Trends in Artificial Intelligence* (pp. 235–246). New York: Springer.
12. Sharma, A., Imoto, S., & Miyano, S. (2012). A top-r feature selection algorithm for microarray gene expression data. *IEEE/ACM Transactions on Computational Biology and Bioinformatics (TCBB), 9*(3), 754–764.
13. Kim, H. J., Choi, B. S., & Huh, M. Y. (2016). Booster in high dimensional data classification. *IEEE Transactions on Knowledge and Data Engineering, 28*(1), 29–40.
14. Kang, S., Kim, D., & Cho, S. (2016). Efficient feature selection-based on random forward search for virtual metrology modeling. *IEEE Transactions on Semiconductor Manufacturing, 29*(4), 391–398.
15. Choi, K. S., Zeng, Y., & Qin, J. (2012). Using sequential floating forward selection algorithm to detect epileptic seizure in EEG signals. In *2012 IEEE 11th International Conference on Signal Processing (ICSP)*, (Vol. 3), IEEE.

16. Apolloni, J., Leguizamón, G., & Alba, E. (2016). Two hybrid wrapper-filter feature selection algorithms applied to high-dimensional microarray experiments. *Applied Soft Computing, 38,* 922–932.
17. Chuang, L.-Y., et al. (2011). A hybrid feature selection method for DNA microarray data. *Computers in Biology and Medicine, 41*(4), 228–237.
18. Lee, C. P., & Leu, Y. (2011). A novel hybrid feature selection method for microarray data analysis. *Applied Soft Computing, 11*(1), 208–213.
19. Hsu, H.-H., Hsieh, C.-W., & Lu, M. D. (2011). Hybrid feature selection by combining filters and wrappers. *Expert Systems with Applications, 38*(7), 8144–8150.
20. Zorarpacı, E., & Özel, S. A. (2016). A hybrid approach of differential evolution and artificial bee colony for feature selection. *Expert Systems with Applications, 62,* 91–103.
21. Lan, L., & Vucetic, S. (2011). Improving accuracy of microarray classification by a simple multi-task feature selection filter. *International Journal of Data Mining and Bioinformatics, 5* (2), 189–208.
22. Wang, X., & Gotoh, O. (2010). A robust gene selection method for microarray-based cancer classification. *Cancer Informatics, 9,* 15–30.
23. Maulik, U., & Chakraborty, D. (2014). Fuzzy preference based feature selection and semi-supervised SVM for cancer classification. *IEEE Transactions on Nanobioscience, 13*(2), 152–160.
24. Liao, B., Jiang, Y., Liang, W., Zhu, W., Cai, L., & Cao, Z. (2014). Gene selection using locality sensitive laplacian score. *IEEE/ACM Transactions on Computational Biology and Bioinformatics, 11*(6), 1146–1156.
25. Liu, Q., Zhao, Z., Li, Y., Yu, X., & Wang, Y. (2013). A novel method of feature selection based on SVM. *Journal of Computers, 8*(8), 2144–2149.
26. Yu, L., Han, Y., & Berens, M. E. (2012). Stable gene selection from microarray data via sample weighting. *IEEE/ACM Transactions on Computational Biology and Bioinformatics, 9*(1), 262–272.
27. Wanga, A., Ana, N., Yanga, J., Chenb, G., Lia, L., & Alterovitzc, G. (2017). Wrapper-based gene selection with Markov blanket. *Computers in Biology and Medicine, 81,* 11–23.
28. He, W., Zhu, X., Cheng, D., Hu, R., & Zhang, S. (2017). Unsupervised feature selection for visual classification via feature representation property. *Neurocomputing, 236,* 5–13.
29. Dadaneh, B. Z., Markid, H. Y., & Zakerolhosseini, A. (2016). Unsupervised probabilistic feature selection using ant colony optimization. *Expert Systems with Applications, 53,* 27–42.
30. Wang, C., Machiraju, R., & Huang, K. (2014). Breast cancer patient stratification using a molecular regularized consensus clustering method. *Methods, 67,* 304–312.
31. Sheikhpour, R., et al. (2017). A survey on semi-supervised feature selection methods. *Pattern Recognition, 64,* 141–158.
32. Aziz, R., Verma, C. K., & Srivastava, N. (2017). Dimension reduction methods for microarray data: A review. *AIMS Bioengineering, 4*(2), 179–197.
33. Hosseinzadeh, F., KayvanJoo, A. M., Ebrahimi, M., & Goliaei, B. (2013). Prediction of lung tumor types based on protein attributes by machine learning algorithms. *Springer Plus, 2,* 238.
34. Herland, M., Khoshgoftaar, T. M., & Wald, R. (2014). A review of data mining using big data in health informatics. *Journal of Big data, 1,* 4.
35. Hira, Z. M., & Gillies, D. F. (2015). A review of feature selection and feature extraction methods applied on microarray data. *Advances in Bioinformatics,* Article ID 198363, pp 1–13.
36. Khalid, S., Khalil, T., & Nasreen, S. (2014). A survey of feature selection and feature extraction techniques in machine learning. In *Science and Information Conference* (pp. 371–378).
37. Masulli, F., Peterson, L. E., & Tagliaferri, R. (2009). Eds., Vol. 6160 of Lecture Notes in Computer Science (pp. 82–96), Berlin, Germany: Springer.
38. Tenenbaum, J. B., de Silva, V., & Langford, J. C. (2000). A global geometric framework for nonlinear dimensionality reduction. *Science, 290*(5500), 2319–2323.

39. Guyon, I., Bitter, H. M., Ahmed, Z., Brown, M., & Heller, J. (2005). Multivariate non-linear feature selection with kernel methods. In *Soft Computing for Information Processing and Analysis* (pp. 313–326).
40. Quiñonero Candela, J., Sugiyama, M., Schwaighofer, A., & Lawrence, N. D. (2009). *Dataset shift in machine learning*. Cambridge, MA: The MIT Press.
41. Moreno-Torres, J. G., Raeder, T., Alaiz-Rodríguez, R., Chawla, N. V., & Herrera, F. (2011). A unifying view of dataset shift in classification. *Pattern Recognition*.
42. Han, X., et al. (2014). Feature subset selection by gravitational search algorithm optimization. *Information Sciences, 281,* 128–146.
43. Xue, B., et al. (2016). A survey on evolutionary computation approaches to feature selection. *IEEE Transactions on Evolutionary Computation, 20*(4), 606–662.
44. Sharbaf, F. V., Mosafer, S., & Moattar, M. H. (2016). A hybrid gene selection approach for microarray data classification using cellular learning automata and ant colony optimization. *Journal on Genomics, 107*(6), 231–238.
45. Solorio-Fernández, S., Carrasco-Ochoa, J. A., & FcoMartínez-Trinidad, J. (2016). A new hybrid filter–wrapper feature selection method for clustering based on ranking. *Journal on Neurocomputing, 214,* 866–880.
46. Lu, H., Chen, J., Yan, K., Jin, Q., Xue, Y., & Gao, Z. (2017). A hybrid feature selection algorithm for gene expression data classification. *Article on Neurocomputing, 256,* 1–7.
47. Zhu, M., & Song, J. (2013). An embedded backward feature selection method for multiple criteria linear programming (MCLP) classification algorithm. *Procedia Computer Science, 17,* 1047–1054.
48. Mishra, S., & Mishra, D. (2015). SVM-BT-RFE: An improved gene selection framework using Bayesian *T*-test embedded in support vector machine (recursive feature elimination) algorithm. *Karbala International Journal on Modern Science, 1,* 86–96.
49. Li, Z. G., Meng, H. H., & Ni, J. (2008). Embedded gene selection for imbalanced microarray data analysis. In *International Multi-symposiums on Computer and Computational Sciences* (pp. 17–24).
50. Bonilla-Huerta, E., Hernandez-Montiel, A., Morales-Caporal, R., & Arjona-Lopez, M. (2016). Hybrid framework using multiple-filters and an embedded approach for an efficient selection and classification of microarray data. *IEEE/ACM Transactions on Computational Biology and Bioinformatics, 13*(1), 12–23.
51. Sheng, L., Pique-Regi, R., Asgharzadeh, S., & Ortega, A. (2009). Microarray classification using block diagonal linear discriminant analysis with embedded feature selection. In *IEEE International Conference on Acoustics, Speech and Signal Processing, 2009* (pp. 1757–1760). ICASSP 2009.
52. Liu, K.-H., Zeng, Z.-H., & Ng, V. T. Y. (2016). A hierarchical ensemble of ECOC for cancer classification based on multi-class microarray data. *Information Sciences, 349,* 102–118.
53. Bolón-Canedo, V., Sánchez-Maroño, N., & Alonso-Betanzos, A. (2012). An ensemble of filters and classifiers for microarray data classification. *Pattern Recognition, 45*(1), 531–539.
54. Mollaee, M., & Mohammad, M. H. (2016). A novel feature extraction approach based on ensemble feature selection and modified discriminant independent component analysis for microarray data classification. *Bio Cybernetics and Biomedical Engineering, 36*(3), 521–529.
55. Seijo-Pardo, B., Porto-Díaz, I., Bolón-Canedo, V., & Alonso-Betanzos, A. (2017). Ensemble feature selection: Homogeneous and heterogeneous approaches. *Knowledge-Based Systems, 118,* 124–139.
56. Das, A. K., Das, S., & Ghosh, A. (2017). Ensemble feature selection using bi-objective genetic algorithm. *Knowledge-Based Systems, 123,* 116–127.
57. Liu, H., Liu, L., & Zhang, H. (2010). Ensemble gene selection by grouping for microarray data classification. *Journal of Biomedical Informatics, 43*(1), 81–87.
58. Ebrahimpour, M. K., & Eftekhari, M. (2017). Ensemble of feature selection methods: A hesitant fuzzy sets approach. *Applied Soft Computing, 50,* 300–312.

# An Investigation on E-Learning Tools and Techniques Towards Effective Knowledge Management

**P. Kalyanaraman, S. Margret Anouncia and V. Balasubramanian**

**Abstract** Knowledge management process has become extremely popular in recent years. The process progresses on the acquisition, organization, retention, storage, dissemination and reuse of knowledge in an organization. Consequently, a suitable strategy is drawn towards a convenient deployment process that aids the need of individual users to dramatically extend the reach of know-how facts. Recently, the process of applying tacit knowledge to solve the business problem has been well formulated with the help of information technology. However, conversion of tacit knowledge to explicit knowledge has become challenging and it requires a powerful leveraging mechanism. One of the organized approaches that are emerging to help knowledge management process is the e-learning. The techniques associated with content management, activity management, support management and assessment of e-learning are observed to be in line with the process of knowledge management. Hence, the practices associated with the e-learning can be very well extended to knowledge management. In order to infiltrate into the idea of mapping the process of e-learning into knowledge management, a study is performed to highlight the commonalities existing between e-learning techniques and its applicability towards knowledge management.

## 1 Introduction

Generally, e-learning is assumed to use electronic technologies to improve the performance of a teaching–learning process. It has emerged with advanced strategy wherein Internet and other forms of information technology play a vital role. The major concern involved in applying the different forms of ICT into this domain lies in the management of knowledge for the effective deployment of e-learning system [1–3]. Rapid progress attained in software technologies and services facilitates the integration of organizational and cultural gaps exercised in the e-learning system [4].

P. Kalyanaraman (✉) · S. Margret Anouncia · V. Balasubramanian
School of Computer Science and Engineering, VIT University, Vellore, India
e-mail: pkalyanaraman@vit.ac.in



S. Margret Anouncia and U. K. Wiil (eds.), *Knowledge Computing and its Applications*, https://doi.org/10.1007/978-981-10-8258-0_15

However, the management of source knowledge for the system leaves greater challenges as the input supports varieties of formats. Thus, the demand for an appropriate knowledge management (KM) process rises to foresee the e-learning system as a pack of well-strategized knowledge. In general, the objective of any KM system is to exploit the existing components, tools and techniques of a domain to foster knowledge sharing and transfer [5]. In view of this motivation, an attempt is made through this investigation to explore the level of significance provided by the e-learning components and tools towards knowledge management.

## 2 E-Learning Components, Features and Tools

Being an integral part of e-learning system, components such as contents, activities, support and assessment decide the course structure, page design, audience, usability and content engagement (http://www.instructionaldesignexpert.com). These components individually or jointly contribute to one or more features that shape up the e-learning system. The content component deals with the design of curriculum and instruction with the help of learning objects such as text, audio, video, diagram, chart, case study and simulation. The activities include administrative tasks such as manage, search, retrieve and reuse of contents. In order to facilitate these activities, support is provided through network connections, service providers, browsers and other Internet tools. In addition, technical support to share multi-format materials, conducting webinars, collaborating with the peers and content administrators are incorporated into the system. The assessment component includes different assessment procedures like goal checks, one-to-one discussion, instructor observation, group presentation and self-assessment that are adopted in the system [6]. These tools are said to work on the features to fulfil the purpose of the e-learning components. However, the process needs an appropriate knowledge management scheme to proceed.

## 3 Knowledge Management Process

Knowledge management is a structured process to identify, acquire, analyse, transfer and store organizations' information, which includes resources, documents, policies and people skills. According to [7], it is addressed as a blend of information technology, people and organizations. Advancement in technology and collaborating tools has contributed to the knowledge management process. Efficient implementation of the knowledge management would bring substantial change in the process flow, user's mindset, technology adoption and deployment along with work culture of the organization. For effective use of information and expertise within an organization, knowledge management should be local and personalized [6]. In order to deploy a proper convincing knowledge management process, it is indeed required to use various KM tools and technologies.

## 4 Knowledge Creation and Acquisition

Knowledge creation deals with extracting new knowledge from data, information, experience and prior knowledge [8]. Tacit knowledge is created through direct or online interactions among the stakeholders. Knowledge can be acquired in various formats that include sources from reports, books, manuals, thesis, proceedings, online chatting, audio, video, case studies and other electronic publications on the Internet. There are three widely adopted techniques for knowledge acquisition such as manual, semi-automated and automated methods. These techniques are performed through structured or semi-structured or unstructured way based on the available resources. Variety of tools such as groupware systems, intranet and extranet technologies, data warehousing, data mining, decision support systems, online analytical processing (OLAP), content management systems, document management systems, simulation tools, artificial intelligence tools and semantic networks are applied for the purpose. The created knowledge is converted into explicit knowledge through various knowledge representation techniques and is maintained.

## 5 Knowledge Organization and Storage

Organization of knowledge is one of the most complex challenges in the knowledge management life cycle. The system organizes its knowledge resources in a suitable and convenient deliverable format so that users can access, share and use the resources very easily. The systematically organized knowledge resources are accumulated as a digital knowledge repository or federated on websites for their preservation.

## 6 Knowledge Retrieval and Dissemination

To facilitate accessing, exploring and retrieving knowledge resources, a good number of tools and techniques, such as browsing, searching, data mining, metadata, knowledge mapping, are being used in KM. It is argued that data mining methods could be a choice to detect unknown patterns in user's behaviour, learning resources and knowledge mastering bottlenecks [9]. Having understood the behaviour, it is easy to adopt a suitable dissemination method that includes email, chat rooms, discussion forums, videoconferencing and other tools of KM as well as e-Learning. Recently, social media and network are being adopted for the knowledge dissemination.

## 7 KM Tools and Technologies

Knowledge management tools and techniques have been widely applied in the process of organizing, structuring and delivering of contents. A KM tool assists to produce new ideas or knowledge to support in dissemination and sharing. Tools such as groupware, intranet and Internet support knowledge sharing, while electronic calendar, database and desktop publishing support knowledge distribution [1]. Systems confined to expert's knowledge modelling, neural models and intelligent agents have been adapted to capture knowledge. Database technologies offer support to store and retrieve knowledge, while mining technology assists in the deriving pattern of knowledge inferred. The determined pattern is henceforth used for classifying and organizing knowledge. Advanced techniques such as social bookmarking and social tagging equally play a vital role in knowledge organization process.

Since most of the task carried for an effective knowledge management is analogous to the process of e-learning, the tasks that are associated with the e-learning may also be scaled up to suit organizational knowledge management [10]. In addition, the e-learning practice leaves an advantage of having evaluated tasks in reality, and hence, it is felt appropriate to integrate e-learning practices with knowledge management techniques to derive at a sophisticated knowledge management strategy.

## 8 Relationship Between E-Learning and KM Process

Knowledge management is inferred to be a process with a key aspect to evolve methods and process that integrates well-defined concepts and facts to support effective functioning of an organization. Therefore, knowledge management process can be viewed as a process of creating, acquiring and utilizing knowledge to enhance organizational performance [11]. On the other hand, e-learning is presumed to be an environment that is supported by collaborative processes in improving and enhancing individual and organizational performance. In the long run, the concept of e-learning encourages lifelong learning which in turn would emphasis on effective organizational learning [12]. Thus, e-learning could be viewed as a cornering stone for knowledge management. The techniques associated with the e-learning eventually help in creating explicit knowledge that could be stored in a knowledge store. Of course, the content that is managed through knowledge management process is the content that is useful for e-learning. Thus, the techniques adopted to support e-learning is indirectly involved in the knowledge management process.

## 9 Integrating E-Learning and KM

According to the knowledge management viewpoint, learning at all levels is important and hence the focus of the team and the organizational learning towards knowledge reuse becomes essential. Also, it is noted that the information chunks of KM system often lack interactivity [13]. Therefore, knowledge management systems must incorporate interactive learning activities for successful learning. Adopting appropriate pedagogical strategy and tailoring of contents suitable for individual needs can make learning more effective.

Since most of the tasks associated with the e-learning are focused on the knowledge capture, organization, delivery and interactions, it is evident that the same can be scaled to develop a knowledge management process [14]. It is noted that the e-learning system has the capability to manage the content through navigation into various course materials, search based on keywords, retrieve documents, grading and re-using contents. Analogously, all these tasks are found to be essential for an effective knowledge management. Hence, the tools adopted for e-learning can be customized to accompany knowledge management process.

In another view, e-learning offers features such as virtual community using tools like discussion boards, emails, chat, browsers that help to interact with peers and administrator to form a sense of a community. This aspect of e-learning is quite useful in the process of knowledge sharing across the demographic regions to formulate an expert forum. The outcome of these forums is recorded, formalized and stored in a knowledge repository. Subsequently, this repository could be used as a knowledge source for decision-making and analysis process of an organization. Eventually, when the repository size becomes larger, a distributed service with the multimodal interactions and delivery could be made possible to enable cross-cultural information sharing. These characteristics are found to be a key focus in the e-learning system. The e-learning system seeks for transparency and accountability by making the outcome and deliverables more measurable which is expected even in knowledge management. Since most of the characteristics of e-learning are applicable to knowledge management, it is evident that the e-learning techniques and technologies could contribute to a good knowledge management significantly. Figure 1 illustrates the integration of e-learning components for knowledge management.

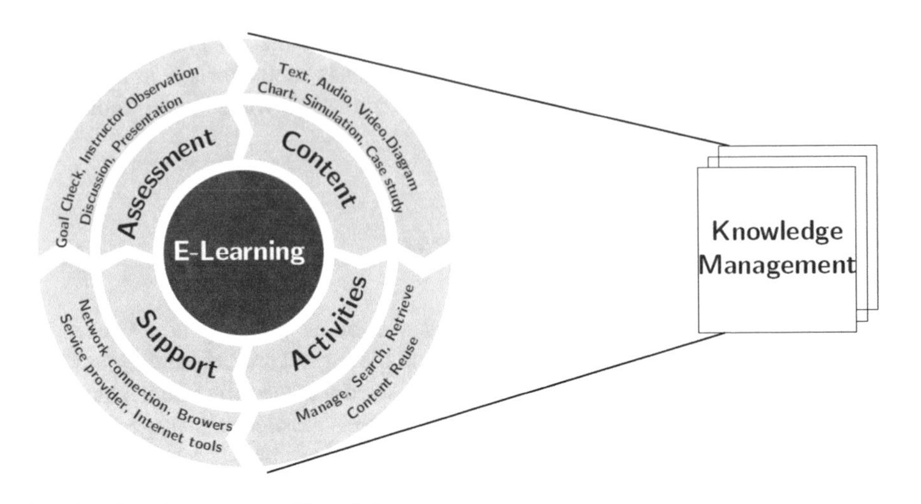


**Fig. 1** E-learning system and knowledge management

## 10 Knowledge Management Through Learning Objects

The working process within an organization with e-learning and knowledge management systems is extremely knowledge concentrated irrespective of the demographic differences and roles. Both e-learning and knowledge management interact with each other for facilitating and developing the capabilities of an organization. Indeed, each of the knowledge management activities depends on the perception of the organization to record the explicit knowledge for reuse. The common method of recording and acquiring such knowledge involves audio, videos, slideshows, documents and Web pages which are processed to offer online resources, databases, periodicals, journals and other materials for reuse. These entities form the basic building blocks of an e-learning system that is called as learning objects (LO), and the process to understand and use this LO is very much branded by learning process of e-learning. Hence, this approach of involving knowledge management activities in an e-learning system would help learners gain a deeper understanding and develop higher cognitive skills towards learning as well training. Though learning and training are intricately linked, the way it is carried out reveals the identity of the process towards knowledge building. However, the collected knowledge is formalized using different techniques, and finally, a knowledge repository is evolved for organizational reuse.

## 11 Knowledge Management Through Repository

Generally, e-learning and knowledge management are adapted to facilitate the learning process adopted in an organization. E-learning services offer support towards ubiquitous learning by utilizing mobile technologies and support social networks (communities) for personalized knowledge management. The convergence of e-learning and knowledge management fosters a constructive, open, dynamic, interconnected communities that are adaptive, user-friendly, socially concerned to access the wealth of knowledge [15]. An effective knowledge management system not only provides a vehicle to share information but also builds a community of learners. As e-learning system caters in the process of building the knowledge repository, the process would drive the knowledge management mechanism in the constant and consistent delivery of relevant information based on user's acquaintance and the environment. The typical process of e-learning system allows learners to acquire information, share to other learners and transfer to knowledge by applying it to organizational problems; further, the e-learning repositories store the acquired knowledge for future reference [1]. The process encourages acquisition, creation, capture and adoption to facilitate knowledge generation which is stored in a repository. A repository is not a large database rather it houses organizational principles which could allow the organization to access stable, reasonable information for the well-being.

Repositories also can complement other document and content management work, such as policy or procedures as that of e-learning system. The repositories prefer to store information that is generated in the form of stories, podcasts and links to the several business processes. Thus, a focus on external information,

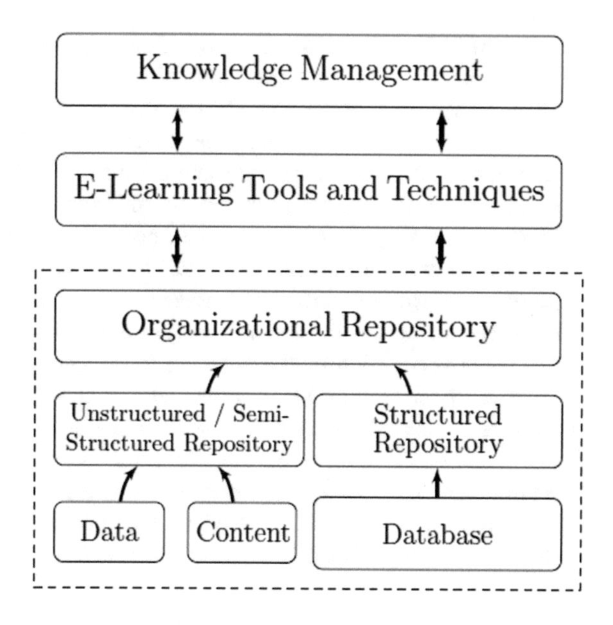


**Fig. 2** Knowledge repository

structured internal information or records of conversations and experiences are encouraged to be stored in a repository to yield a better knowledge management [16]. Links to several repositories and peer-to-peer work on them made them suit well for problem solving, tacit knowledge creation and sharing which forms a prime concern of knowledge management. Hence, it is evident that repositories can be a predominant resource for a good knowledge management system (Fig. 2).

## 12 Purpose of E-Learning Components Towards Knowledge Management

As stated in earlier sections, a good knowledge management is possible through a set of tasks, supporting tools and technologies that resemble close to e-learning process. In another view, it is admissible that the tasks of e-learning process help to gain a good knowledge management. Especially, the components of e-learning when thought of for knowledge management, it serves significantly towards organizational learning, pattern analysis and subsequently for personalization of the knowledge. Consequently, good, realistic knowledge dissemination among the stakeholders becomes active to yield more usability.

## 13 Organizational Learning

By using knowledge management techniques, organizations look for creating or discovering valuable knowledge that is accessible and used appropriately in a convenient form to gain maximum organizational performance. In order to cater the need, the process needs to be fostered with innovation to organize strategic assets. Further, these organizations should capitalize the existing organizational knowledge to improve and build up intellectual assets that gain a competitive advantage. Unfortunately, knowledge management is linked to organizational perspectives; on the contrary, e-learning focuses on an individual perspective. In other words, achieving the integration between the knowledge management and e-learning is more accountable for the learning outcomes. Most of the e-learning systems offer solutions irrespective of dynamic interactions in an organization where learning takes place in a specific context based on real-world need. Yet, the approach neglects delivery of information leading to no learning.

Hence, understanding users/individuals learning styles and delivering content according to their skill/capabilities will ensure learning towards the organization goals. Organizational resources may be used as learning materials when KM and e-learning are integrated, annotated and made into smaller learning units with specific objectives. This is hindered by the specifics of e-learning materials with their personalized content, internal connections, links and references. Annotation with metadata about the connection to other objects, technical prerequisites and training styles is also considered to be challenging. Therefore, using tools associated with groupware, data mining, data warehousing, decision support systems, OLAP, document management, content management, semantics and simulation would ensure the effective learning.

## 14 Pattern Analysis

One of the major concerns in a knowledge management system is the knowledge dissemination mechanism. In order to adopt a suitable dissemination process in an organization, user's capability towards learning and reuse needs to be studied. Pattern analysis is a process that helps to analyse the individual behaviour to detect and manage interactions among the individuals and organization. Subsequently, this analysis can lead to personalization process, by which an individual can aspire to extract the convenient knowledge for the organization development. One of the techniques that are widely adopted for extracting a different pattern of information is the data mining [17].

Individual information such as experience, domain expertise, career objectives, tasks handled along with interest would contribute substantially to predict future behaviour of different classes of employed professionals. Also, additional information about an individual like personal interests and hobbies is also found to be valuable to discover hidden patterns. Usage of mining in the knowledge domain involves recording information of individual profiles, Web access information, skill acquisition details, work compliments and evaluation results to track individual activities, and hence, real-time recommendations can be fostered for individuals and organizations. A well-organized, structured knowledge management aids in the process of skill development, planning and decision-making processes.

## 15 Personalization

In general, personalization is the process of increasing user satisfaction through addressing their needs effectively and efficiently. The process becomes easy and tangible if the source of action, i.e. knowledge is augmented with proper organization [18]. The process of knowledge management influences this augmentation significantly than any other techniques. The process aids for both individual learning and personalized learning. The individual learning in an organization expects the process to provide multiple services in building individual's experience rather than imposing a particular method of learning. In other words, the amount of data and information available for the learners must be adapted to the needs of workers and the environment. Providing the right learning object to a learner requires additional information regarding the learner's behaviour and their environment. Various methodologies have been adopted by researchers to establish the knowledge requirement of the learner and to guide them appropriately. As the knowledge management system provides limited guidance, e-learning components on the other hand attempt to provide overwhelming guidance that prevents the notion of self-directed learning, free navigation and content selection. The e-learning tools and techniques could be exploited to provide the structured knowledge to the stakeholders. Mining the domain related to recording information of learner's profile, Web access information, collaborative activities and their performance can track learning activities and identify access patterns and user behaviours [19]. The behaviour analysis thus carried out helps to sync personalization of the maintained knowledge. The personalized knowledge would be applied in the decision-making process suitably.

# 16 Relationship Between Knowledge Management and E-Learning

Comparing the functionalities of knowledge management process and e-learning process, a relationship between these two domains is explored. The relationship as inferred from the study is tabulated in Table 1.

**Table 1** Relationship between Knowledge Management and E-Learning

| Knowledge management | E-learning |
|---|---|
| It is a process of creating, capturing and using knowledge | It is a process of using multimedia technologies for creating, capturing and organizing the knowledge |
| It offers reusable objects | It generates reusable objects |
| It makes use of e-learning process as a tool for knowledge management | It helps knowledge management through creating explicit knowledge from tacit knowledge and offers intricacy for knowledge sharing |
| Uses the content management system as a subset of the process | It focuses on creation of structured knowledge |
| It uses the information of enterprise and human resources for planning and skill development | It provides information for enterprise and human resources |

## 17 Conclusion

A study on the process of e-learning and knowledge management is performed to understand the process structure and the functionalities offered by these domains. From the study, it is clear that there exists an intersecting point that makes use of functionalities of both the processes. The e-learning techniques such as content management, activity management and support management are found as an appropriate mechanism for knowledge structuring. Hence, the process as such is employed to be a tool for effective knowledge management. The study revealed the commonalities between these two domains, and hence a comparison is brought out among these fields. Consequently, the study helped to understand the process of converting tacit knowledge into explicit knowledge through e-learning technologies for an effective knowledge management.

## References

1. Islam, S., Kunifuji, S., Miura, M., & Hayama, T. (2011). Adopting knowledge management in an e-learning system: Insights and views of KM and EL research scholars. *Knowledge Management E-Learning, 3*(3), 375–398.
2. Judrups, J. (2015). Analysis of knowledge management and e-learning integration approaches.In *{ICEIS} 2015—Proceedings of 17th International Conference on Enterprise Information Systems* (Vol. 2, pp. 451–456). Barcelona, Spain, April 27–30, 2015, no. Dalkir.
3. Radwan, N. M. (2015). Investigating knowledge management in e-learning environment. *International Journal of e-Education, e-Business, e-Management e-Learning, 5*(3), 136–143.
4. Mihalca, R., Uţă, A., Andreescu, A., & Întorsureanu, I. (2008). Knowledge Management in E-Learning Systems. *Revista Informatica Economic nr, 2*(46), 60–65.
5. Mendes, M. E. S., Martinez, E., & Sacks, L. (2002). Knowledge-based content navigation in e -learning applications. In *Proceedings of 2002 London Communication Symposium LCS*.
6. Owayid, A., Alrawi, K., & Shaalan, K. (2013). Strategic change in knowledge management and e-learning: enhancing workplace learning. *European Journal of Scientific Research, 100,* 230–240.
7. Tochtermann, K. (2002) Personalization in knowledge management. In *International Symposium on Metainformatics* (pp. 29–41).
8. Sabherwal, R., & Sabherwal, S. (2005). Knowledge management using information technology: Determinants of short-term impact on firm value. *Decision Sciences, 36*(4), 531–567.
9. Ponce, D. (2003). What can e-learning learn from knowledge management? (pp. 1–4). April, 2003.
10. Murugaboopathi, V. S. G., & Harish, K. (2012). Knowledge management through e-learning. *International Journal of Advanced Research Computer Science Software Engineering, 2*(9), 300–305.
11. Stacey, R. (2000). The emergence of knowledge in organization. *Emergence, A Journal of Complex Issues Organizations and Management, 2*(4), 23–39.
12. Zahid, A., & Alam, R. (2015). Knowledge management process—perspective on e-learning uses & effectiveness, *4*(9), 136–139.
13. Yacci, M. (2005). The promise of automated interactivity. In *Biennial Conference on Professional Knowledge Management/Wissensmanagement* (pp. 214–221).

14. Yordanova, K. (2007). Integration of Knowledge management and E-learning—common features. In *International Conference on Computer System and Technology—CompSysTech 2007* (pp. 94:1–94:6).
15. Naeve, A., Lytras, M., Nejdl, W., Balacheff, N., & Hardin, J. (2006). Advances of the semantic web for e-learning: expanding learning frontiers. *British Journal of Educational Technology, 37*(3), 321–330.
16. Muhire, A. (2012) E-learning and knowledge management : The development of an e-learning system for organisational training e-learning and knowledge management.
17. Abd Elaal, S. A. E. A. (2013) E-learning using data mining. In *21th International Conference on Chinese–Egyptian Research Journal.*
18. Tochtermann, K. (2003). Personalization in knowledge management. *Metainformatics, 2641,* 29–41.
19. Du, Z., Fu, X., Zhao, C., Liu, Q., & Liu, T. (2013). Proceedings of the 2012 International Conference on Information Technology and Software Engineering (Vol. 212, pp. 11–19).

Printed in the United States
By Bookmasters